A School Algebra Complete

A SCHOOL ALGEBRA COMPLETE

BY

FLETCHER DURELL, PH.D.

MATHEMATICAL MASTER IN THE LAWRENCEVILLE SCHOOL

AND

EDWARD R. ROBBINS, A.B.

MATHEMATICAL MASTER IN THE WILLIAM PENN CHARTER SCHOOL

NEW YORK

MAYNARD, MERRILL, & CO.

44–60 EAST TWENTY-THIRD STREET

A COMPLETE SERIES OF

Text-Books in Mathematics

PRICE

FIRST LESSONS IN NUMBERS (Durell & Robbins). 88 pages, $0.25

The development of numbers to 100, attractively illustrated

THE ELEMENTARY PRACTICAL ARITHMETIC (Durell & Robbins). 202 pages .... .40

Begins with the development of numbers and closes with the subject of Interest, covering the more useful subjects of arithmetic.

THE ADVANCED PRACTICAL ARITHMETIC (Durell & Robbins). 367 pages.... .65

Covers the courses of the state normal schools, meets the requirements for admission to colleges, and is also especially adapted to the more practical demands of the rural schools.

A MENTAL ARITHMETIC. 173 pages.... .35

Its success is already fully attested by its extensive use in Pennsylvania and its adoption for exclusive use in all the public schools of Tennessee.

MENSURATION. 82 pages .... .50

A valuable subject, treated in a masterly manner by a practical teacher.

A GRAMMAR SCHOOL ALGEBRA (Durell & Robbins). 287 pages.... .80

This volume contains only so much of the subject as pupils in grammar schools are likely to study.

A SCHOOL ALGEBRA (Durell & Robbins). 373 pages...... 1.00

This volume covers the requirements for admission to the classical course of colleges.

A SCHOOL ALGEBRA COMPLETE (Durell & Robbins). 463 pages.... 1.25

This book contains, in addition to the subjects usually treated in a school algebra, the more advanced subjects required for admission to universities and scientific schools.

PLANE GEOMETRY (Durell). 320 pages.... *In press.*

Leads to original demonstrations.

SOLID GEOMETRY (Durell). 204 pages.... *In press.*

A companion to Plane Geometry and on the same plan.

Write for descriptive circulars and special introductory prices.

MAYNARD, MERRILL, & CO., Publishers, 44-60 East 23d Street, New York.

PREFACE.

THE principal object in writing this School Algebra has been to simplify principles and make them attractive, by showing more plainly, if possible, than has been done heretofore, the practical or common-sense reason for each step or process. Thus, at the outset it is shown that new symbols are introduced into algebra not arbitrarily, but for the sake of definite advantages in representing numbers. The fundamental laws of algebra governing the use of symbols derive their importance in like manner from the economies which they make possible in dealing with the symbols for numbers. Each successive process is taken up for the sake of the economy or new power which it gives as compared with previous processes.

It is hoped that this treatment not only makes each principle clearer to the pupil, but also gives increased unity to the subject as a whole. It is also believed that this treatment of algebra is better adapted to the practical American spirit, and gives the study of the subject a larger educational value.

While seeking to develop the theory of the subject in this manner, it has been deemed best to keep in close touch with the best current practice of teachers in other respects. For instance, the order of topics in text-books most used at present has been followed.

Great care has been taken in the selection and gradation of a large number of examples. It is hoped that they have been so graded that any example may be considered the last of a series of progressive steps, provided the teacher wishes to limit the work at any particular point. Frequent reviews have been provided for, especially in the all-important subjects of Factoring, Fractions, Exponents, and Radicals.

This volume contains, besides the specified requirements in algebra for admission to the classical course of colleges, the more advanced subjects required by universities and scientific schools—to wit, Permutations and Combinations, Undetermined Coefficients, The Binomial Theorem, Continued Fractions, and Logarithms.

The authors will sincerely appreciate the courtesy, if their friends and fellow-teachers will kindly advise them of any discovered errors.

FLETCHER DURELL,
EDWARD R. ROBBINS.

LAWRENCEVILLE, N. J.,
December 23, 1897.

CONTENTS.

CHAPTER XV.

CHAPTER XVI.

CHAPTER XVII.

CHAPTER XVIII.

CHAPTER XIX.

CHAPTER XX.

SCHOOL ALGEBRA COMPLETE.

CHAPTER I.

ALGEBRAIC SYMBOLS.

1. First Source of New Power in Algebra. Let the following problem be proposed for solution:

James, John, and William together have 120 marbles. John has three times as many marbles as James, and William has twice as many as John. How many marbles has each boy?

The solution of the problem is facilitated by the use of some symbol, as x, for one of the unknown numbers. Thus,

Let x	= number of marbles which James has,				
then $3x$	=	"	"	"	John has,
$6x$	=	"	"	"	William has.
Hence, $x + 3x + 6x$	=	"	"	"	all have together.
But 120	=	"	"	"	all have together.

Hence, $x + 3x + 6x = 120$

that is, $10x = 120$

$x = 12$,	number of marbles which James has,				
$3x = 36$,	"	"	"	John	"
$6x = 72$,	"	"	"	William	"

This solution of the problem illustrates the first new principle in algebra—viz. that a more extended use of symbols than is practised in arithmetic gives increased ease and power in the investigation of properties of number. In the above example, a symbol being used for one unknown number, the other unknown numbers may be expressed in terms of this symbol; then the relation of all the unknown numbers to the known number is expressed in a form which is readily reduced to another form so simple that from it the value of the first unknown, and afterward the values of the other unknown numbers are at once perceived.

2. Definition of Algebra. Hence, *Algebra*, in its first conception, is that branch of mathematics which treats of the properties of number (or quantity expressed by number) by the extended use of symbols.

Algebra may be briefly described as generalized arithmetic.

3. Three Classes of Symbols. Three principal kinds of symbols are used in algebra:

I. SYMBOLS OF QUANTITY.
II. SYMBOLS OF OPERATION.
III. SYMBOLS OF RELATION.

I. SYMBOLS OF QUANTITY.

4. Symbols for Known Quantities. Known quantities are represented in arithmetic by figures; as, 2, 3, 27, etc. They are represented in the same way in algebra, but also in another more general way—viz. by the first letters of the alphabet; as, a, b, c, etc.

The advantage in the use of letters to represent known numbers lies in the fact that a letter may stand for any known number, and a result be obtained by the use of letters which is true for all numbers.

5. Symbols for Unknown Quantities. Unknown quantities in algebra are usually denoted by the last letters of the alphabet; as, x, y, z, u, v, etc.

The advantage in the use of a distinct symbol for an unknown quantity is stated in Art. 1.

6. Symbols for Groups of Similar Quantities. Groups of similar quantities are usually represented by groups of similar symbols; as,

(1) By the same letter with different *accents;* for example, a', a'', a''', etc., read "a prime," "a second," "a third," etc.

(2) By the same letter with different *subscript figures;* as, a_1, a_2, a_3, etc., read "a sub-one," "a sub-two," etc.

The advantages of these ways of representing groups of similar quantities are obvious.

7. Sign of Continuation. After a group of similar quantities a series of dots is often written to indicate that the group is continued indefinitely.

Ex. a_1, a_2, a_3, ...

This series of dots is called the sign of continuation, and reads "and so on."

8. Positive and Negative Quantity. Negative quantity is quantity exactly opposite in quality or condition to quantity taken as positive.

If distance east of a certain point is taken as positive, distance west of that point is called negative.

If north latitude is positive, south latitude is negative.

If temperature above zero is taken as positive, temperature below zero is negative.

If in business matters a man's assets are his positive possessions, his debts are negative quantity.

Positive and negative quantity are distinguished by the signs + and − placed before them. Thus, \$50 assets are denoted by + \$50, and \$30 debts by − \$30. We denote 12° above zero by + 12°, and 10° below zero by − 10°.

The use of the symbols + and − for this purpose, as well as to indicate the operations of addition and subtraction (see Arts. 9 and 10), will be justified later on (see Arts. 31, 32).

II. Symbols of Operation.

9. The **Sign of Addition** is $+$, and is called "plus." Placed between two quantities, it indicates that the quantity after the plus sign is to be added to the quantity before it.

Thus, $a+b$ is read "a plus b," and indicates that the quantity b is to be added to the quantity a.

10. The **Sign of Subtraction** is $-$, and is called "minus." Placed between two quantities, it indicates that the second quantity is to be subtracted from the first.

Thus, $a-b$ is read "a minus b," and means that b is to be subtracted from a.

11. The **Sign of Multiplication** is $\times$, and reads "times" or "multiplied by," or simply "into." Placed between two quantities, it indicates that the one is to be multiplied by the other.

Thus, $a \times b$ reads "a multiplied by b," and means that a and b are to be multiplied together.

The multiplication of literal quantities (and sometimes of arithmetical numbers) may also be indicated more simply by a dot placed between the quantities. The multiplication of literal quantities is indicated most simply of all by the omission of any symbol between the quantities.

Thus, instead of $a \times b$, we may write $a \cdot b$, or ab.

12. Factors. The factors of a number are the numbers which multiplied together produce the given number.

For example, the factors of 14 are 7 and 2; the factors of abc are a, b, and c.

13. Coefficients. In case a numerical factor occurs in a product, it is written first, and is called a *coefficient.* Hence,

A **Coefficient** is a number prefixed to a given quantity to show how many times the given quantity is taken.

For example, in $5xy$, 5 is called the coefficient. When the coefficient is 1, the 1 is not written, but is understood.

Thus, xy means $1xy$.

When a number of factors are multiplied together, any one of them or the product of any number of them may be regarded as the coefficient.

Thus, in $5abx$, $5ab$ is sometimes regarded as the coefficient.

14. Powers and Exponents. A **Power** is the product of a number of equal factors.

The expression for a power is abbreviated by the use of the exponent.

An **Exponent** is a small figure or letter written above and to the right of a quantity to indicate how many times the quantity is taken as a factor.

Thus, for $xxxx$, or four x's multiplied together, we write x^4, the exponent in this case being 4.

An exponent is thus in effect a symbol of operation.

When the exponent is unity it is omitted. Thus, x is used instead of x^1, and means x to the first power.

15. The **Sign of Division** is $\div$, and reads "divided by." Placed between two quantities, it indicates that the quantity to the left of the sign is to be divided by the quantity to the right of it.

Thus, $a \div b$ means that a is to be divided by b.

Division may also be indicated by placing the quantity to be divided above a horizontal line and the divisor below the line. Thus, for $a \div b$ we may write $\frac{a}{b}$.

The expression $\frac{a}{b}$ is often read "a over b."

16. The **Radical Sign** is $\sqrt{}$, and means that the root of the quantity following it is to be extracted. The degree of the root is indicated by a small figure placed above the radical sign. For the square root the figure or index of the root is omitted.

Thus, $\sqrt{9}$ means "square root of 9."

$\sqrt[3]{a}$ means "cube root of a."

III. Symbols of Relation.

17. The **Sign of Equality** is $=$, and reads "equals" or "is equal to."

When placed between two quantities it indicates that they are equal to each other.

Thus, $a = b$ means that a and b are equal quantities.

18. The **Signs of Inequality** are $>$, which reads "is greater than," and $<$, which reads "is less than."

Thus, $a > b$ means that a is greater than b.

$c < b$ means that c is less than b.

It is to be observed that in both the signs of inequality the opening is toward the greater quantity.

19. The **Signs of Aggregation** are the parenthesis ( ), the brackets [ ], the braces { }, and the vinculum $\overline{\quad\quad}$. Any one of these indicates that all the quantities inclosed by it are to be treated as a single quantity; that is, subjected to the same operation.

Thus, $5(2a - b + c)$ means that all the quantities inside the parenthesis—viz. $2a$, $-b$, $+c$—are each to be multiplied by 5.

Again, $(a + 2b)(a + 2b + c)$ means that the sum of the quantities in the first parenthesis is to be multiplied by the sum of those in the second parenthesis.

The sign of aggregation ordinarily used is the parenthesis; the other symbols of aggregation are used in cases where confusion might result if several parentheses were used together.

20. The **Sign of Deduction** is $\therefore$, and is read "therefore" or "hence."

It is used to show the relation between succeeding propositions.

ALGEBRAIC EXPRESSIONS.

21. An **Algebraic Expression** is an algebraic symbol or combination of algebraic symbols, representing some quantity.

For example, $5x^2y - 6ab + 7\sqrt[3]{ax}$.

In algebra the terms "number," "quantity," and "algebraic expression" represent aspects of the same thing, and may be used interchangeably to express these aspects as occasion may require.

22. A **Term** is a part of an algebraic expression contained between a plus or minus sign and the next plus or minus sign (neither of the two plus or minus signs being inside a parenthesis).

Ex. 1. $5x^2y - 6ab + 7\sqrt[3]{ax}$.

This algebraic expression contains three terms—viz. $5x^2y$, $-6ab$, and $7\sqrt[3]{ax}$.

Ex. 2. $5x + a \div b + c$.

This expression also contains three terms, $5x$, $a \div b$, and c.

Ex. 3. $7ax^2 + 5(a + b) - c^3$.

Since the parenthesis, $(a + b)$, is treated as a single quantity, three terms occur in this expression also—viz. $7ax^2$, $5(a + b)$, $-c^3$.

23. A **Monomial** is an algebraic expression of only one term; as, $5x^2y$, or c.

24. A **Polynomial** is an algebraic expression containing more than one term.

Ex. $3ab - c + 2x + 5y^2$.

A monomial is sometimes called a *simple expression*, and a polynomial a *compound expression*.

25. A **Binomial** is an algebraic expression of two terms; as, $2a - 3b$.

A **Trinomial** is an algebraic expression of three terms; as, $2a - 3b + 5c$.

26. Positive and Negative Terms. Terms preceded by the plus sign are called *positive*, and those preceded by the minus sign are called *negative*. If no sign is written before a term (as at the beginning of an expression), the plus sign is understood.

In the expression $5x^2y - 6ab + 7$ the positive terms are $5x^2y$ and 7. The negative term is $-6ab$.

27. Similar and Dissimilar Terms. Similar (or like) terms are those which have the same literal factors with the same exponents (the coefficients and signs of similar terms may be unlike).

Ex. $7ab^2 - 5ab^2$ are similar terms.

Dissimilar (or unlike) terms are unlike either in their letters or the exponents of the letters.

Ex. $5ab^2$, $5ab^3$ are dissimilar terms.

28. Reading Algebraic Expressions. It is important at this point that the student acquire the power of reading algebraic expressions—*i. e.* of converting algebraic symbols into ordinary language, and conversely of converting language expressing relations of quantity into algebraic symbols.

Ex. 1. Read the algebraic expression, $5a^3 - 6(x + y) + 7c^2$.

Expressed in ordinary language, this expression becomes, "five times a cube, minus six times the quantity x plus y, plus seven times c square."

Ex. 2. Express in algebraic symbols the following: six times a cube, plus five times b square, minus three times the sum of a and b.

We obtain $6a^3 + 5b^2 - 3(a + b)$.

EXERCISE 1.

Read and copy the following algebraic expressions:

1. $9x - 11y^3$.
2. $2ab + 7b^2c$.
3. $(a+b)^2 - c^3$.
4. $3(x - 2y) - z^5$.
5. $2 + 3(x - 4)$.
6. $3x(x - 1) - x^8$.
7. $5x - 2x(1 + 3x)$.
8. $7x^3y - 8x^2y^2 + 9xy^3$.
9. $\frac{1}{x} + \frac{a - y}{3y} = 2(x^2 + ay)^2$.
10. $\frac{5 - 3(x^3 + 1)}{5(x^3 + 1) - 3} > \frac{2x}{y^2} - \frac{y}{3x^4}$.
11. $(x - 2a)(7x + 11a) < \frac{a^2 - (x^2 + 1)^2}{(a^2 - x^2 + 1)^2}$.
12. $\frac{(m + n)^5}{3(x^2 - y)^4} = 5(m^2 - 3x)^2(y^3 - 2n^2z)$.

Express in algebraic symbols—

13. Five times a plus seven times b.

14. Six a square equals twice the quantity a minus b.

15. Four times the quantity a square minus nine b is less than the square of the quantity seven a plus b cube.

16. The product of x minus ten y square, and x cube plus y times z, equals two a times x to the eighth power.

17. The quantity nine x plus two y divided by three z, is equal to nine x, plus two y divided by three z cube.

18. Five a cube, plus six b square over the square of the quantity x minus two y cube, is greater than five times the quantity a cube plus b square, over the cube of the quantity x plus two y to the fourth power.

19. Four x minus a fraction whose numerator is x plus y square, and whose denominator is the square of the quantity x plus y, equals one, minus x square over y.

29. The **Numerical Value** of an algebraic expression is obtained by substituting for each letter in the expression the value which it represents, and performing the operations indicated.

Thus, if $a=1$, $b=2$, $c=3$.

Ex. 1. Find the numerical value of $7ab-c^2$.

$$\begin{aligned} 7ab-c^2 &= 7\times1\times2-3^2\\ &= 14-9\\ &= 5, \textit{ Ans.}\end{aligned}$$

Ex. 2. Find numerical value of $\frac{9b}{c}+5ab^2-7(a^3+2b)^2+3c^2$.

$$\begin{aligned} \text{We obtain} &= \frac{9\times2}{3}+5\times1\times2^2-7(1^3+2\times2)^2+3\times3^2\\ &= 6+20-175+27\\ &= -122, \textit{ Ans.}\end{aligned}$$

EXERCISE 2.

Find the numerical value of each of the following when $a=5$, $b=3$, $c=1$, $x=6$:

1. $2a$.
2. $3x$.
3. ax.
4. $3ac$.
5. b^2.
6. a^2c.
7. c^2x.
8. $3ab^2$.
9. a^3b.
10. $7abc^2$.
11. $a+2c$.
12. $3b-x$.
13. x^2-ax.
14. $2x-4bc$.
15. a^3-bx^2.
16. $3(a+c)$.
17. $x(a-b)$.
18. $2b(x-c^2)$.
19. $4(a-3c)^2$.
20. $2x(2a-3b)^2$.
21. $3+2(x-a)$.
22. $5x-3(2b+c)$.
23. $2(x^2-a^2)+3ac$.
24. $7x(5a-4x)-cx^2$.
25. $3x(x-3)^2-9x$.
26. $(x-1)(x-3)+x(x-a)$.
27. $3(2x-5c)-a(2b^2-3x)$.
28. $a(3x-ab)^2+x(a^2b^2-x^2)$.
29. $(5b+x)(x-b+a-5c^2)$.
30. $3x(x-a)(x^2-ab^2+2ac^2)$.
31. $a^2b^2c-b^2c^2x+c^2x^2-(2x+a-b)^2$.

If $a=\frac{1}{2}$, $b=\frac{2}{3}$, $x=2$, $y=\frac{3}{5}$, find values of—

32. $6a$.
33. by.
34. abx.
35. a^2x^2.
36. $3ab^2$.
37. $x-2b$.
38. $2x+5y$.
39. $6ab-b^2y$.

40. $b(10y-3b)$.

41. $3x(4a+3b)$.

42. $ax+5x(3b-y)$.

43. $3a+a(3x-10y)$.

44. $5x-3(by-ab)$.

45. $5x(by-a^2)-bx$.

46. $6(a+b)^2+10(y-a)^2$.

If $a=4$, $b=\frac{3}{2}$, $c=0$, $x=1$, $y=9$, find values of—

47. $\sqrt{y}$.

48. $\sqrt[3]{2ax}$.

49. $\sqrt{ax^2y}$.

50. $\sqrt[3]{a^2by}$.

51. $2\sqrt{ab^2y}$.

52. $3b\sqrt{6b^3x^3}$.

53. $\sqrt{a^2+y}$.

54. $x\sqrt{7x^2-ab}$.

55. $ab\sqrt[3]{y^2-a^2-x}$.

56. $3cx\sqrt{a^2+bc^2}$.

57. $ab\sqrt[3]{2a^3-2bx^3}$.

58. $a+\sqrt{x^2y^2}$.

59. $3a-\sqrt{4b^2+3y}$.

60. $5x+x\sqrt{9x^2+a^2}$.

61. $3x\sqrt{aby-5x}-b^2c^2$.

62. $(\sqrt{7c+4y})(\sqrt{11a+5x})$.

63. $\sqrt{ab^2+bc^2+cx^2+xa^2}$.

64. $(a-2b)\sqrt{a^3+4x^2y}$.

Find the numerical value of—

65. $(x+3)(2x-1)-5x$ when $x=2$.

66. $3(2x-1)+6x+7$ when $x=3$.

67. $3(6x+1)^2+2(x-3)-5x$ when $x=4$.

68. $8x(2x+1)-3(x+1)^2$ when $x=1$.

69. $3x(2x+1)^2+5x-(x+2)^2$ when $x=2$.

70. $2x(x+3)(x+2)-(x+4)^2$ when $x=5$.

71. $8x^3-5x(x+2)^2-6x^2$ when $x=3$.

72. $3x(x+1)^2-6x-7(x+2)$ when $x=4$.

73. $7x+5(4x+1)-3x(x+2)^2$ when $x=1$.

74. $3x^2(x+1)^2-(x+2)^2(x-1)^2$ when $x=2$.

75. $(3x+y)^3-2(x-2y)(x+2y)+5(2x-3y)^3$ when $x=3$, $y=2$.

CHAPTER II.

METHODS OF USING ALGEBRAIC SYMBOLS.

30. Second Source of New Power in Algebra. The second general source of new power in algebra lies in certain standard ways or methods in which the symbols of algebra are used. These ways are termed the Laws of Algebra. There are two divisions of the primary laws of algebra:

I. Laws for + and − Signs.

II. Laws for Grouping and Arrangement of Symbols of Quantity.

I. Laws for + and − Signs.

31. First Law for + and − taken together. As was explained in Arts. 8, 9, 10, the signs + and − are employed for two purposes—first, to express positive and negative quantity; and second, to indicate the operations of addition and subtraction. We are able to put these signs to this double use because, as used in both of these ways, the signs are governed by the same laws.

Thus, if the distance to the right of O be regarded as positive, and therefore the distance to the left of O as negative,

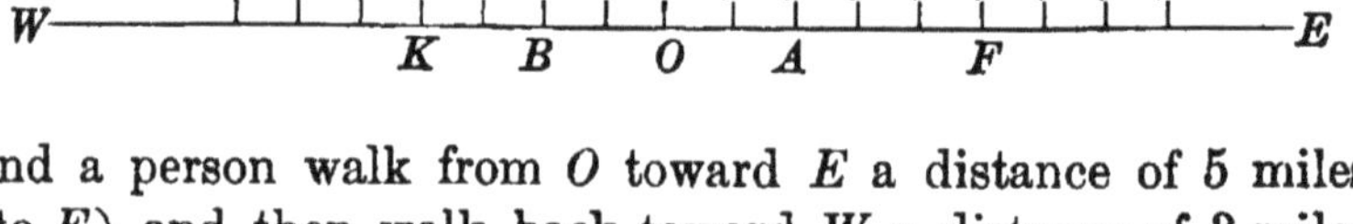

−7 −6 −5 −4 −3 −2 −1 +1 +2 +3 +4 +5 +6 +7 +8
W K B O A F E

and a person walk from O toward E a distance of 5 miles (to F), and then walk back toward W a distance of 3 miles (to A), the distance travelled by him may be expressed as the sum of a positive quantity and a negative quantity; that is,

$$(\text{positive distance } OF) + (\text{negative distance } FA),$$

$$\text{or, } +5 + (-3) = 5 - 3 = 2.$$

The position arrived at may also be determined in another way—viz. by deducting (that is, using the operation of subtraction) 3 miles from 5 miles. We obtain

$$5 - (+3) = 5 - 3 = 2.$$

Hence, we see that adding negative quantity is the same in effect as subtracting positive quantity; therefore, in the expression

$$5 - 3$$

the minus sign used may be considered either a sign of the *quality* of 3, or as a sign of *operation* to be performed on 3.

Hence, we are able to use the signs + and − to cover two meanings, as stated above.

Whichever of these two meanings be assigned, we see that $+(-3) = -3$; also, $-(+3) = -3$. Hence,

LAW 1. *The signs + and − applied in succession to a quantity are equivalent to the single sign −.*

Or in symbols,

$$+(-a) = -a; \qquad -(+a) = -a.$$

32. Second Law for + and − taken together. Again, if in the above illustration a person walk in the negative direction from O—*i. e.* toward W—a distance of 4 miles to K, and then reverse his direction and go 2 miles, he will be at B; or the distance travelled is expressed as

$$-4 - (-2) = -4 + 2 = -2;$$

that is, the two minus signs in $-(-2)$ taken together give +.

So also we see that deducting a certain sum from a man's debts is the same in effect as adding this sum to his assets;

or, in general, that a double reversal of the quality of any quantity gives the original quality. Hence,

LAW 2. *The sign – applied twice to a given positive quantity gives a + result.*

Or in symbols, $-(-a) = +a$.

It is also evident that $+(+a) = +a$.

By these laws any succession of + and – signs applied to a quantity can be at once reduced to a single + or – sign.

Ex. $-[+(-a)] = -[-a] = +a$.

These laws therefore enable us to use negative quantity with as great freedom as we use positive quantity, and hence are an important source of power. They also open the way to a free use of the second group of the laws of algebra.

II. Laws of Arrangement and Grouping.

33. Formal Statement of Laws of Arrangement. The laws which govern the arrangement and grouping of the symbols for quantity in algebra are—

A. **The Commutative Law.**

1. *For Addition*, $a + b = b + a$.
2. *For Multiplication*, $ab = ba$.
3. *For Division*, $a \div b \times c = a \times c \div b$.

B. **The Associative Law.**

1. *For Addition*, $a + b + c = a + (b + c) = (a + b) + c$.
2. *For Multiplication*, $abc = a(bc) = (ab)c$.

C. **The Distributive Law.**

1. *For Multiplication*, $a(b + c) = ab + ac$. Hence, inversely, $ab + ac = a(b + c)$.
2. *For Division*, $\frac{b + c}{a} = \frac{b}{a} + \frac{c}{a}$.

34. Meaning of Commutative Law for Addition. The meaning of these laws is best shown by examples. If it is required to combine a group of 7 objects and another group of 5 objects into a single group, we may either count the 7 objects first, and then count on the 5 objects afterward, or the 5 objects first and the 7 afterward; that is, groups of objects may be counted together into a single group in any order we please. The algebraic symbols representing different groups in like manner may be arranged in any order we please; that is, briefly,

$$a + b = b + a.$$

An example of the advantage in this quality of algebraic symbols is that similar terms in an expression may, by rearranging the terms, be brought together, and then counted into a single term by the use of the Distributive Law for Multiplication (inverse form).

Thus, the terms of the algebraic expression,

$$7a^2b - 5xy^3 + 6xy^3 + 3a^2b + 4xy^3 - 2a^2b,$$

by the use of the Commutative Law may be arranged thus,

$$7a^2b + 3a^2b - 2a^2b - 5xy^3 + 6xy^3 + 4xy^3.$$

By the Distributive Law for Multiplication (inverse form) the first three terms may be combined into a single term, and the last three into another term, giving

$$8a^2b + 5xy^3.$$

Thus, by use of these laws 6 terms are reduced to 2 terms.

The symbols used to represent number in arithmetic cannot be changed about in this manner. Thus, the number 234 cannot be written 324. If, however, we employ the + sign, the symbols used for the number may be put in a commutative form; as,

$$\begin{aligned} 234 &= 200 + 30 + 4 \\ &= 30 + 4 + 200 \\ &= \text{etc.} \end{aligned}$$

The arithmetical form, 234, has the advantage of greater brevity than the algebraic form, 200 + 30 + 4, but the disadvantage of less flexibility.

35. Meaning of Commutative Law for Multiplication.

To illustrate the Commutative Law for Multiplication, we recognize that if we have 15 objects, the number of the objects is the same whether they be arranged in 5 rows of 3 objects each or 3 rows of 5 objects each—

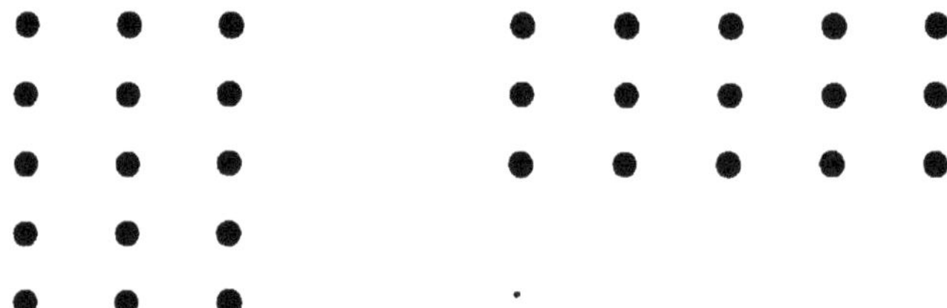

So, in general,

$$ab = ba.$$

The advantages resulting from this property of algebraic symbols are illustrated by the fact that we are enabled by it to have a standard order for the arrangement of the literal factors in a term—viz. the alphabetical order. Thus, instead of writing $7c^2xa$, or $7axc^2$, or $7xac^2$, we write the literal factors in the alphabetical order, $7ac^2x$.

When the factors in each term of an expression are thus arranged, it is much easier to recognize similar terms.

36. Meaning of the Distributive Law for Multiplication. Let it be required to reduce the expression,

$$5(a - b + c) + 2(a + b - c) + 3(a + b + c),$$

to its simplest form.

Applying the Distributive Law, the expression becomes

$$5a - 5b + 5c + 2a + 2b - 2c + 3a + 3b + 3c.$$

Using the Commutative Law, we obtain

$$5a + 2a + 3a - 5b + 2b + 3b + 5c - 2c + 3c.$$

Hence, by the Distributive Law,

$$10a + 6c.$$

In other cases the Distributive Law enables us to perform work, part by part, which would be difficult if not impossible in the undivided form.

In general, therefore, these laws enable us to arrange and group the parts of an algebraic expression to the best advantage according to the work to be done. They are, therefore, to be considered, from one standpoint, as economic methods which govern the use of algebraic symbols.

It will be a useful exercise for the student to determine which of the fundamental laws for grouping and arranging algebraic symbols are used in the following illustrative examples:

Ex. 1. $6(x+y)+3(x-y+z)+2(x+2y-z).$

$$= 6x+6y+3x-3y+3z+2x+4y-2z$$
$$= 6x+3x+2x+6y-3y+4y+3z-2z$$
$$= 11x+7y+z$$

Ex. 2. $\dfrac{12a^3b^3+9a^3b^2+6a^2b^3}{3ab}.$

$$= \frac{12a^3b^3}{3ab}+\frac{9a^3b^2}{3ab}+\frac{6a^2b^3}{3ab}$$
$$= 4a^2b^2+3a^2b+2ab^2$$
$$= ab(4ab+3a+2b)$$

Ex. 3. $(2x+3y)(3x+4y).$

$$= 2x(3x+4y)+3y(3x+4y)$$
$$= 6x^2+8xy+9xy+12y^2$$
$$= 6x^2+17xy+12y^2$$

CHAPTER III.

ADDITION AND SUBTRACTION.

ADDITION.

37. Addition, in algebra, is the combination of several algebraic expressions, representing numbers, into a single equivalent expression.

38. Addition of Similar Terms. If the question be asked, How many books are

$$3 \text{ books} + 7 \text{ books} + 4 \text{ books?}$$

the answer is, 14 books.

In like manner, if the question be asked, How many a^2b^3's are

$$3a^2b^3 + 7a^2b^3 + 4a^2b^3?$$

the answer is, $14a^2b^3$.

The simplification is obtained by the use of the Distributive Law for Multiplication (Art. 36).

Thus, similar terms are added by adding the coefficients of the terms and setting the result before the literal part common to the terms.

If some of the similar terms are negative, the sum of the coefficients is taken, respect being had to their signs. The sum thus taken is called the *algebraic sum* of the coefficients.

For example, add the similar terms

$$8a^2x - 7a^2x - 6a^2x + 10a^2x - a^2x.$$

The sum of the plus coefficients is $+18$, the sum of the negative coefficients is -14, the algebraic sum of $+18-14$ is $+4$; hence, the sum of all the given similar terms is $+4a^2x$.

39. Addition of Dissimilar Terms. If the terms to be added are dissimilar, the addition of them can be indicated only.

Thus, b added to a gives $a + b$; also, $a^3 - 3a^2b$ added to $3a^2 - b^3$ gives $a^3 - 3a^2b + 3a^2 - b^3$.

Simplifications are possible only where there are similar terms.

40. General Method of Addition. The most convenient general method for addition is shown in the following examples:

Ex. 1. Add $4x^2 + 3x + 2$, $3x^2 - 4x - 3$, $-2x^2 - x - 5$.

Arranging similar terms in the same column, and adding each column separately, we obtain

$$\begin{array}{rrr} 4x^2 & +\ 3x & +\ 2 \\ 3x^2 & -\ 4x & -\ 3 \\ -\ 2x^2 & -\ \ x & -\ 5 \\ \hline 5x^2 & -\ 2x & -\ 6, \end{array} \text{ \textit{Sum.}}$$

Ex. 2. Add $2a^3 - 5a^2b + 4ab^2 + a^2b^3$, $4a^2b + 2a^3 - ab^4 - 3ab^2$, $a^2b - a^3 + 2ab^2$.

Proceeding as in Ex. 1,

$$\begin{array}{rrrrr} 2a^3 & -\ 5a^2b & +\ 4ab^2 & +\ a^2b^3 & \\ 2a^3 & +\ 4a^2b & -\ 3ab^2 & & -\ ab^4 \\ -\ a^3 & +\ \ a^2b & +\ 2ab^2 & & \\ \hline 3a^3 & & +\ 3ab^2 & +\ a^2b^3 & -\ ab^4, \end{array} \text{ \textit{Sum.}}$$

In the second column the algebraic sum of the coefficients is $-5 + 4 + 1$, which $= 0$; and as zero times a number is zero, the sum of the second column is zero, which need not be set down in the result.

Hence, the general process for addition may be stated as follows:

Arrange the terms to be added in columns, similar terms in the same column; in each column take the sum of the + coefficients, and also the sum of the − coefficients;

Subtract the less sum from the greater, prefix the sign of the greater, and annex the common letters with their exponents.

41. Collecting Terms. It is often required to add together the similar terms which occur in a single polynomial. This is called *collecting terms.*

Ex. Simplify $2x + 7ab + 5 - 3ab + 2ab + 3x$

Collecting terms, we obtain $5x + 5 + 6ab$.

EXERCISE 3.

Find the sum of—

1.	2.	3.	4.	5.
-11	4	$8x$	$-x$	$-7x$
6	-10	$-6x$	$-3x$	$12x$

6.	7.	8.	9.	10.
$2a$	$-x^2$	$7xy$	a^2b	$7x^2y^2$
$5a$	$3x^2$	$-10xy$	$5a^2b$	$-10x^2y^2$
$-12a$	$5x^2$	$2xy$	$-3a^2b$	x^2y^2

11. $3ax$, $-2ax$, $5ax$, ax, $-3ax$.

12. $5x^2$, $12x^2$, $-10x^2$, x^2, $-16x^2$, $3x^2$, $-x^2$.

13. $7a^2b^2$, $-12a^2b^2$, $-a^2b^2$, $-4a^2b^2$, $5a^2b^2$, $6a^2b^2$.

14.	15.	16.
$3x - 2y$	$5x^2 + 7$	$a^2 - ax + 4x^2$
$-2x + 3y$	$x^2 - 10$	$3a^2 + 2ax - 5x^2$
$x - y$	$-7x^2 + 1$	$-a^2 - ax - x^2$

17. $a - 2b$, $3a + 4b$, $a + 5b$, $-5a - b$, $a - 5b$.

18. $3x^2 + y^2$, $2x^2 - 7y^2$, $-4x^2 - 5y^2$, $x^2 + 3y^2$, $-3y^2$.

19. $3ax^3 - 5by^3$, $2ax^3 + 4by^3$, $2by^3 - 4ax^3$, $by^3 - ax^3$.

20. $x^2 - xy + 3y^2,\ 2x^2 + 2xy - 2y^2,\ x^2 + y^2,\ 3x^2 - xy.$

21. $mn - 3n^2 + m^2,\ m^2 + 2n^2 - 3mn,\ m^2 - n^2,\ mn - 2m^2.$

22. $x^2 + y^2 - 2z^2,\ 3x^2 - y^2 + 2z^2,\ z^2 - 2x^2,\ x^2 - z^2.$

23. $2x^2 - xy,\ 3xy - 5y^2,\ 3y^2 - 3x^2,\ x^2 + 2y^2 - 2xy.$

24. $7x + y + 5z - 10xy,\ 2y - 3z + 13xy - 4xz,\ 5z - 6x - 4xz + 2xy,\ -3y + 9z + 7x - xz,\ 21xz - 16z + x - 5xy.$

25. $x^3 + 3x^2y + 3xy^2 + y^3,\ x^3 - 3x^2y + 3xy^2 - y^3,\ 2x^2y - 2xy^2 + y^3,\ x^3 - y^3,\ x^2y - 4x^3 - xy^2 - y^3,\ y^3 + x^3 - x^2y + xy^2.$

Collect similar terms in the following:

26. $2x - 3y - 5x + 4z + 4y + z - 2y - x - 3z + 2x - 3y.$

27. $3xy - 5ax + 3y^2 - 2xy - 3x^2 + 4ax - 2y^2 + 3ax - 2xy.$

28. $x - 3y + 2z + 2y - 2x - z - 3x - 4z - 2x + z + 2x.$

29. $2x - 1 + 5y - 2 + 3x + 2 + 3y - 3 - 2x + 1 - x - 3y.$

30. $3a^2b - 2a^2c + 3a^3 - 5a^2b - a^3 - 3a^2c + a^2b + 6a^2c - 2a^3.$

31. $5x^3 - 3x + 4 - 2x^2 - 6x^3 + 4x - 7 - x^2 + x^3 + 3x^2 - x + 5 + 3x^2 - 6x - x^2 + 4x - 2x^2 + 2x.$

32. $2x^n - 5x^m + 3x^2 - x^n - 7x + 3x^2 - 3 + 2x^m - 5x^2 + 5 + 3x^m$

SUBTRACTION.

42. Subtraction, in arithmetic, is the process of finding the difference between two numbers, and subtraction in algebra includes this work. But inasmuch as negative quantity is dealt with in algebra as well as positive quantity, the word difference takes a broader meaning, and we need a broader definition of subtraction which will cover both positive and negative quantity.

If the quantity to be subtracted be named the **Subtrahend,** and the quantity from which the subtrahend is taken be named the **Minuend,** and the result obtained be named the **Difference,** it is evident that for both positive and negative quantity, the Difference added to the Subtrahend will give the Minuend. Hence,

Subtraction, in algebra, is the process of finding a quantity which, added to a given quantity (the subtrahend), will produce another given quantity (the minuend).

Thus, if we subtract $3ab$ from $10ab$, we obtain $7ab$, for $7ab$ added to $3ab$ (subtrahend) gives $10ab$ (minuend).

43. Signs in Subtraction. From Art. 31 it is clear that subtracting a positive quantity is the same as adding a negative quantity of the same absolute magnitude; and from Art. 32, that subtracting a negative quantity is the same as adding a positive quantity of the same absolute magnitude.

Hence, in subtraction, the most convenient way to govern the signs is to change (mentally) the signs of the terms in the subtrahend.

$$\begin{array}{r} 7b \\ 4b \\ \hline 3b \end{array}$$

Thus, to subtract $4b$ from $7b$, we change $4b$ mentally to $-4b$, and add $+7b$ and $-4b$, and obtain the result $3b$.

Again, to subtract $-2b$ from $7b$, we mentally change the sign of $-2b$, and add the result $+2b$ to $+7b$, and obtain $+9b$.

$$\begin{array}{r} 7b \\ -2b \\ \hline +9b \end{array}$$

(Some concrete illustration of the reason for changing the sign of a negative term of the subtrahend to plus in the process of subtraction should be frequently recalled by the student. For example, subtracting a \$10 debt from a man's possessions is the same in effect as adding \$10 to them.)

44. General Method for Subtraction. Accordingly, the most convenient general method in subtraction is to—

Place the terms of the subtrahend under the terms of the minuend, similar terms in the same column.

Change the signs of the terms in the subtrahend mentally; proceed as in addition.

Ex. 1. From $5x^3-2x^2+x-3$ subtract $2x^3-3x^2-x+2$.

We obtain

$$\begin{array}{r} 5x^3-2x^2+\ \ x-3 \\ \underline{2x^3-3x^2-\ \ x+2} \\ 3x^3+\ \ x^2+2x-5, \end{array} \textit{ Difference,}$$

since the coefficient of x^3 is $5-2$, or 3, of x^2 is $-2+3$, or 1, etc.

Ex. 2. Subtract $2a^4-3a^3b-6a^2b^2-2ab^3+2b^4$ from $a^4+5a^3b-6a^2b^2-3ab^3$.

We obtain

$$\begin{array}{r} a^4+5a^3b-6a^2b^2-3ab^3\qquad \\ \underline{2a^4-3a^3b-6a^2b^2-2ab^3+2b^4} \\ -a^4+8a^3b\qquad\qquad\ -\ \ ab^3-2b^4 \end{array}$$

The coefficient of a^2b^2 is $-6+6$, or 0. The coefficient of b^4 is $0-2$, or -2.

EXERCISE 4.

	1.	2.	3.	4.	5.	6.
From	$7ab$	$5x$	x	$5x$	$-3x^2$	$-7xy$
Take	$3ab$	$9x$	$2x$	$-3x$	$-4x^2$	$3xy$

	7.	8.	9.	10.
From	$3x^2-4x$	$3x-9$	$2x^3-5$	$5x^2+4x-3$
Take	$2x^2+\ x$	$5x+1$	$-x^3+2$	$-x^2-3x+5$

11. From $3a+2b-3c-d$ take $2a-2b+c-2d$.
12. From $7-3x+2x^2$ take $15-4x-5x^2$.
13. From $x^2-y^2-z^2+8$ take $2x^2+y^2-2z^2+10$.
14. From $5xy-3xz+5yz+x^2$ take $4xz-2xy-x^2$.
15. From $2-x+x^2+x^4$ take $3+x-x^2-x^3-2x^4$.
16. Subtract $10x^2y+3x^2y^2-13xy^2$ from $x^2y-xy^2+2x^2y^2$.
17. Subtract $3-2ab+3ac-4cd$ from $5-ac+8cd-5ad$.
18. Subtract $1+x-x^2+x^3-x^4$ from $2-x-x^2-x^3+x^5$.
19. Subtract $a+2b-3c+4d$ from $m+2b+d-x+a$.

20. Subtract $3x^4 - 2x^2 + 5x - 7$ from $3x^3 + 2x^2 - x - 7$.

21. Subtract $-x^5 - 2x^4 + x^2 + 5$ from $x^5 - x^3 + x^2 - 2x + 5$.

22. Subtract $3x^m - 3x^n + x - 3$ from $x^m + x^n - x^2 + x - 1$.

If $A = x^3 - 3x^2 + 1$, $B = 2x^3 - 5x - 3$, $C = 3x^3 + x^2 + 3x$, find the values of—

23. $A + B + C$.

24. $B - A + C$.

25. $A + B - C$.

26. $A - B + C$.

USE OF PARENTHESIS.

45. I. Removal of Parenthesis. Addition and subtraction may be indicated briefly by the use of the parenthesis.

Thus, the expression

$$2a + 3b - 5c + (3a - 2b + 3c)$$

indicates that $3a - 2b + 3c$ is to be added to $2a + 3b - 5c$.

The process of addition thus indicated by the parenthesis may be performed in the usual way by placing similar terms in the same column, etc. But expressions like the above occur so frequently in algebra that it is found more convenient to simplify them simply by setting down the terms to be added in succession (omitting the parenthesis) and collecting similar terms.

In accordance with this method we obtain,

$$2a + 3b - 5c + 3a - 2b + 3c$$
$$= 5a + b - 2c.$$

Similarly, the expression

$$2a + 3b - 5c - (3a - 2b + 3c)$$

indicates that $3a - 2b + 3c$ is to be subtracted from $2a + 3b - 5c$.

The most convenient way of making the subtraction is to

change the signs of terms of the subtrahend (dropping the parenthesis which contains them) and to collect terms.

Accordingly we obtain

$$2a + 3b - 5c - 3a + 2b - 3c$$
$$= -a + 5b - 8c.$$

Addition or subtraction performed in this way is called *removing a parenthesis.* The special rule to be observed in removing a parenthesis is that—

When a parenthesis preceded by a + sign is removed, the signs of the terms inclosed by the parenthesis remain unchanged. But—

When a parenthesis preceded by a minus sign is removed, the signs of the terms inclosed by the parenthesis are changed, the + signs to −, and the − signs to +.

46. The **Sign of the First Term within a Parenthesis** is usually + understood, it being the custom to put a plus term first in an algebraic expression if possible. Owing to the absence of this + sign, the beginner frequently makes the mistake of using the sign of the parenthesis as the sign of the first term within it. This error may be obviated at first by writing out the sign of the first term in the parenthesis in full, till the fact of its existence is firmly realized.

$$\text{Thus, } 5a - (+3a - b)$$

plainly reduces to $5a - 3a + b$, the $-3a$ being obtained by changing the + sign before $3a$ to −. This is equally true when the + sign is understood, as in $5a - (3a - b)$

$$= 5a - 3a + b = 2a + b.$$

In both cases the minus sign before the parenthesis belongs to the parenthesis, indicates subtraction, and disappears with the parenthesis.

47. Parenthesis within Parenthesis. Using the parenthesis as a general name for the signs of aggregation, as brace, bracket, vinculum, it is evident that several parentheses may occur one within another in the same algebraic expression. The best general method of removing several parentheses occurring thus, is as follows:

Remove the parentheses one at a time, beginning with the innermost;

On removing a parenthesis preceded by a minus sign, change the signs of the terms inclosed by the parenthesis;

Collect the terms of the result.

Ex. Simplify $5x - y - [4x - 6y + \{-3x + y + 2z - (2x - z)\}]$.

$$\begin{aligned} &= 5x - y - [4x - 6y + \{-3x + y + 2z - 2x + z\}] \\ &= 5x - y - [4x - 6y - 3x + y + 2z - 2x + z] \\ &= 5x - y - 4x + 6y + 3x - y - 2z + 2x - z \\ &= 6x + 4y - 3z. \end{aligned}$$

EXERCISE 5.

Remove parentheses and collect similar terms:

1. $3a + (2a - b)$.
2. $2x - (x - 1)$.
3. $x + (1 - 2x)$.
4. $3x - (1 + 3x)$.
5. $x - (-x - 1)$.
6. $x + 2y - (2x - y)$.
7. $x - [2x + (x - 1)]$.
8. $5x + (1 - [2 - 4x])$.
9. $2 - \{1 - (3 - a) - a\}$.
10. $2x - [-x - (x - 1)]$.
11. $2y + \{-x - (2y - x)\}$.
12. $a - \{-a - (-a - 1)\}$.
13. $[x^2 - (x^2y - z^2) - z^2] + (x^2y - x^2)$.
14. $1 - \{1 - [1 - (1 - x) - 1] - 1\} - x$.
15. $x - [-\{-(-x - 1) - x\} - 1] - 1$.
16. $1 - \{2 + [-3 - (-4 - \overline{5 - 6}) - 7]\}$.
17. $a - \{a + [b - (a + b + c - \overline{a + b + d}) - c]\}$.
18. $x - \{2x^2 + (3x^3 - 3x - [x + x^2]) + [2x - (x^2 + x^3)]\}$.
19. $x^4 - [4x^3 - [3x^2 - (2x + 2)] + 3x] - [x^4 + (3x^3 + 2x^2 - 3x - 1)]$.

20. $x - \overline{x - y} - \{-x - [-(x - y) - (x + y) - x] - (x - y)\}$.

21. $-[-2x - \{-(-2x - 1) - 2x\} - 1] - 2x$.

22. $x - [x + (x - y) - \{x + (y - x) - 2y\} + y] - y + x$.

23. $25x - [12 + \{3x - 7 - (-12x - \overline{5 + 15x}) - (3 + 2x)\}] + 7 - (3x + 5) + (2x - 3) + x + 8$.

48. II. Insertion of Parenthesis. It is plain that the process of removing a parenthesis may be reversed; that is, that terms may be inclosed in a parenthesis.

Inverting the statements of Art. 45,

Terms may be inclosed in a parenthesis preceded by the plus sign, provided the signs of the terms remain unchanged;

Terms may be inclosed in a parenthesis preceded by the minus sign, provided the signs of the terms be changed.

Ex. $a - b + c + d - e = a - b + (c + d - e)$,
or, $= a - b - (-c - d + e)$.

EXERCISE 6.

In each of the following insert a parenthesis, inclosing the last three terms; each parenthesis to be preceded by a minus sign:

1. $x^3 - 3x^2 + 3x - 1$.
2. $a - b + c + d$.
3. $1 + 2a - a^2 - 1$.
4. $1 - a^2 - 2ab - b^2$.
5. $x^4 + 4x - x^2 - 4$.
6. $a^2b^2 - 2cd - c^2 - d^2$.
7. $4x^4 - 9x^2 + 12xy - 4y^2$.
8. $a^2 - 2a + 1 - 9 - 6x - x^2$.
9. $x^4 - 4x^3 + 4x^2 + 4x - 4 - x^2$.

It is often useful to collect the coefficients of a letter into a single coefficient.

Let it be required to collect the coefficients of x, y, and z in the expression,

$$3x - 4y + 5z - ax - by - cz - bx + ay + az.$$

The complete coefficient of x is $(3 - a - b)$; of y, $(-4 - b + a)$ or $-(4 + b - a)$; of z, $(5 - c + a)$.

Hence, the same expression may be written,

$$(3 - a - b)x - (4 + b - a)y + (5 - c + a)z.$$

In like manner collect the coefficients of x, y, and z—

10. $mx - ny + 3z + 2x + nz - 4y$.
11. $x - y - 2z - ax + by - az - bx - ay + cz$.
12. $-7x + 12y - 10z - 2ax + 3bz - cy + 2bx - 6dy$.
13. $abx - bcy - cdz + acx - ady - acz - aby + adz$.
14. $5y - 3acx - 5cdz - 4abx - 3cdy + 2cx - 4z - 5ax$.

Collect coefficients of x^3, x^2, and x—

15. $3x^3 + x - 2x^2 - ax^3 - 5 + ax^2 - 2ax - cx^3 - cx^2 - cx$.
16. $-x^2 - x - ax^2 + x^3 - ax + bx^2 - ax^2 - 3bx - 2bx^3 + 3a$.
17. $a^2x^3 - ax - a - b^3x^3 - 2b^2x^2 + 3bx - a^3x^3 - cx^2 + 3cx - c$.

EXERCISE 7.

SPECIAL REVIEW.

Add—

1. $2x^4 - 5x^3 - 3x^2 + 2x - 5$, $2x^3 - 3x^4 - 2x + 2x^2 - 6$, $3x^2 + x^4 - 3x^3 + 7 - x$, and $2 + 3x^3 + 2x^4 - 4x - 2x^2$.

2. $5xyz + 3x^2yz - 3xy^2z - 3xyz^2$, $5xy^2z - 3x^2yz - 4xyz$, and $7xyz^2 - xyz - x^2yz + xy^2z$.

3. $3\sqrt{2} - 5\sqrt{3} + 8$, $5\sqrt{3} - 2\sqrt{2} - 7$, $3\sqrt{3} - 4\sqrt{2} - 2$.

4. $2(x + y) - 3(x + z) + 2(y + z)$, $4(x + z) - 3(x + y) - 5(y + z)$, and $4(x + y) - (x + z) + 4(y + z)$.

Subtract—

5. $2ab - 3bc + d$ from $1 - 3ab - bc + x$.
6. $c - d^2 + x - 10y$ from $3x - a + c$.

7. $19ab - c - 4x + \sqrt{y}$ from $12ab - 3c + c^2 - \sqrt{y}$.

8. $3 - 2\sqrt{x} + 5x - x^2 - x^4$ from $2\sqrt{x} + x^3 - 1$.

Find value of—

9. $3x - (x-2)^2 + 2(x+1)(4-x) - \sqrt{5x+1}$, when $x = 3$.

10. $6x^2 - 3x(x + \frac{1}{3}) + \sqrt{3x^2 - 5x + 2}$, when $x = 1$. When $x = \frac{2}{3}$.

11. $6x - 2(4x^2 - 2x - 5) + x(x + \frac{1}{2})(5 - 2x)$, when $x = 2$. When $x = \frac{3}{2}$.

Simplify and collect—

12. $3x - \{-2x + [-4x - (x-2) - x] - x\} - 1$.

13. $9x - \{-8x - [7x + (-6x + 1) - 5x] - 4x\} - (3x + 1) - 2x$.

14. $x^2 - (y^2 - x^2) - [(x^2 + z^2) - \{(x^2 - z^2) + (y^2 - z^2) - (x^2 + z^2)\} - x^2]$.

Bracket coefficients of like powers of x—

15. $x^5 - x^3 + 2 - 3x^4 - ax^3 + ax^5 - cx^4 - 2ax^2 + 3cx^3 - 2cx^5 - 5x^2$.

16. $1 - x - x^2 - x^3 + 2a - 2ax + 2ax^2 - 2ax^3 - 3bx + 3bx^2 + 3bx^3 + cx$.

17. From the sum of $a^2 - 7ab + 3b^2$ and $2a^2 - 6b^2 + 7a^2b^2$, take the sum of $4a^2b^2 - 3a^3 + 2a^2 - b^2$ and $3ab - 2b^2 + a^2$.

18. What must be added to $x^2 - x + 1$ that the sum may be x^3? That the sum may be $3x$? 15? 0?

19. What must be subtracted from $2x^2 - 3x + 1$ that the remainder may be x^3? $x^2 + 10$? 7? $a - x + 1$?

If $A = 4x^3 - 2x^2y + 3xy^2 + y^3$, $C = 3x^3 - x^2y + 2y^3$,
$B = 4x^3 - x^2y - xy^2 - 3y^3$, $D = x^3 - 2xy^2 + y^3$.

Find the values of—

20. $A - B + C - D$.

21. $A - [B - (D + C)]$.

22. $A - (B + C) + D$.

23. $B + \{A - [C - D]\}$.

24. $B - \{-A - [-B - (-C) - D] - C\} - (C - B)$.

CHAPTER IV.

MULTIPLICATION.

49. Multiplication is the process of finding the result of taking one quantity as many times as there are units in another quantity.

The **Multiplicand** is the quantity to be multiplied.

The **Multiplier** is the quantity showing how many times the multiplicand is to be taken.

The **Product** is the result of the multiplication. By definition of "factors" in Art. 11 it is seen that the multiplier and multiplicand are factors of the product.

Thus, if x is the multiplicand and y the multiplier, the product is xy, and the factors of xy are x and y.

MULTIPLICATION OF MONOMIALS.

50. Multiplication of Coefficients. To multiply $4a$ by $3b$, we evidently take the product of all the factors of the multiplier and multiplicand, and thus get $4 \times a \times 3 \times b$, or, rearranging factors as we are enabled to do by the Commutative Law,

$$4 \times 3 \times a \times b = 12ab.$$

Hence, in multiplying two monomials we multiply their coefficients together to produce the coefficient of the product.

51. Multiplication of Literal Factors or Law of Exponents. To multiply a^3 by a^2:

$$\text{Since } a^3 = a \times a \times a$$

$$\text{and } a^2 = a \times a$$

$$\therefore a^3 \times a^2 = a \times a \times a \times a \times a = a^5.$$

This may be expressed in the form

$$a^3 \times a^2 = a^{3+2} = a^5,$$

$$\text{or, in general, } a^m \times a^n = a^{m+n},$$

where m and n are positive whole numbers.

Hence, in multiplying the literal factors of a monomial, we add the exponents of each letter that occurs in both multiplier and multiplicand.

$$\text{Ex. } 4a^2bc^3 \times 3a^3b^2x = 12a^5b^3c^3x.$$

52. Law of Signs. The law of signs in multiplication follows directly from the general law of signs as stated in Art. 31.

To proceed by way of illustration:

(1) + \$100 taken 5 times gives + \$500,

or, in general, + quantity taken a + number of times gives a + result.

(2) \$100 of debts—that is, − \$100, taken 5 times, gives − \$500,

or, in general, − quantity taken a + number of times, gives − quantity as a result.

(3) \$100 deducted 5 times, or \$100 × − 5, gives as total amount of deduction − \$500,

or, in general, + quantity taken a − number of times, gives − quantity as a result.

(4) Deducting \$100 of debts 5 times from a man's possessions is the same as adding \$500 to his assets; that is, − \$100 × − 5 = + \$500,

or, in general, − quantity taken a − number of times gives + quantity as a result.

Thus, we see from (1) and (4) that

either + × +, or − × −, gives +,

and from (2) and (3), that

$$\text{either } - \times +, \text{ or } + \times -, \text{ gives } -;$$

or, in brief, that in multiplication

Like signs give plus, unlike signs give minus.

53. Multiplication of Monomials. Combining the results of Arts. 50, 51, 52, the process of multiplying one monomial by another may be expressed as follows:

Multiply the coefficients together for a new coefficient;

Annex the literal factors, adding the exponents of each letter that occurs in both multiplier and multiplicand;

Determine the sign of the result by the rule that like signs give +, unlike signs give −.

Ex. 1. Multiply $5a^2bx^3$ by $-6ab^3y^2$.

The product is $-30a^3b^4x^3y^2$.

Ex. 2. Multiply $5a^{n+3}$ by $2a^{n-1}$.

Since $n+3$ and $n-1$, added, give $2n+2$,
the product is $10a^{2n+2}$

MULTIPLICATION OF A POLYNOMIAL BY A MONOMIAL.

54. Since, by the Distributive Law, Art. 33,

$$a(b+c)=ab+ac,$$

it follows that to multiply any polynomial by a monomial we proceed thus:

Multiply each term of the multiplicand by the multiplier, and set down the results as a new polynomial.

Ex. Multiply $2a^3-5a^2b+3ab^2$ by $-3ab^2$.

$$\begin{array}{l} 2a^3 - 5a^2b + 3ab^2 \\ -3ab^2 \\ \hline -6a^4b^2 + 15a^3b^3 - 9a^2b^4, \textit{ Product.} \end{array}$$

EXERCISE 8.

	1.	2.	3.	4.	5.	6.
Multiply	-5	$-3a$	$3ab$	$30x^2y^2$	$4x$	$-5x$
By	4	-2	-5	-1	$-2x$	$-3x$

	7.	8.	9.	10.	11.	12.
Multiply	$3ax$	$-6xy^2$	$7ax$	$-5a^2b$	$6c^2d$	$-2x^2yz$
By	$-4ax$	$-7xy^2$	$-3ay$	$-4cd^2$	$-3cd^2$	$-8xy^2z^3$

	13.	14.	15.	16.	17.
Multiply	$11cdx^2$	$-4x^n$	$5x^ny^m$	$3x^ny^{n-1}$	$x^{2n}y^{n-3}$
By	$-3c^2d$	$3x^n$	$-7x^ny^m$	$-x^2y^{n+1}$	$-x^ny^{m+3}$

	18.	19.	20.	21.
Multiply	$2a+3x$	$3x-2y$	$4x^2y-xy^2$	$7ax-4by$
By	$3ax$	$-5xy$	$2xy$	$-3abxy$

Multiply—

22. $8ac^2-3m^2n$ by $5an$.
23. $m-m^2-3m^3$ by $-7m^2n$.
24. $8x^2y-5xy^2-y^3$ by $3xy$.
25. $2x^n-3x^{n-1}$ by x^3.
26. $3x^{n+1}+7x^n$ by $-4x$.
27. $2x^{2n}-5x^ny$ by $3x^ny^m$.
28. ax^n-7by^n by x^ny^n.
29. $5x^{1-n}-3x^{2-n}$ by $4x^n$.
30. $2x^{n+3}-3x^{n+2}-x^{n+1}-x^n$ by $5x^{n-2}$.
31. $x^ny+3x^{n+1}y^2-4x^{n+2}y^3$ by $-2x^{n-1}y^n$.

MULTIPLICATION OF A POLYNOMIAL BY A POLYNOMIAL.

55. Arranging the Terms of a Polynomial. The multiplication of polynomials is greatly facilitated by arranging the terms in each polynomial according to the powers of some letter, the powers being taken either in the ascending or descending order.

In arranging the terms of a polynomial according to the ascending powers of a letter, the term containing the lowest power of the letter is placed first; the term containing the next higher power of this letter is placed next, etc.

Ex. $5x^2 + 3 - x + x^4 - 7x^3$, arranged according to the ascending powers of x, becomes

$$3 - x + 5x^2 - 7x^3 + x^4.$$

In arranging the terms of an expression according to the descending powers of a letter, the term containing the highest power of the letter is placed first; the term containing the next higher power is placed next, etc.

Ex. $a^4 + b^4 - 4a^2b^2 - 5a^3b$, arranged according to the descending powers of a, becomes

$$a^4 - 5a^3b - 4a^2b^2 + b^4.$$

When arranging two polynomials for purposes of multiplication the same letter should be used, and the same order, either ascending or descending, in both polynomials.

56. Multiplication of Polynomials. The terms of each polynomial having been arranged, we proceed to multiply each term of the multiplicand by each term of the multiplier, and take the sum of the results. The reason for this is made clear by taking two polynomials, $a + b$ and $c + d$, and forming their product by use of the Distributive Law:

$(a + b)(c + d) = a(c + d) + b(c + d)$, by Distributive Law.
$= ac + ad + bc + bd$, by a second use of this law.

We see that a similar result is obtained, no matter how many terms occur in each polynomial.

Therefore, to multiply two polynomials

Arrange the terms of the multiplier and multiplicand according to the ascending or descending powers of the same letter;

Multiply each term of the multiplicand by each term of the multiplier;

Add the partial products thus formed.

Ex. 1. Multiply $2x - 3y$ by $3x + 5y$.

The terms as given are arranged in order.

The most convenient way of adding partial products is to set down similar terms in columns, thus:

$$\begin{array}{lll}
2x & - 3y & \\
3x & + 5y & \\
\hline
6x^2 & - 9xy & \\
 & + 10xy & - 15y^2 \\
\hline
6x^2 & + xy & - 15y^2, \textit{ Product.}
\end{array}$$

(The two middle lines are marked *Partial products.*)

Ex. 2. Multiply $2x - x^3 + 1 - 3x^2$ by $2x + 3 - x^2$.

Arrange the terms in both polynomials according to the ascending powers of x. (Why is the ascending order chosen rather than the descending?)

$$\begin{array}{llllll}
1 & + 2x & - 3x^2 & - x^3 & & \\
3 & + 2x & - x^2 & & & \\
\hline
3 & + 6x & - 9x^2 & - 3x^3 & & \\
 & + 2x & + 4x^2 & - 6x^3 & - 2x^4 & \\
 & & - x^2 & - 2x^3 & + 3x^4 & + x^5 \\
\hline
3 & + 8x & - 6x^2 & - 11x^3 & + x^4 & + x^5, \textit{ Product.}
\end{array}$$

Let the student also multiply the two polynomials together with their terms in the order as first given, and hence discover the advantage of arranging the terms in order before multiplying.

Ex. 3. Multiply $3ab - 4b^2 + 2a^2$ by $-2b^2 + 3a^2 - 5ab$.

Arrange the terms in each polynomial according to the descending powers of a.

$$\begin{array}{lllll}
2a^2 & + 3ab & - 4b^2 & & \\
3a^2 & - 5ab & - 2b^2 & & \\
\hline
6a^4 & + 9a^3b & - 12a^2b^2 & & \\
 & - 10a^3b & - 15a^2b^2 & + 20ab^3 & \\
 & & - 4a^2b^2 & - 6ab^3 & + 8b^4 \\
\hline
6a^4 & - a^3b & - 31a^2b^2 & + 14ab^3 & + 8b^4, \textit{ Product.}
\end{array}$$

Ex. 4. Multiply $a^2+b^2+c^2+2ab-ac-bc$ by $a+b+c$.

Arranging the terms according to powers of a,

$$\begin{array}{l} a^2+2ab-ac+b^2-bc+c^2 \\ a+b+c \\ \hline a^3+2a^2b-a^2c+ab^2-abc+ac^2 \\ \quad +a^2b \quad +2ab^2-abc \quad +b^3-b^2c+bc^2 \\ \quad\quad +a^2c \quad +2abc-ac^2 \quad +b^2c-bc^2+c^3 \\ \hline a^3+3a^2b+3ab^2+b^3+c^3, \textit{ Product.} \end{array}$$

Ex. 5. Multiply $x^{3m}+x^{2m}+x^m+1$ by x^m+1.

$$\begin{array}{l} x^{3m}+x^{2m}+x^m+1 \\ x^m+1 \\ \hline x^{4m}+x^{3m}+x^{2m}+x^m \\ \quad +x^{3m}+x^{2m}+x^m+1 \\ \hline x^{4m}+2x^{3m}+2x^{2m}+2x^m+1, \textit{ Product.} \end{array}$$

57. Degree of a Term. Homogeneous Expressions. The degree of a term is determined by the number of literal factors which the term contains; hence, the degree of a term is also equal to the sum of the exponents of the literal factors.

Ex. $7a^3bc^2$ is a term of the sixth degree, since the sum of the exponents $3+1+2=6$.

A polynomial is said to be *homogeneous* when all its terms are of the same degree.

Ex. $5a^2b-b^3+ab^2$ is a homogeneous polynomial, since each of its terms is of the 3d degree.

58. Multiplication of Homogeneous Polynomials. If two monomials be multiplied together, the degree of the product must equal the sum of the degrees of the multiplier and multiplicand.

For instance, in Ex. 3. Art. 56, the multiplicand and multiplier are both homogeneous, and each is of the second degree, and their product is seen to be homogeneous and of the fourth degree.

The fact that the product of two homogeneous expressions

must also be homogeneous affords a partial test of the accuracy of the work. For if, for instance, in the above example, a term of the 5th degree, such as $5a^3b^2$, had been obtained in the product, it would have been at once evident that a mistake had been made in the work. The student should make use of this principle in testing the results obtained in examples 8, 9, 10, 12, 14, 15, 20, 21, 22, 23, 24, 25, of the following exercise.

EXERCISE 9.

Multiply—

1. $x - 4$ by $2x + 1$.
2. $x - 3$ by $3x + 2$.
3. $2x + 5$ by $x - 7$.
4. $3x - 4y$ by $4x - 3y$.
5. $7x^2 - 5y^2$ by $4x^2 + 3y^2$.
6. $5xy + 6$ by $6xy - 7$.
7. $4a^2 - b^2c$ by $8a^2c + 2ab^2c^2$.
8. $11x^2y - 7xy^2$ by $3x^2 + 2y^2$.
9. $a^2 - ab + b^2$ by $a + b$.
10. $x^3 + x^2y + xy^2 + y^3$ by $x - y$.
11. $4x^3 - 3x^2 + 2x - 1$ by $2x + 1$.
12. $2x^2 - 3xy + 2y^2$ by $3x - 5y$.
13. $x^3 - 3x^2 + 2x - 1$ by $2x^2 + x - 3$.
14. $3x^2y - 4xy^2 - y^3$ by $x^2 - 2xy - y^2$.
15. $x^3 - 3x^2y + 3xy^2 - y^3$ by $x^2 - 2xy + y^2$.
16. $4x^3 - 3x^2 + 5x - 2$ by $x^2 + 3x - 3$.
17. $x^4 - 3x^2 + 5$ by $x^2 - x - 4$.
18. $x^3 - 3xy + y^3$ by $x^3 - 3xy - y^3$.
19. $x^2 - 7x + 2$ by $x^2 - 7x - 2$.
20. $a^2 - ab + b^2$ by $a^2 + ab + b^2$.
21. $4x^2 + 9y^2 - 6xy$ by $4x^2 + 9y^2 + 6xy$.
22. $x^4 - 7x^2y^2 + 6xy^3 - y^4$ by $x^3 - 2xy^2 + y^3$.
23. $x^3 - 6ax^2 + 12a^2x - 8a^3$ by $-x^2 - 4ax - 4a^2$.
24. $a^2 + b^2 + x^2 + 2ab - ax - bx$ by $a + b + x$.
25. $ab + cd + ac + bd$ by $ab + cd - ac - bd$.

26. $x^n + 2x^{n-1} + 3x^{n-2} - 2$ by $x - 2$.

27. $x^{n+1} - 3x^n + 4x^{n-1} - 5x^{n-2}$ by $x^n + 2x^{n-1}$.

28. $x^{n-4} - 2x^{n-3} + 3x^{n-2} - 4x^{n-1} + 5x^n$ by $2x^2 + 3x + 1$.

59. Multiplication indicated by the Parenthesis. Simplifications. The parenthesis is useful in indicating multiplications or combinations of multiplications.

Thus, $(a - b + 2c)^2$ means that $a - b + 2c$ is to be multiplied by itself.

$(a - b + 2c)^3$ means that $a - b + 2c$ is to be taken as a factor three times and multiplied.

To multiply out such a power is termed to *expand* the power.

Again, $(a - b)(a - 2b)(a + b - c)$ means that the three factors, $a - b$, $a - 2b$, $a + b - c$, are all to be multiplied together.

Also, $(a - 2x)^2 - (a + 2x)(a - 2x)$ means that $a + 2x$ is to be multiplied by $a - 2x$, and the product subtracted from the product of $a - 2x$ by itself.

To *simplify* an expression in which multiplications are indicated in any of the above manners, means to perform the operations indicated and to collect terms.

Ex. Simplify $3(x - 2y)(x + 2y) - (x - 2y)^2$.

$$3(x - 2y)(x + 2y) = 3x^2 - 12y^2$$
$$(x - 2y)^2 = \underline{x^2 - 4xy + 4y^2}$$

Subtracting the second expression from the first, we obtain

$$2x^2 + 4xy - 16y^2.$$

EXERCISE 10.

Simplify by removing parentheses and collecting terms:

1. $x^2 - x(1 + x)$.
2. $(x - 2)(2x + 4)$.
3. $3x(x - 2) - 2x(x - 3)$.
4. $(x - 5)^2 - (x + 5)^2$.
5. $3x - 2x(1 + x + x^2)$.
6. $x - (x - 1)(x + 2)$.

7. $3(x-3)(x+1)+9$.
8. $(a-2b)(3a+4b)-3a^2$.
9. $(a+2b-3c)(a-2b+3c)$.
10. $(x-y+z)^2-x(x-2y+2z)$.
11. $2x^2-3(x-1)^2+(x-2)^2$.
12. $3x^2+x(1-x)(2+x)+x^3$.
13. $2-3(x-2)^2-2(3-2x)(1+x)$.
14. $a^2-[x(a-x)-a(x-a)]-x^2$.
15. $(x-1)(x-2)-(x-2)(x-3)+(x-3)(x-4)$.
16. $3(x-y)^2-2\{(x+y)^2-(x-y)(x+y)\}+2y^2$.
17. $x(x-y-z)-y(z-x-y)-z(z-y-x)-y^2$.
18. $3[(a+2b)x+2my]-5[(m-c)y+bx]-4[(x-a)a+cy]$.
19. $26ab-(9a-8b)(5a+2b)-(4b-3a)(15a+4b)$.
20. $6x^4-2x^3+x^2-2(x^2+x-1)(3x^2-x+1)+(3x^2+2)(2x-1)$.

If $a=3$, $x=-2$, $y=-5$, find the values of—

21. $2ax$.
22. x^3y.
23. $3x^2+ay$.
24. $xy-ax^2$.
25. $y^2+3x(x-y)$.
26. $4x^2-ax(4x-y)$.
27. $3x-5(2x+3)$.
28. $2(x^3+y)-ay+ax^3$.

29. $2(1-2x)^2+(x+y)(a^2+x)$.
30. $(x-1)^2-3(x+1)(x+2)-x(x^2-2)(y-2x)$.
31. $3a(a-2x)-\{a-(a-1)(x+1)-(a+x)^2\}+5ax$.

CHAPTER V.

DIVISION.

60. Division is the inverse of multiplication, and may be defined as the process of finding one factor when the product and other factor are given.

The **Dividend** is the product of the two factors, and hence is the quantity to be divided by the given factor.

The **Divisor** is the given factor.

The **Quotient** is the required factor.

Thus, if it is required to divide $10xy$ by $5x$, it is meant that we must find a quantity which, multiplied by $5x$, will produce $10xy$. The factor $5x$ is the divisor, $10xy$ is the dividend, and the other factor, or required quotient, is evidently $2y$.

61. General Principle. Division being the inverse of multiplication, the methods of division are obtained by inverting the processes used in multiplication.

DIVISION OF MONOMIALS.

62. Division of Coefficients. In the division of one monomial by another the coefficient of the quotient is obtained by dividing the coefficient of the dividend by the coefficient of the divisor.

For, by multiplication, Art. 50,

Coeff. of dividend (*i. e.* of product) = coeff. of divisor factor × coeff. of quotient factor.

∴ Dividing these equals by coeff. of divisor factor,

$$\frac{\text{Coeff. of dividend}}{\text{Coeff. of divisor}} = \text{Coeff. of quotient.}$$

63. Index Law for Division. If a^5 is to be divided by a^2, we have

$$\frac{a^5}{a^2}=\frac{a\times a\times a\times a\times a}{a\times a}=a\times a\times a=a^3.$$

Or, in general, $$\frac{a^m}{a^n}=a^{m-n},$$

where m and n are positive whole numbers.

Hence, in general, the exponent of a literal factor in the quotient is obtained by subtracting the exponent of this letter in the divisor from the exponent of the same letter in the dividend.

64. Law of Signs in Division. This law is obtained by inverting the processes of multiplication.

Thus, in multiplication, if a and b stand for any positive quantities (see Art. 52),

$$\left.\begin{aligned} +a\times+b&=+ab\\ +a\times-b&=-ab\\ -a\times+b&=-ab\\ -a\times-b&=+ab \end{aligned}\right\}\ \text{Hence, by definition of division,}\ \left\{\begin{aligned} +ab\div+b&=+a\ldots(1)\\ -ab\div-b&=+a\ldots(2)\\ -ab\div+b&=-a\ldots(3)\\ +ab\div-b&=-a\ldots(4) \end{aligned}\right.$$

From (1) and (2) we see that *the division of like signs gives* +. From (3) and (4) we see that *the division of unlike signs gives* −. Hence, the law of signs is the same in division as in multiplication.

65. Division of Monomials in General. Combining the results obtained in Arts. 62, 63, 64, we have the following general process for the division of one monomial by another:

Divide the coefficient of the dividend by the coefficient of the divisor;

Obtain the exponent of each literal factor in the quotient by sub-

tracting the exponent of each letter in the divisor from the exponent of the same letter in the dividend;

Determine the sign of the result by the rule that like signs give plus, and unlike signs give minus.

Ex. 1. Divide $27a^3b^4x^3$ by $-9a^2bx^3$.

$$\frac{+27a^3b^4x^3}{-9a^2bx^3} = -3ab^3, \textit{ Quotient,}$$

since the factor x^3 in the divisor cancels x^3 in the dividend.

Ex. 2. Divide a^{2m-3} by a^{m-1}.

$$\frac{a^{2m-3}}{a^{m-1}} = a^{m-2}, \textit{ Quotient.}$$

DIVISION OF A POLYNOMIAL BY A MONOMIAL.

66. Since $\frac{a+b}{c} = \frac{a}{c} + \frac{b}{c}$ (Distributive Law, Art. 33), the process of dividing a polynomial by a monomial may be stated as follows:

Divide each term of the dividend by each term of the divisor, and connect the results by the proper signs.

Ex. 1. Divide $12a^3x - 10a^2y + 6a^4z^2$ by $2a^2$.

$$\frac{12a^3x - 10a^2y + 6a^4z^2}{2a^2} = \frac{12a^3x}{2a^2} - \frac{10a^2y}{2a^2} + \frac{6a^4z^2}{2a^2} = 6ax - 5y + 3a^2z^2, \textit{ Quotient.}$$

Or, we may arrange the work more conveniently thus:

$$\begin{array}{l} 2a^2\underline{)12a^3x - 10a^2y + 6a^4z^2} \\ \quad\;\; 6ax - \quad 5y + 3a^2z^2, \textit{ Quotient.} \end{array}$$

Ex. 2. Divide $6a^{3n+3} - 4a^{2n+2} - 2a^{5n-3}$ by $2a^{n-1}$.

$$\begin{array}{l} 2a^{n-1}\underline{)6a^{3n+3} - 4a^{2n+2} - 2a^{5n-3}} \\ \qquad\;\; 3a^{2n+4} - 2a^{n+3} - a^{4n-2}, \textit{ Quotient.} \end{array}$$

EXERCISE 11.

Divide—

1. $15a$ by $-5a$.
2. $-3x^3$ by x.
3. $8a^2x^2$ by $-4ax^2$.
4. $-30x^3y^2$ by $-6x^2y$.
5. $-7xz^3$ by $7z^3$.
6. $21xy^2z$ by $-3xz$.
7. $18bc^3d^3$ by $-9c^3d$.
8. $-33x^5y^6z^7$ by $11xy^4z^5$.
9. $28x^2y^2z^3$ by $-14xy^2z^3$.
10. $-m^3n$ by $-m^2$.
11. $6x^{3n}$ by $-3x^{2n}$.
12. $7x^ny^{n+1}$ by $-x^ny^n$.
13. $-15x^{2n-1}y^{3n}$ by $5x^{n-1}y^{2n}$.
14. $x^3 - 3x^2$ by $-x$.
15. $20x^2 - 8xy$ by $4x$.
16. $4ab^2 - 6a^2bc$ by $-2ab$.
17. $x^3 - x^2 + x$ by x.
18. $-3x^3 + 7x^2 - x$ by $-x$.
19. $15x^3y - 10x^2y^2 - 5xy^3$ by $5xy$.
20. $-m - m^2 + m^3 - m^4$ by $-m$.
21. $14x^3y^2z - 21xy^2z^3 + xyz$ by $-xyz$.
22. $9x^{3n} - 6x^{2n} + 12x^n$ by $-3x^n$.
23. $-4x^{2n+1} + 10x^{2n+2} - 6x^{n+2}$ by $-2x^{2n}$.
24. $x^{n+3} - 2x^{n+2} + 3x^{n+1} + x^n$ by x^{n-1}.
25. $8x^{m+2} - 16x^{m+1} - 4x^m - 12x^{m-1}$ by $-4x^{m-2}$.
26. $9x^{3n-2} - 6x^{2n-1} + 12x^{2n} - 3x^{2n+1}$ by $3x^{n-1}$.

DIVISION OF A POLYNOMIAL BY A POLYNOMIAL.

67. General Method. The work of dividing one polynomial by another is performed to the best advantage if we first arrange the polynomials according to the ascending or descending powers of some one letter, and then, in effect, separate the dividend into partial dividends (by the Distributive Law, Art. 33), which are then divided in succession by the divisor.

Thus, in order to divide $6x^4 + 7x^3 - 3x^2 + 11x - 6$ by $2x^2$

$+3x-2$, if we divide the first term of the dividend, $6x^4$, by the first term of the divisor, $2x^2$, and multiply the quotient obtained, $3x^2$, by the entire divisor, we obtain the first partial dividend. If we subtract this from the entire dividend and proceed in like manner with the remainder, we have a process like the following:

$$\begin{array}{l|l}
6x^4+7x^3-3x^2+11x-6 & 2x^2+3x-2, \textit{ Divisor.} \\
\underline{6x^4+9x^3-6x^2} & 3x^2-\ x+3, \textit{ Quotient.} \\
\quad -2x^3+3x^2+11x-6 & \\
\quad \underline{-2x^3-3x^2+\ 2x} & \\
\qquad\qquad 6x^2+\ 9x-6 & \\
\qquad\qquad \underline{6x^2+\ 9x-6} &
\end{array}$$

The partial dividends into which the entire dividend is separated are,

$$\begin{array}{r}
6x^4+9x^3-6x^2 \\
-2x^3-3x^2+2x \\
6x^2+9x\ -6.
\end{array}$$

These are divided in succession by the divisor, and give the partial quotients, $3x^2$, $-x$, $+3$, which combined form the polynomial quotient, $3x^2-x+3$.

Hence, the process of dividing one polynomial by another may be formally stated as follows:

Arrange the terms of both divisor and dividend according to the ascending or descending powers of some one letter;

Divide the first term of the dividend by the first term of the divisor, and set down the result as the first term of the quotient;

Multiply the entire divisor by the first term of the quotient, and subtract the result from the dividend;

Continue the division, regarding each remainder as a dividend, till the remainder is zero, or a quantity which cannot be divided.

Ex. 1. Divide $4x + 4x^5 - x^3$ by $2x + 2x^3 - 3x^2$.

Arrange the terms according to the descending powers of x.

$$\begin{array}{l}
4x^5 \qquad - x^3 \qquad + 4x \;\big|\; 2x^3 - 3x^2 + 2x, \textit{ Divisor.} \\
4x^5 - 6x^4 + 4x^3 \qquad\qquad 2x^2 + 3x + 2, \textit{ Quotient.} \\
\qquad 6x^4 - 5x^3 \\
\qquad 6x^4 - 9x^3 + 6x^2 \\
\qquad\qquad 4x^3 - 6x^2 + 4x \\
\qquad\qquad 4x^3 - 6x^2 + 4x
\end{array}$$

Ex. 2. Divide $31ab^3 - 20b^4 - 10a^2b^2 + 6a^4 - a^3b$ by $3a^2 - 5b^2 + 4ab$.

$$\begin{array}{l}
6a^4 - a^3b - 10a^2b^2 + 31ab^3 - 20b^4 \;\big|\; 3a^2 + 4ab - 5b^2, \textit{ Divisor.} \\
6a^4 + 8a^3b - 10a^2b^2 \qquad\qquad 2a^2 - 3ab + 4b^2, \textit{ Quotient.} \\
\quad - 9a^3b \qquad + 31ab^3 \\
\quad - 9a^3b - 12a^2b^2 + 15ab^3 \\
\qquad + 12a^2b^2 + 16ab^3 - 20b^4 \\
\qquad + 12a^2b^2 + 16ab^3 - 20b^4
\end{array}$$

Ex. 3. Divide $x^3 + y^3 + z^3 + 3x^2y + 3xy^2$ by $x + y + z$.

Arranging terms according to the descending powers of x,

$$\begin{array}{l}
x^3 + 3x^2y + 3xy^2 + y^3 + z^3 \;\big|\; x + y + z \\
x^3 + x^2y + x^2z \qquad\qquad x^2 + 2xy - xz + y^2 + z^2 - yz \\
\quad + 2x^2y - x^2z + 3xy^2 + y^3 + z^3 \\
\quad + 2x^2y \qquad + 2xy^2 \qquad + 2xyz \\
\qquad - x^2z + xy^2 - 2xyz + y^3 + z^3 \\
\qquad - x^2z - xz^2 - xyz \\
\qquad\quad xy^2 + xz^2 - xyz + y^3 + z^3 \\
\qquad\quad xy^2 \qquad\qquad + y^3 + y^2z \\
\qquad\qquad + xz^2 - xyz \qquad - y^2z + z^3 \\
\qquad\qquad + xz^2 \qquad\qquad + yz^2 + z^3 \\
\qquad\qquad\quad - xyz - y^2z - yz^2 \\
\qquad\qquad\quad - xyz - y^2z - yz^2
\end{array}$$

Ex. 4. Divide $a^{m+3} - 4a^{m+2} - 27a^{m+1} + 42a^m$ by $a^m + 3a^{m-1} - 6a^{m-2}$.

$$\begin{array}{l}
a^{m+3} - 4a^{m+2} - 27a^{m+1} + 42a^m \;\big|\; a^m + 3a^{m-1} - 6a^{m-2} \\
a^{m+3} + 3a^{m+2} - 6a^{m+1} \qquad\qquad a^3 - 7a^2 \\
\quad - 7a^{m+2} - 21a^{m+1} + 42a^m \\
\quad - 7a^{m+2} - 21a^{m+1} + 42a^m
\end{array}$$

EXERCISE 12.

Divide—

1. $3x^2 + 7x + 2$ by $x + 2$.
2. $6x^2 + 7x + 2$ by $3x + 2$.
3. $12x^2 + xy - 20y^2$ by $3x + 4y$.
4. $6x^2 - xy - 12y^2$ by $2x - 3y$.
5. $3x^2 + x - 14$ by $x - 2$.
6. $6x^2 - 31xy + 35y^2$ by $2x - 7y$.
7. $12a^2 - 11ac - 36c^2$ by $4a - 9c$.
8. $-15x^2 + 59x - 56$ by $3x - 7$.
9. $44x^2 - xy - 3y^2$ by $11x - 3y$.
10. $a^2 - 4b^2$ by $a - 2b$.
11. $x^3 - y^3$ by $x - y$.
12. $27x^3 + 8$ by $3x + 2$.
13. $9x^2 - 49$ by $3x + 7$.
14. $125 - 64x^3$ by $5 - 4x$.
15. $8a^3x^3 + y^6$ by $2ax + y^2$.
16. $2x^3 - 9x^2 + 11x - 3$ by $2x - 3$.
17. $35x^3 + 47x^2 + 13x + 1$ by $5x + 1$.
18. $6a^3 - 17a^2x + 14ax^2 - 3x^3$ by $2a - 3x$.
19. $4y^4 - 18y^3 + 22y^2 - 7y + 5$ by $2y - 5$.
20. $c^5 + c^4x + c^3x^2 + c^2x^3 + cx^4 + x^5$ by $c + x$.
21.* $11x - 8x^2 + 5x^3 - 20 + 2x^4$ by $x + 4$.
22. $4x + 6x^5 + 3x^2 - 11x^3 - 4$ by $3x^2 - 4$.
23. $-x^3y - 11xy^3 - 2x^2y^2 + 6x^4 - 6y^4$ by $2x - 3y$.
24. $4y^3 + 6x^6 - 13x^4y$ by $3x^2 - 2y$.
25. $x^4 - 16y^4$ by $x - 2y$.
26. $x^5 + 32y^5$ by $x + 2y$.
27. $x^6 - y^6$ by $x + y$.
28. $256x^8 - y^8$ by $4x^2 - y^2$.
29. $9x - 18x^3 + 8x^4 - 13x^2 + 2$ by $4x^2 + x - 2$.
30. $10 - x^3 - 27x^2 + 12x^4 - 3x$ by $x + 4x^2 - 2$.

---

* **Arrange dividend in descending powers of x.**

31. $22x^2 - 13x^3 + 10x^5 - 18x^4 + 5x - 6$ by $x + 5x^3 - 2$.

32. $14x^2y^3 - 16x^3y^2 + 6x^5 + y^5 + 5x^4y - 6xy^4$ by $3x^2 + y^2 - 2xy$.

33. $5a^4b - 3a^3b^2 - a^2b^3 + 3a^5 - 4b^5$ by $a^2 + 3ab + 2b^2$.

34. $x^3 - y^3 + z^3 - xyz - 2x^2z + 2yz^2$ by $x - y - z$.

35. $c^3 + d^3 + n^3 - 3cdn$ by $c + d + n$.

36. $y^6 - 2y^3 + 1$ by $y^2 - 2y + 1$.

37. $2x^6 + 1 - 3x^4$ by $1 + 2x + x^2$.

38. $6x^2y^2 - 6y^2z^2 - 6x^2z^2 - 13xyz^2 - 5xy^2z$ by $3xy + 2yz + 3xz$.

39. $x^5 - 39x + 15 - 2x^3$ by $3x^2 + 6x + x^3 + 15$.

40. $4x^6 - 9x^4 + 25 - 14x^3 - x^2$ by $2x^3 - x - 5 + 3x^2$.

41. $6x^{2n+1} - 13x^{2n} + 6x^{2n-1}$ by $3x^{n+1} - 2x^n$.

42. $12x^{4n} + 13x^{3n} - x^n$ by $3x^n + 1$.

43. $4x^{n+3} + 5x^{n+2} - x^{n+1} - x^n + x^{n-1}$ by $x^2 + 2x + 1$.

44. $6x^{n+1} - 5x^n - 6x^{n-1} + 13x^{n-2} - 6x^{n-3}$ by $2x^2 - 3x + 2$.

68. Division of Polynomials having Fractional Coefficients. Polynomials having fractional coefficients are divided by the same methods that are used for those having integral coefficients.

Ex. 1. Divide $\frac{1}{4}a^3 - \frac{1}{2}a^2b + \frac{25}{72}ab^2 - \frac{1}{6}b^3$ by $\frac{1}{2}a^2 - \frac{1}{3}ab + \frac{1}{4}b^2$.

$$\begin{array}{l|l}
\frac{1}{4}a^3 - \frac{1}{2}a^2b + \frac{25}{72}ab^2 - \frac{1}{6}b^3 & \frac{1}{2}a^2 - \frac{1}{3}ab + \frac{1}{4}b^2, \textit{ Divisor.} \\
\underline{\frac{1}{4}a^3 - \frac{1}{6}a^2b + \frac{1}{8}ab^2} & \frac{1}{2}a - \frac{2}{3}b, \textit{ Quotient.} \\
\quad - \frac{1}{3}a^2b + \frac{2}{9}ab^2 - \frac{1}{6}b^3 & \\
\quad \underline{- \frac{1}{3}a^2b + \frac{2}{9}ab^2 - \frac{1}{6}b^3} &
\end{array}$$

Ex. 2. Divide $0.2x^4 - 0.01x^3y - 0.44x^2y^2 + xy^3 - 1.92y^4$ by $0.5x^2 - 0.4xy + 1.2y^2$.

$$\begin{array}{l|l}
0.2x^4 - 0.01x^3y - 0.44x^2y^2 + xy^3 - 1.92y^4 & 0.5x^2 - 0.4xy + 1.2y^2 \\
\underline{0.2x^4 - 0.16x^3y + 0.48x^2y^2} & 0.4x^2 + 0.3xy - 1.6y^2 \\
\quad + 0.15x^3y - 0.92x^2y^2 + xy^3 & \\
\quad \underline{+ 0.15x^3y - 0.12x^2y^2 + 0.36xy^3} & \\
\qquad - 0.8x^2y^2 + 0.64xy^3 - 1.92y^4 & \\
\qquad \underline{- 0.8x^2y^2 + 0.64xy^3 - 1.92y^4} &
\end{array}$$

EXERCISE 13.

Multiply—

1. $\frac{1}{2}x^2 + \frac{2}{3}x - \frac{1}{3}$ by $\frac{3}{2}x - 2$.
2. $\frac{3}{5}x^2 - 2x + \frac{2}{3}$ by $\frac{5}{2}x + \frac{5}{6}$.
3. $\frac{4}{3}x^3 - \frac{3}{2}x^2 + \frac{2}{3}x - \frac{3}{4}$ by $\frac{3}{2}x + \frac{2}{3}$.
4. $1.2x^2 + 1.5x + 6.4$ by $2.4x - 3$.
5. $3.6x^3 - 2.8x^2 + 7.2x - 0.32$ by $1.5x + 0.25$.
6. $4.5x^2 - 2.8xy + 5.6y^2$ by $1.5x^2 + 1.2xy + 2.4y^2$.

Divide—

7. $\frac{1}{2}x^3 - \frac{53}{8}x + 2$ by $\frac{3}{4}x + \frac{9}{2}$.
8. $\frac{3}{5}x^3 + \frac{7}{15}x^2 - \frac{59}{36}x + \frac{2}{3}$ by $\frac{3}{4}x - \frac{2}{3}$.
9. $\frac{1}{3}x^4 + \frac{11}{15}x^3y + \frac{79}{20}x^2y^2 + \frac{45}{4}y^4$ by $\frac{5}{6}x^2 - \frac{2}{3}xy + \frac{5}{2}y^2$.
10. $4.5x^3 - 7.1x^2 - 0.4x + 0.24$ by $2.5x + 0.5$.
11. $0.25x^4 - 1.8x^3 + 3.24x^2 - 12.25$ by $0.5x^2 - 1.8x - 3.5$.
12. $4.8x^4 + 0.18x^3y - 8x^2y^2 + 115.19xy^3 - 275.2y^4$ by $1.6x^2 - 2.5xy + 12.8y^2$.

Perform the operations indicated—

13. $[1 - x^3 + 2y(4y^2 + 3x)] \div (1 - x + 2y)$.
14. $[1 - 2x^2(x^4 - x^3 + 1) - 3x^4] \div [1 - x(2x^2 + 1)]$.
15. $(1 + x^8 + x^{16}) \div [(1 - x + x^2)(1 - x^4 + x^8)]$.
16. $[x^3 + (4ab - b^2)x - (a - 2b)(a^2 + 3b^2)] \div (x + 2b - a)$.
17. $\{x^4 + (3 - b)x^3 + (c - 3b - 2)x^2 + (2b + 3c)x - 2c\} \div (x^2 + 3x - 2)$.
18. $\{x^4(y - z) + y^4(z - x) + z^4(x - y)\} \div \{x^2(y - z) + y^2(z - x) + z^2(x - y\}$.
19. $\{(a^2 - 3ab)x^2 + (2a^2 + 4ab + 3b^2)x - (2ab + 5b^2)\} \div (ax - b)$.

CHAPTER VI.

SIMPLE EQUATIONS.

69. An **Equation** is the statement of the equality of two algebraic expressions.

An equation, therefore, consists of the sign of equality and an algebraic expression on each side of it.

Ex. $3x - 1 = 2x + 3$.

70. Members of an Equation. The algebraic expression to the left of the sign of equality is called the *first member* of the equation; the expression to the right of the sign of equality is called the *second member*.

Thus, in the equation $3x - 1 = 2x + 3$,

the first member is, $3x - 1$; the second member is, $2x + 3$.

The members of an equation are sometimes called *sides* of the equation.

71. Use of an Equation. An equation expresses the relation of at least one unknown quantity to certain given or known quantities, the object in the use of the equation being to determine the value of the unknown quantity in terms of the known.

Thus, in the above example x represents the unknown quantity, and -1, 2, 3 are known quantities.

72. A **Numerical Equation** is one in which all the known quantities are expressed as arithmetical numbers.

Ex. $3x - 1 = 2x + 3$.

73. A **Literal Equation** is one in which at least some of the known quantities are represented by letters.

Ex. $ax + 2b = 3c - dx$.

a, b, c, and $-d$ represent known quantities.

74. Degree of Equation having One Unknown Quantity. If an equation contain but one unknown quantity, the degree of the equation (after the equation has been reduced to its simplest form) is determined by the exponent of the highest power of the unknown quantity in the equation.

Exs. $2x + 1 = 5x - 8$ is an equation of the first degree.

$ax = b^2 + cx$ is of the first degree.

$4x^2 - 5x = 20$ " second "

$3x^2 - x^3 = 6x + 8$ " third "

An equation of the first degree is also called a **Simple Equation.**

75. The **Root** of an equation is the number which, substituted for the unknown quantity in the equation, *satisfies* the equation; that is, reduces the two members of the equation to identical numbers.

Ex. If in the equation, $3x - 1 = 2x + 3$,

we substitute 4 in the place of x,

we obtain $$12 - 1 = 8 + 3$$
$$11 = 11,$$

and the equation is satisfied. Hence, 4 is the root of the given equation.

76. The **Solution** of a simple equation is the process of finding the value of its root.

In a simple equation the relation between the unknown quantity and certain known quantities is given, but in a more or less complex form, from which the value of the unknown quantity in terms of the known is difficult to perceive. But since the algebraic symbols which represent these quantities may be rearranged and combined by the methods of Chapter II., as well as by the axioms of Art. 77, the complex relation first given can be reduced to a simple one, whence the value of x can be at once perceived.

The processes of reducing an equation to its simplest form have been systematized, and take certain standard forms.

77. Axioms. Besides the principles given in Chapter II., which govern the use of algebraic symbols for quantity and enable us to use these symbols to the best advantage, there are other principles which are true of quantity in general, and therefore of algebraic quantity. These principles, like those of Chapter II., are of great value in enabling us to use algebraic expressions to the best advantage.

They are the so-called axioms:

1. *Things equal to the same thing, or to equals, are equal to each other.*
2. *If equals be added to equals, the sums are equal.*
3. *If equals be subtracted from equals, the remainders are equal.*
4. *If equals be multiplied by equals, the products are equal.*
5. *If equals be divided by equals, the quotients are equal.*
6. *Like powers or like roots of equals are equal.*
7. *The whole is equal to the sum of its parts.*

78. Application of Axioms to the Members of an Equation. Since the two members of an equation are equal quantities, it follows from the axioms of Art. 77 that—

The members of an equation may be increased, diminished, multiplied, or divided by the same quantity, and the results will be equal.

Ex. 1. If $8x = 24$,

dividing both members by 8 (Ax. 5),

$$x = 3.$$

Ex. 2. If $x^2 + 3x = x^2 + 12$,

subtracting x^2 from each member (Ax. 3),

$$3x = 12.$$

Hence $$x = 4 \text{ (Ax. 5)}.$$

79. Transposition of Terms. If we take the equation

$$x + b = a,$$

and we subtract b from each member (Ax. 3),

$$x + b - b = a - b,$$
$$\text{or } x = a - b.$$

This result might have been obtained at once from $x + a = b$ in a mechanical way, by transferring $+ b$ from the left-hand member to the right-hand member, at the same time changing its sign to minus.

So, if we take the equation,

$$x - b = a,$$

and add b to each member, we obtain

$$x - b + b = a + b,$$
$$\text{or } x = a + b.$$

This result also might have been obtained at once by transferring $- b$ to the opposite member, at the same time changing its sign.

This process occurs so often in simplifying an equation that we abbreviate into the mechanical form and call it **Transposition.**

Any term of an equation may be transposed from one side of an equation to the other, provided the sign of the term be changed.

The main object of transposition of terms is to get all the terms containing the unknown quantity on the left-hand side of the equation, and all the known terms on the right-hand side.

This is the most important single step in simplifying an equation. This importance is indicated by the fact that the word algebra means transposition (*al gebre*, Arabic words meaning "the transposition").

80. Clearing of Parentheses. If an equation contain quantities in parentheses, it is necessary first of all to remove the parentheses by performing the operations indicated by them.

81. Changing Signs of all Terms. The signs of all the terms of an equation may be changed. This may be regarded either as the result of multiplying both members of the equation by -1, or as the result of transposing all the terms of the equation.

Ex. For $-ax = -b + c$,

write $$ax = b - c.$$

82. General Process of Solving a Simple Equation. This may now be stated as follows:

Clear the equation of parentheses by performing the operations indicated by them;

Transpose the unknown terms to the left-hand side of the equation, the known terms to the right-hand side;

Collect terms;

Divide both members by the coefficient of the unknown quantity.

Ex. 1. Solve the equation, $3x - 7 = 14 - 4x$.

Transposing the terms -7 and $-4x$, $3x + 4x = 14 + 7$

Collecting terms, $7x = 21$

Dividing by 7, $x = 3$, *Root.*

Verification. Substituting 3 for x in $3x - 7 = 14 - 4x$

$$3 \times 3 - 7 = 14 - 4 \times 3$$
$$9 - 7 = 14 - 12$$
$$2 = 2.$$

Ex. 2. Solve $x(x-2)=x(x+4)-3(x-3)$. . . . . . . (1)

Removing parentheses, $x^2-2x=x^2+4x-3x+9$

Transposing terms, $x^2-x^2-2x-4x+3x=9$

Collecting terms, $-3x=9$ . . (2)

Dividing by -3, $x=-3$, *Root.*

Verification. Substituting -3 for x in equation (1),

$$-3(-3-2)=-3(-3+4)-3(-3-3)$$
$$-3(-5)=-3(1)-3(-6)$$
$$15=-3+18$$
$$15=15.$$

The value of the processes employed in the solution of an equation—viz. transposition, etc.—is realized when we compare the ease with which the value of x is perceived from equation (2), with the difficulty in assigning immediately in equation (1) a value for x which will satisfy that equation.

EXERCISE 14.

Solve the following equations:

1. $3x=15$.
2. $2x=-6$.
3. $-13x=26$.
4. $-6x=-12$.
5. $3x=-5$.
6. $-2x=11$.
7. $-5x=-13$.
8. $4x=-1$.
9. $-7x=0$.
10. $4x-5=1+3x$.
11. $2x-7=8+5x$.
12. $2x-3(x-3)+2=0$.
13. $7(2-3x)=2(7-8x)$.
14. $x^2-x(x+5)=x+12$.
15. $3-2(3x+2)=7$.
16. $2x-(x+5)=4x$.
17. $(x-1)(x+3)=(x-4)(x+2)$.
18. $3x-4x+10+5x=0$.
19. $x(2x+1)-2x(x+3)=7$.

20. $3(x-1)(x+1)=x(3x+4)$.

21. $4(x-3)^2=(2x+1)^2$.

22. $8(x-3)-(6-2x)=2(x+2)-5(5-x)$.

23. $5x-(3x-7)-\{4-2x-(6x-3)\}=10$.

24. $x+2-[x-8-2\{8-3(5-x)-x\}]=0$.

25. $2(x+1)(2x-1)+2\{x-(x+3)(2x-1)\}=-32$.

26. $2x(x-5)-\{x^2+(3x-2)(1-x)\}=(2x-4)^2$.

27. $(x+1)^2-2\{(x-1)^2-3(x+2)^2\}=3(x+4)(2x-4)-(x^2-5)$.

28. $8x^2+13x-2\{x^2-3[(x-1)(3+x)-2(x+2)^2]\}=3$.

SOLUTION OF PROBLEMS.

83. The **General Method** of stating and solving will first be illustrated by an example similar to that given in Art. 1.

James and John together have $18. If James has twice as many dollars as John, how many dollars has each boy?

By solution of the problem is meant, of course, finding the value of the unknown quantity or quantities of the problem. The first thing to be done, therefore, is to determine what are the unknown quantities or numbers. In the given problem there are evidently two unknown quantities to be determined—first, the number of dollars which James has; second, the number of dollars which John has.

The method of procedure is then to state the relation of the unknown quantities to the known quantities in the form of an equation, which is afterward solved.

As a rule, when there are two unknown quantities in a problem, it is more convenient to represent the smaller of them by x.

Let	$x =$	number of dollars John has.
Therefore,	$2x =$	" " James has.
Hence,	$x + 2x =$	" " both have.
But	18 also $=$	" " "

Hence, by use of Axiom 1, Art. 77,

$$x + 2x = 18,$$

$$3x = 18,$$

$x = 6$, number of dollars John has.

$2x = 12$, " " James has.

It is to be noticed particularly that we *let x equal a definite number, not a vague quantity.*

We do not let $x =$ the *money* which John has,

nor $x =$ *what* John has,

but let $x =$ *number of dollars* which John has.

In solving problems the student will find it necessary to study each problem carefully by itself, as no rule or method can be found which will cover all cases.

The following general directions will, however, be found of service:

By study of the problem determine what the unknown quantity or quantities are whose values are to be obtained;

Let x equal one of these expressed as a number;

State all the other unknown quantities which are either to be determined or to be utilized in the process of the solution, in terms of x;

Obtain an equation by the use of a principle, such as the whole is equal to the sum of its parts, or things equal to the same things are equal to each other;

Solve the equation, and find the value of each of the unknown quantities.

Ex. 1. James and John together have $24, and James has 8 dollars more than John; how many dollars has each?

Let $x =$ number of dollars which John has.

Since James has 8 dollars more than John,

$x + 8 =$ number of dollars which James has.

$\therefore x + (x + 8) =$ " " " both have.

But $24 =$ " " " " "

Hence, by Axiom 1,

$$x + x + 8 = 24$$

$$2x = 16$$

$x = 8$, number of dollars which John has.

$x + 8 = 16$, " " " James has.

Ex. 2. The sum of two numbers is 36, and three times the greater number exceeds four times the less number by 10. Find the numbers.

Let $x =$ the less number.

If the sum of the numbers is 36 (as the line AB), and one part is x (AC), then the other part will be $36 - x$ (CB).

36

A C B

x $36 - x$

$\therefore \quad 36 - x =$ the greater number.

$\therefore 3(36 - x) =$ three times the greater number.

$4x =$ four " " less "

But (three times the greater number) − (four times the less) = 10.

$\therefore \quad 3(36 - x) - 4x = 10.$

Hence, $108 - 3x - 4x = 10$

$$-7x = -98$$

$x = 14$, the less number.

$36 - x = 22$, the greater number.

Let the student verify these results.

Problems may frequently be solved in more than one way. Thus, Ex. 1 above may be stated and solved as follows:

Let $x =$ number of dollars which John has.

Then $24 - x =$ " " " James has.

Also, (number of James's dollars) − (number of John's dollars) = 8;

that is,

$$(24 - x) - x = 8$$

$$24 - x - x = 8$$

$$-2x = -16$$

$$x = 8, \text{ number of John's dollars.}$$

$$24 - x = 16, \text{ " James's "}$$

EXERCISE 15.

ORAL.

1. A has x marbles, and B has twice as many; how many has B? How many have both?

2. There are 100 pupils in a school, of which x are boys; how many are girls?

3. If I have x dollars, and you have three dollars more than twice as many, how many have you? How many have we together?

4. Two boys together solved a examples: the one did x; how many did the other solve?

5. The difference between two numbers is 15, and the less is x; what is the greater? What is their sum?

6. If n is a whole number, what is the next larger number? The next less?

7. Write three consecutive numbers, the least being x. Write them if the greatest is y.

8. John has x dollars, and James has seven dollars less than three times as many; how many has James?

9. If I am x years old now, how old was I ten years ago? a years ago? How old will I be in c years?

10. A man bought a horse for x dollars, and sold it so as to gain a dollars; what did he receive for it?

11. A man sold a horse for $200, and lost x dollars; what did the horse cost?

12. If a yard of cloth cost m dollars, what will x yards cost?

13. If a boy ride a miles an hour, how far will he ride in c hours?

14. A bicyclist rides x yards in y seconds; how far will he ride in one second? In n seconds?

15. How many hours will it require to walk x miles at a miles an hour?

16. A man has a dollars, and b quarters; how many cents has he?

17. How many dimes in x dollars and y halves?

18. I have x dollars in my purse and y dimes in my pocket; if I give away fifty cents, how much have I remaining?

19. By how much does 30 exceed x?

20. Express the sum of the squares of two consecutive even numbers if the larger is x.

21. A gentleman is out x hours, of which he rides a hours at the rate of eight miles an hour, and walks the rest of his time at the rate of three miles an hour; how far did he ride? How far did he walk?

EXERCISE 16.

1. A boy has three times as many marbles as his brother, and together they have 48; how many has each?

2. A and B pay $100 taxes; if A pays $22 more than B, what does each pay?

3. John solved a certain number of examples, and William did 12 less than twice as many; both solved 96. How many did each solve?

4. Three boys earned together $98; if the second earned $11 more than the first, and the third $28 less than the other two together, how many dollars did each earn?

5. A man walked 15 miles, rode a certain distance in a coach, and then took a boat for twice as far as he had previously traveled; altogether, he went 120 miles. How far did he ride by boat?

6. Find three consecutive numbers whose sum is 84.

7. The sum of two numbers is 92, and the larger is 3 less than four times the less; find the numbers.

8. The sum of three numbers is 50: the first is twice the second, and the third is 16 less than three times the second; find the numbers.

9. A farmer paid $94 for a horse and cow; what did each cost, if the horse cost $13 more than twice as much as the cow?

10. Distribute $485 among A, B, and C so that B and C each get twice as much as A.

11. Divide the number 35 into two parts, such that three times the smaller shall be equal to twice the larger.

12. Find five consecutive numbers whose sum shall be 3 less than six times the least.

13. The difference between two numbers is 6, and if 3 be added to the larger the sum will be double the less; find the numbers.

14. Three men rent a store for $500: the first is to pay twice as much as the second, and the second $60 more than the third; how much does each pay?

15. Divide $4500 among two sons and a daughter so that each son gets $150 less than twice the daughter's share.

16. A father is four times as old as his son, and the difference of their ages is 24 years less than the sum; how old is each?

17. A man is twice as old as his daughter, who is 5 years younger than her brother, and the combined ages of all three are 109 years; what is the age of each?

18. A father is now twice as old as his son; 21 years ago he was three times as old. How old are they now?

19. Find two numbers whose difference is 14, such that the greater exceeds twice the less by 3.

20. Find three consecutive odd numbers whose sum is 63.

21. The greater of two numbers is 5 more than the less, and five times the less exceeds three times the greater by 3; find the numbers.

22. A man had five sons, to whom he gave $56, giving to each $5 less than twice the amount his next younger brother received; what did each receive?

23. It is required to divide 75 into two such parts that three times the greater exceeds seven times the less by 15.

24. The difference of the squares of two consecutive numbers is 43; find the numbers.

25. The difference of the squares of two consecutive even numbers is 60; find the numbers.

26. The joint ages of father and son are 64 years: if the age of the son were doubled, he would still be four years younger than his father; find the age of each.

27. Two bicyclists, A and B, start respectively from New York and Philadelphia, 90 miles apart, and ride toward each other; A rides 8 and B, 12 miles per hour. How long and how far will A ride before meeting B?

28. A man walks to the top of a mountain at the rate of 2 miles an hour, and back down at 4 miles an hour; if he is out 6 hours, how far is it to the top of the mountain?

29. How far into the country will a man go who rides out at the rate of 9 miles an hour, walks back at 6 miles an hour, and is gone 10 hours?

30. A boy was engaged to work 50 days at 75 cents each day for the days he worked, and to forfeit 25 cents every day he was idle. On settlement he received $25.50; how many days did he work?

31. Five years ago A was twice as old as B, but 10 years ago he was three times as old as B; how old is each now?

32. A man is 30 years older than his son, and 10 years ago he was three times as old; what is the age of each?

33. Twenty yards of silk and 30 yards of cloth cost $99, and the silk cost three times as much per yard as the cloth; how much did each cost per yard?

34. A merchant paid a bill of $72 with dollar, two-dollar, and five-dollar bills, paying the same number of each; how many of each did he use?

35. How can $2.25 be paid in five- and ten-cent pieces so as to use the same number of each?

36. How can $5.95 be paid in dimes and quarters, using the same number of each?

37. A purse contains $10.50 in dollar bills and quarters, but there are twice as many quarters as bills; how many are there of each?

38. Twenty coins, dimes and half dollars, make together $8.80; how many are there of each?

39. A gentleman gave $525 to his son and daughter, so that for every dime the daughter received the son got a quarter; how much did each receive?

40. A person was desirous of giving 30 cents apiece to some beggars, but found he had not money enough by 80 cents; he gave them, therefore, 20 cents each, and had 30 cents remaining. Required the number of beggars.

41. A sum of money is divided among 3 persons, A, B, and C, so that A and B have $79, B and C have $70, and A and C $75; how much has each?

42. A woman sold 12 new baskets for $3. For a part she got 20 cents each, and for the rest 32 cents each; how many of each grade did she sell?

43. A certain flag-pole is 69 feet long, and has 12 feet of its length in water: the part in air is 3 feet more than five times the length of the part in earth; what is the length of the part above water?

44. B had $5 more than three times as much as A, but he gave A $9, and now he has a dollar less than twice A's sum; how much did each have at first?

45. There is a fish whose tail weighs 9 pounds; his head weighs as much as his tail and half his body; and his body weighs as much as his head and his tail. What is the weight of the whole fish?

46. A set out from a town, P, to walk to Q, 45 miles distant, an hour before B started from Q toward P. A walked at the rate of 4 miles an hour, but rested 2 hours on the way; B walked at the rate of 3 miles an hour. How many miles did each travel before they met?

EXERCISE 17.

REVIEW.

Add—

1. $\frac{2}{3}x^2 + \frac{1}{3}xy - \frac{3}{4}y^2$, $2y^2 - x^2 - \frac{2}{3}xy$, $\frac{2}{3}x^2 - xy - \frac{5}{4}y^2$.
2. $\frac{3}{8}x^3 - x^2 + \frac{1}{2}x$, $\frac{1}{4}x^2 - \frac{3}{4}x + \frac{2}{3}$, $\frac{1}{4}x - 1 - \frac{1}{2}x^3$, $\frac{1}{4}x^3 + \frac{3}{4}x^2$.
3. $1.5x^2 - 3.2xy + 0.6y^2$, $3.2x^2 - 3.1y^2 + 1.5xy$, $1.7xy - y^2 - 0.7x^2$.
4. $1.6 + 1.1x^3 - 2.2x^2$, $5.1x^2 - 0.7x - 2.3x^3$, $1 + 0.2x^3 - 2.9x^2 + 1.6x$.

Subtract—

5. $x^3 - \frac{2}{5}x^2 + \frac{3}{4}x - 5\frac{1}{2}$ from $\frac{3}{2}x^3 - \frac{3}{5}x^2 + \frac{1}{2}x + \frac{2}{3}$.
6. $2.7x^3 - 0.4x^2y - 1.3xy^2 + y^3$ from $3x^3 - 1.1x^2y + 2xy^2 - 1.5y^3$.

Multiply—

7. $\frac{1}{2}x^2 - \frac{3}{2}x + 4$ by $\frac{2}{3}x + 2$.
8. $1.5x - 0.4y$ by $2.4x + 1.5y$.
9. $\frac{3}{2}x^2 - ax + \frac{2}{3}a^2$ by $\frac{3}{4}x^2 + \frac{1}{2}ax + \frac{1}{3}a^2$.
10. $\frac{5}{3}x^2 - \frac{2}{3}x + 1$ by $\frac{3}{2}x^2 + \frac{3}{5}x - 1$.
11. $3.2a^2 - 2.3ab + 5.2b^2$ by $1.5a + 2.5b$.
12. $0.4x^2 - 1.8xy - 2.8y^2$ by $0.5x - 1.5y$.
13. $4.5x^2y - 1.2x^2y^2 - 5.4xy^3$ by $0.4x^2y - 0.5xy^2$.
14. $3.2x^2 - 4.5xy + 1.8y^2$ by $1.5x - 3.5y$.

Divide—

15. $6x^3 - 4y^3 - 2x^2 - 16x^2y + 14xy^2 + 4xy - 2y^2$ by $3x - 2y - 1$.
16. $x^7 - 5x^5y^2 + 13x^4y^3 - 2x^3y^4 - 16x^2y^5 + 9y^7$ by $x^2 + 2xy - 3y^2$.
17. $2 - x$ by $1 + x$ to five terms in the quotient.
18. $1 - x + x^2$ by $1 + x + x^2$ to five terms.
19. $x^6 - 15$ by $x^2 + x - 1$ to five terms.
20. 1 by $1 - 2x - 3x^2$ to five terms.
21. $\frac{27}{8}x^3 - \frac{9}{2}x^2y + 2xy^2 - \frac{8}{27}y^3$ by $\frac{3}{2}x - \frac{2}{3}y$.
22. $\frac{9}{16}x^4 - \frac{7}{8}x^3y + \frac{19}{36}x^2y^2 + \frac{1}{6}xy^3$ by $\frac{3}{2}x + \frac{1}{3}y$.
23. $36x^2 + \frac{1}{9}y^2 + \frac{1}{4} - 4xy - 6x + \frac{1}{3}y$ by $6x - \frac{1}{3}y - \frac{1}{2}$.
24. $2.4x^3 - 0.12x^2y + 4.32y^3$ by $1.5x + 1.8y$.

25. $8.4x^4 - 1.6x^3 - 10.3x^2 + 10.2x - 3.9$ by $2.4x^2 + 1.6x - 2.6$.

26. $5x^4 - 6.85x^3y + 3.09xy^3 - 0.36y^4$ by $2.5x - 0.3y$.

Simplify—

27. $6z + [4z - \{8x - (2z + 4x) - 22x\} - 7x] - [7x + \{14z - (4z - 5x)\}]$.

28. $a^2(b - c) - b^2(a - c) + c^2(a - b) - (a - b)(a - c)(b - c)$.

29. $2x - [-3(x - y) + \{(x - 2y) - 2(x + 3y) - x\} - 2(y - 2x)]$.

30. $6\{a - 2[b - 3(c + d)]\} - 4\{a - 3[b - 4(c - d)]\}$.

31. $1 - 2\{-[-(x - y)] - x\} + 2\{-2[-1 - (x - y)] - 1\}$.

32. $8a^2(a - b) - (3b - 2a)[a(2b - 3a) - (a + b)^2] + b(b + a)^2$.

Solve and verify—

33. $(2x + 1)(x - 3) + 7 = x - 2(x - 4)(2 - x)$.

34. $7x - 2(x - 1)(2 - x) - 17 = x(3x + 7) - (x + 1)^2$.

35. $2x^2 - 3x - 2(2x + 1)^2 + (2x - 3)(3x + 2) = 8$.

36. $3x^2 - \{5x - [4 - (x - 1)(2x - 3) - 7x] + (x - 3)^2\} = 0$.

37. $5x + 1 - 2\{2x - 3[x - (x + 1)(x + 3)] - 3(x + 2)^2\} = 0$.

38. What is the dividend when the quotient is $x^3 + 2x^2 + 7x + 20$, the remainder $62x + 59$, and the divisor $x^2 - 2x - 3$?

39. What is the divisor if the quotient is $x^3 + 3x$, the dividend $x^5 - 8$, and the remainder $9x - 8$?

40. If $x = -\frac{2}{3}$ and $y = -\frac{3}{2}$, find the value of

$$(3x - 2y)^2(9x^2 + 4y^2) - 6(y - x)\sqrt{6xy(x + 2y^2 + \tfrac{1}{3})}.$$

CHAPTER VII.

CASES OF ABBREVIATED MULTIPLICATION AND DIVISION.

ABBREVIATED MULTIPLICATION.

84. Value of Abbreviated Multiplication. In certain cases of multiplication, by observing the character of the expressions to be multiplied, it is possible to write out the product at once, without the labor of the actual multiplication. Almost all the multiplication of binomials, and that of many trinomials, will fall under these cases, and by the use of the abbreviated methods at least three-fourths of the labor of multiplication in them will be saved. The student should therefore master them as thoroughly as he has done the multiplication table in arithmetic.

85. I. Square of the Sum of Two Quantities.

Let $a + b$ be the sum of any two algebraic quantities.

By actual multiplication,

$$\begin{array}{l} a + b \\ \underline{a + b} \\ a^2 + ab \\ \underline{\quad\;\; + ab + b^2} \\ a^2 + 2ab + b^2, \textit{ Product.} \end{array}$$

Or, in brief, $(a + b)^2 = a^2 + 2ab + b^2$,

which, stated in general language, is the rule—

The square of the sum of two quantities equals the square of the first, plus twice the product of the first by the second, plus the square of the second.

Ex. $(2x+3y)^2 = 4x^2 + 12xy + 9y^2$, *Product.*

Since the square of $2x$ is $4x^2$,

twice the product of $2x$ and $3y$ is $12xy$,

and the square of $3y$ is $9y^2$.

86. II. Square of the Difference of Two Quantities.

By actual multiplication,

$$\begin{array}{l} a-b \\ \underline{a-b} \\ a^2-ab \\ \underline{\quad -ab+b^2} \\ a^2-2ab+b^2, \textit{ Product.} \end{array}$$

Or, in brief, $(a-b)^2 = a^2 - 2ab + b^2$,

which, stated in general language, is the rule—

The square of the difference of two quantities equals the square of the first, minus twice the product of the first by the second, plus the square of the second.

Ex. $(2x-3y)^2 = 4x^2 - 12xy + 9y^2$, *Product.*

87. III. Product of the Sum and Difference of Two Quantities.

By actual multiplication,

$$\begin{array}{l} a+b \\ \underline{a-b} \\ a^2+ab \\ \underline{\quad -ab-b^2} \\ \quad a^2-b^2, \textit{ Product.} \end{array}$$

Or, in brief, $(a+b)(a-b) = a^2 - b^2$,

which, stated in general language, is the rule—

The product of the sum and difference of two quantities is the difference of their squares.

Ex. $(2x+3y)(2x-3y) = 4x^2 - 9y^2$, *Product.*

EXERCISE 18.

Write by inspection the values of—

1. $(n+y)^2$.
2. $(c-x)^2$.
3. $(2x-y)^2$.
4. $(3x-2y)^2$.
5. $(5x+1)^2$.
6. $(x^2+1)^2$.
7. $(x-y^2)^2$.
8. $(1-7y^3)^2$.
9. $(3x^4+5x^3)^2$.
10. $(6x^2y-11y^2z^3)^2$.
11. $(x+z)(x-z)$.
12. $(y-3)(y+3)$.
13. $(3x-y)(3x+y)$.
14. $(7x+4y)(7x-4y)$.
15. $(x^2-2)(x^2+2)$.
16. $(ax^2-b^2y)(ax^2+b^2y)$.
17. $(1-11x^3)(1+11x^3)$.
18. $(2x^n+5y^m)(2x^n-5y^m)$.
19. $(5x^n-3y^nz^m)^2$.
20. $(4x^3y^5z^{2n}+9y^{3n})^2$.

88. Special Case under III. In applying any of the above abbreviated methods of multiplication we may have parentheses containing two or more terms used as a single quantity. This is of especial importance in obtaining the product of a sum and difference.

Ex. 1. Multiply $x+(a+b)$ by $x-(a+b)$.

We have

$$[x+(a+b)][x-(a+b)] = x^2-(a+b)^2, \text{ by III.}$$
$$= x^2-(a^2+2ab+b^2), \text{ by I.}$$
$$= x^2-a^2-2ab-b^2, \textit{ Product.}$$

It is frequently necessary to re-group the terms of trinomials in order that the multiplication may be performed by the above method.

Ex. 2. Multiply $x+y-z$ by $x-y+z$.

$$(x+y-z)(x-y+z) = [x+(y-z)][x-(y-z)]$$
$$= x^2-(y-z)^2, \text{ by III.}$$
$$= x^2-(y^2-2yz+z^2), \text{ by II.}$$
$$= x^2-y^2+2yz-z^2, \textit{ Product.}$$

EXERCISE 19.

Write by inspection the product of—

1. $[(a+b)+3][(a+b)-3]$.
2. $[2x-1-y][2x-1+y]$.
3. $[4-(x+1)][4+(x+1)]$.
4. $[a+(b-2)][a-(b-2)]$.
5. $(2x+3y+1)(2x-3y-1)$.
6. $(x^2+3x-2)(x^2+3x+2)$.
7. $(4-x-y)(4+x+y)$.
8. $(3x^2-2x+1)(3x^2+2x-1)$.
9. $(x^2-xy+y^2)(x^2+xy+y^2)$.
10. $(a^2+a+1)(a^2-a+1)$.
11. $(2x^2-3x-5)(2x^2+3x-5)$.
12. $(2x^2+5xy-y^2)(2x^2-5xy-y^2)$.
13. $(ax^2-bx+2c)(ax^2+bx-2c)$.
14. $(x^2+xy-y^2)(x^2-xy-y^2)$.
15. $(6a^2-3a-2)(6a^2+3a-2)$.
16. $[(a+b)-(c-1)][(a+b)+(c-1)]$.
17. $[(x^2+y^2)+(x^2y^2+1)][(x^2+y^2)-(x^2y^2+1)]$.
18. $(x-y+z-1)(x+y+z+1)$.
19. $(x^3-2x^2-x-2)(x^3+2x^2+x-2)$.

Simplify—

20. $(3a-1)^2+(2-3a)(2+3a)$.
21. $(2x-7y)(2x+7y)-4(x-2y)^2+13y(5y-x)$.
22. $(3x^2+5)^2+x^2(10-3x)(10+3x)-(5+13x^2)^2$.
23. $(a-c+1)(a+c-1)-(a-1)^2+2(c-1)^2$.
24. $(x+y-xy)(x-y-xy)+x^2y-(x-y^2)(x+y^2)$.
25. $(x^2-x-3)(x^2+x-3)+(3+x+x^2)(x-x^2+3)$.

89. IV. Square of any Polynomial.

By actual multiplication,

$$
\begin{array}{llllll}
a+b & +c & & & & \\
a+b & +c & & & & \\
\hline
a^2+ab & +ac & & & & \\
\quad +ab & & +b^2+bc & & & \\
 & +ac & & +bc & +c^2 & \\
\hline
a^2+2ab & +2ac & +b^2 & +2bc & +c^2, &
\end{array}
$$

or, in brief, $(a+b+c)^2 = a^2+b^2+c^2+2ab+2ac+2bc.$

In like manner we obtain

$$(a+b+c+d)^2 = a^2+b^2+c^2+d^2+2ab+2ac+2ad+2bc + 2bd + 2cd.$$

It is seen that each of these results consists (1) of the square of each term of the polynomial, and (2) of other terms formed by taking twice the product of each term into each term which follows it. It is also perceived that this method of forming the square of a polynomial will hold good no matter how many terms there are in the polynomial. For if we examine the above process of the multiplication of $a+b+c$ by itself, we see that not only is each term multiplied by itself, but also a of the multiplicand is multiplied by b of the multiplier; and, *vice versâ*, b of the multiplicand is multiplied by a of the multiplier, and so for any other pair of terms, and that this list exhausts the partial products. Hence, in general,

The square of any polynomial equals the sum of the squares of all the terms, together with twice the product of each term into each term which follows it.

Ex. $(a-2b+c-3x)^2 = a^2+4b^2+c^2+9x^2-4ab+2ac - 6ax - 4bc + 12bx - 6cx.$

EXERCISE 20.

Write by inspection the values of—

1. $(2x + y + 1)^2$.
2. $(x - 2y + 2z)^2$.
3. $(3x - 2y - 5)^2$.
4. $(2a - b + 3c)^2$.
5. $(x - 2y - 3z)^2$.
6. $(4x + 3y - 1)^2$.
7. $(x^2 - x + 1)^2$.
8. $(2a^2 + 5a - 3)^2$.
9. $(x - y + z - 1)^2$.
10. $(2x + 3y - 4z - 5)^2$.
11. $(3x^3 - 4x^2 + x - 2)^2$.
12. $(2a - b^3)(2a + b^3)$.
13. $(2a^2b - 5bc^2)^2$.
14. $(a - b + 7)(a + b - 7)$.
15. $(2x^2 - 7x + 11)(2x^2 + 7x + 11)$.
16. $(5x^2 - 6x - 3)^2$.
17. $(x^2 + 3x - 5)(x^2 - 3x - 5)$.
18. $(3a^2 - 5ab + b^2)(3a^2 - 5ab - b^2)$.
19. $(7x^2 - 5x + 4)(7x^2 + 5x - 4)$.
20. $[(a^2 + 2b^2) + (2ab - 1)][(a^2 + 2b^2) - (2ab - 1)]$.

90. V. Product of Two Binomials of the Form $x + a$, $x + b$.

By actual multiplication,

$$\begin{array}{l} x + 5 \\ \underline{x + 3} \\ x^2 + 5x \\ \underline{\quad + 3x + 15} \\ x^2 + 8x + 15 \end{array} \qquad \begin{array}{l} x - 5y \\ \underline{x + 3y} \\ x^2 - 5xy \\ \underline{\quad + 3xy - 15y^2} \\ x^2 - 2xy - 15y^2 \end{array} \qquad \begin{array}{l} x + a \\ \underline{x + b} \\ x^2 + ax \\ \underline{\quad + bx + ab} \\ x^2 + (a + b)x + ab. \end{array}$$

By comparing each pair of binomials with their product, the following relation is observed:

The product of two binomials of the form $x + a$ and $x + b$ consists of three terms:

The first term is the square of the first term of either binomial;

The last term is the product of the second terms of the binomials;

The middle term consists of the first term of the binomials with a coefficient equal to the algebraic sum of the second terms of the binomials.

Ex. 1. Multiply $x-8$ by $x+7$.

$$-8+7=-1, \qquad -8\times 7=-56$$

$$\therefore (x-8)(x-7)=x^2-x-56, \textit{ Product.}$$

Ex. 2. Multiply $(x-6a)(x-5a)$.

$$(-6a)+(-5a)=-11a, \qquad (-6a)\times(-5a)=+30a^2$$

$$\therefore (x-6a)(x-5a)=x^2-11ax+30a^2.$$

In Case V., also, a parenthesis containing two or more terms may be used instead of a single quantity. In all cases the student should thoroughly acquire the ability to use a parenthesis as a single quantity. By so doing many of the difficulties of algebra are at once overcome.

Ex. 3. Multiply $x+y+6$ by $x+y-2$.

$$(x+y+6)(x+y-2)=[(x+y)+6][(x+y)-2]$$

$$=(x+y)^2+4(x+y)-12, \textit{ Product.}$$

EXERCISE 21.

Write by inspection the product of—

1. $(x+2)(x+5)$.
2. $(x-5)(x-3)$.
3. $(x-7)(x+4)$.
4. $(x-4)(x+8)$.
5. $(x+1)(x-7)$.
6. $(x^2-2)(x^2-3)$.
7. $(x^2+3)(x^2+1)$.
8. $(a+3x)(a-10x)$.
9. $(x-7y)(x+y)$.
10. $(ab+c)(ab+3c)$.
11. $(xy-7z^2)(xy+3z^2)$.
12. $(bc^2-5a^2)(bc^2+7a^2)$.
13. $(x^2-yz)(x^2-4yz)$.
14. $(xy-4ab)(xy+ab)$.
15. $(x^2+7)(x^2-7)$.
16. $(m^2-8n^2)(m^2-8n^2)$.
17. $(a+2b-1)(a-2b+1)$.
18. $(a-bc-4)(a+bc-4)$.

19. $(a^2 - 2ax - 3x^2)^2$.

20. $(a - 1)(a + b)$.

21. $(a - 2x)(a + y)$.

22. $(a + 3b)(a - 3c)$.

23. $(x + 2y - 7)(x + 2y + 2)$.

24. $(xy^2z^3 - 10)(xy^2z^3 + a)$.

25. $(a^2bc - x^2yz)(a^2bc + 5x^2yz)$.

26. $(ab + bc - ac)(ab - bc + ac)$.

27. $[(a + b) - 3][(a + b) + 4]$.

28. $(m - 5 + 4n)(m - 5 - 8n)$.

91. VI. Product of Two Binomials whose Corresponding Terms are Similar.

By actual multiplication,

$$\begin{array}{r} 2a - 3b \\ 4a + 5b \\ \hline 8a^2 - 12ab \qquad \\ + 10ab - 15b^2 \\ \hline 8a^2 - \ \ 2ab - 15b^2 \end{array}$$

It is seen that the middle term of this product may be obtained directly from the two binomials by taking the algebraic sum of the cross products of their terms. Thus,

$$(+2a)(+5b) + (-3b)(+4a) = 10ab - 12ab = -2ab.$$

Hence, in general,

The product of any two binomials of the given form consists of three terms:

The first term is the product of the first terms of the binomials;

The third term is the product of the second terms of the binomials;

The middle term is formed by taking the algebraic sum of the cross products of the terms of the binomials.

Ex. Multiply $(10x + 7y)(8x - 11y)$.

For the middle term,

$$(10x)(-11y) + (7y)(8x) = -110xy + 56xy = -54xy.$$

$$\therefore (10x + 7y)(8x - 11y) = 80x^2 - 54xy - 77y^2, \textit{ Product.}$$

EXERCISE 22.

Write by inspection the product of—

1. $(2x+3)(x+2)$.
2. $(2x+5)(x-2)$.
3. $(3x-1)(x-2)$.
4. $(5x+1)(x-1)$.
5. $(x+3y)(3x-8y)$.
6. $(2x-7y)(3x+10y)$.
7. $(3a^2-5b)(3a^2-5b)$.
8. $(x^2-2yz)(5x^2+3yz)$.
9. $(3x-y^2)^2$.
10. $(x^2-3xy+4y^2)^2$.
11. $(5x^2-7y)(5x^2+7y)$.
12. $(11x^2y^2-3)(7x^2y^2+2)$.
13. $(2x^3+ay)(3x^3+2ay)$.
14. $(a^2-a+7)(a^2+a-7)$.
15. $(x^2-3x+1)(x^2-3x-1)$.
16. $(5x^2-9x-8)^2$.
17. $(3x^2+5xy-4y^2)^2$.
18. $(4a^2-3b)(5a^2+4b)$.
19. $[(8x^3+x)-(4x^2-1)][(8x^3+x)+(4x^2-1)]$.

ABBREVIATED DIVISION.

92. Value of Abbreviated Division. In certain cases much of the labor of division may be saved by performing the division operation in a typical case, noting the relation between the quantities divided and the quotient, and formulating this relation into a mechanical rule.

93. I. Division of the Difference of Two Squares.

Either by actual division, or by inverting the relation of Art. 87, we obtain

$$\frac{a^2-b^2}{a+b}=a-b, \qquad \text{and} \qquad \frac{a^2-b^2}{a-b}=a+b.$$

Hence, in general language,

The difference of the squares of two quantities is divisible by the sum of the quantities, and also by the difference of the quantities, the quotients in the respective cases being the difference of the quantities and the sum of the quantities.

Ex. 1. $\dfrac{4x^2 - 9y^2}{2x - 3y} = 2x + 3y$, *Quotient.*

Ex. 2. $\dfrac{x^2 - (a+b)^2}{x + (a+b)} = x - (a+b)$, *Quotient.*

EXERCISE 23.

Write by inspection the quotient of—

1. $\dfrac{a^2 - x^2}{a - x}$.

2. $\dfrac{9 - 4x^2}{3 - 2x}$.

3. $\dfrac{x^2 - 81y^2}{x + 9y}$.

4. $\dfrac{25x^2 - 36y^4}{5x - 6y^2}$.

5. $\dfrac{16x^4 - 49y^4}{4x^2 + 7y^2}$.

6. $\dfrac{25x^{10} - y^{12}}{5x^5 - y^6}$.

7. $\dfrac{a^2b^4 - 36c^6d^8}{ab^2 + 6c^3d^4}$.

8. $\dfrac{(x+1)^2 - a^2}{(x+1) + a}$.

9. $\dfrac{a^2 - (b - 2c)^2}{a - (b - 2c)}$.

10. $\dfrac{4x^4 - (y^2 + 1)^2}{2x^2 + (y^2 + 1)}$.

11. $\dfrac{(a-b)^2 - (c-1)^2}{(a-b) + (c-1)}$.

12. $\dfrac{1 - (a+b-c)^2}{1 + (a+b-c)}$.

94. II. and III. Division of Sum or Difference of Two Cubes.

By actual division we can obtain,

$$\frac{a^3 + b^3}{a + b} = a^2 - ab + b^2, \qquad \text{and} \qquad \frac{a^3 - b^3}{a - b} = a^2 + ab + b^2.$$

Hence, in general language,

The **sum** *of the cubes of two quantities is divisible by the sum of the quantities, and the quotient is the square of the first quantity, minus the product of the two quantities, plus the square of the second quantity;* also,

The **difference** *of the cubes of two quantities is divisible by the*

difference of the quantities, and the quotient is the square of the first quantity, plus the product of the two quantities, plus the square of the second.

Ex. 1. $$\frac{8x^3-27y^3}{2x-3y}=\frac{(2x)^3-(3y)^3}{2x-3y}$$
$$=(2x)^2+(2x)(3y)+(3y)^2$$
$$=4x^2+6xy+9y^2, \textit{ Quotient.}$$

Ex. 2. $$\frac{(a-b)^3+27}{(a-b)+3}=(a-b)^2-3(a-b)+9$$
$$=a^2-2ab+b^2-3a+3b+9.$$

EXERCISE 24.

Write the quotient of—

1. $\dfrac{a^3+8}{a+2}$.
2. $\dfrac{x^3-1}{x-1}$.
3. $\dfrac{27x^3-64}{3x-4}$.
4. $\dfrac{1+8x^6}{1+2x^2}$.
5. $\dfrac{125-x^9}{5-x^3}$.
6. $\dfrac{27a^6+y^{12}}{3a^2+y^4}$.
7. $\dfrac{x^6+y^6}{x^2+y^2}$.
8. $\dfrac{(a-1)^3-x^3}{(a-1)-x}$.
9. $\dfrac{c^3+(1-x)^3}{c+(1-x)}$.
10. $\dfrac{8-(x+y)^3}{2-(x+y)}$.
11. $\dfrac{x^6y^6-(xy-1)^3}{x^2y^2-(xy-1)}$.
12. $\dfrac{27x^6+125y^9}{3x^2+5y^3}$.
13. $\dfrac{(a-1)^3-x^6}{(a-1)-x^2}$.
14. $\dfrac{8x^3+(x^2-1)^3}{2x+x^2-1}$.
15. $\dfrac{8(x-y)^3-z^3}{2(x-y)-z}$.
16. $\dfrac{27(x^2+1)^3+125z^{12}}{3x^2+3+5z^4}$.

95. IV., V., and VI. Division of Sum or Difference of any Two Like Powers.

By actual division we can obtain,

$$\frac{a^4 - b^4}{a + b} = a^3 - a^2b + ab^2 - b^3, \textit{ Quotient.}$$

$$\frac{a^4 - b^4}{a - b} = a^3 + a^2b + ab^2 + b^3, \textit{ Quotient.}$$

But $a^4 + b^4$ is not divisible by either $a + b$ or $a - b$:

$$\frac{a^5 + b^5}{a + b} = a^4 - a^3b + a^2b^2 - ab^3 + b^4,$$

$$\frac{a^5 - b^5}{a - b} = a^4 + a^3b + a^2b^2 + ab^3 + b^4.$$

Hence, in like manner,

The difference of any two like even powers of two quantities is divisible by the sum of the quantities, and also by their difference;

The sum of two like odd powers of two quantities is divisible by the sum of the quantities;

The difference of any two like odd powers of two quantities is divisible by the difference of the quantities.

For the quotient in all these cases—

(1) *The number of terms in a quotient equals the degree of the powers whose sum or difference is divided;*

(2) *The terms of each quotient are homogeneous (since the exponent of a decreases by 1 in each term, and that of b increases by 1 in each term).*

(3) *If the divisor is a difference, the signs of the quotient are all plus; if the divisor is a sum, the signs of the quotient are alternately plus and minus.*

The last statement forms a general rule for signs of a quotient in all the cases of abbreviated division, including I., II., and III.

Ex. $\frac{32x^5 + y^5}{2x + y} = \frac{(2x)^5 + y^5}{2x + y}$

$= (2x)^4 - (2x)^3y + (2x)^2y^2 - (2x)y^3 + y^4$

$= 16x^4 - 8x^3y + 4x^2y^2 - 2xy^3 + y^4$, *Quotient.*

EXERCISE 25.

Write the quotient of—

1. $\frac{a^3 - 27b^3}{a - 3b}$.
2. $\frac{a^4 - 16b^4}{a - 2b}$.
3. $\frac{x^5 + 1}{x + 1}$.
4. $\frac{x^5 - 243}{x - 3}$.
5. $\frac{x^7 + y^7}{x + y}$.
6. $\frac{x^9 - 1}{x - 1}$.
7. $\frac{x^{12} + 1}{x^4 + 1}$.
8. $\frac{81 - x^8}{3 + x^2}$.
9. $\frac{x^4 - y^8}{x + y^2}$.
10. $\frac{x^{10} - y^{15}}{x^2 - y^3}$.

Write all the exact binomial divisors for each of the following:

11. $1 + 64x^3$.
12. $9x^2 - 25y^4$.
13. $x^4 - 81$.
14. $x^5 - 32$.
15. $1 + x^7$.
16. $x^6 - 27y^6$.
17. $x^6 - y^6$.
18. $x^{10} + y^{10}$.
19. $1 - (x - y)^4$.
20. $x^8 - y^8$.
21. $x^{12} - 64y^6$.
22. $x^{2n} - y^{2n}$.

For what values of n will—

23. $a^n - b^n$ be divisible by $a - b$? By $a + b$?
24. $a^n + b^n$ be divisible by $a - b$? By $a + b$?

CHAPTER VIII.

FACTORING.

96. THE **Factors** of an expression (see Art. 12) are the quantities which, multiplied together, produce the given expression.

Factoring is the process of separating an algebraic expression into its factors.

97. Illustration of Value of Factoring. If a fraction,

$$\frac{x^2-8x+15}{2x^2-13+21}, \text{ be given,}$$

and it is known that

$$x^2-8x+15=(x-3)(x-5),$$

and that $$2x^2-13x+21=(2x-7)(x-3),$$

for the original fraction we may write,

$$\frac{(x-3)(x-5)}{(x-3)(2x-7)},$$

then cancel out the factor $x-3$, common to both numerator and denominator, and obtain the simple fraction,

$$\frac{x-5}{2x-7}.$$

This is an illustration of the usefulness of a knowledge of factoring in enabling us to simplify work and save labor.

98. A **Prime Quantity** in algebra is one which cannot be divided by any quantity except itself and unity.

EXS. a, b, a^2+b^2, 17.

99. Perfect Square and Perfect Cube. When an expression is separable into two equal factors, the expression is called a *perfect square*, and each of the factors is called the *square root* of the expression.

Ex. $9a^2x^4 = 3ax^2 \cdot 3ax^2$.

$\therefore 3ax^2$ is the square root of $9a^2x^4$.

Also, $x^2 - 4x + 4 = (x-2)(x-2)$, and is therefore a perfect square, with $x - 2$ for its square root.

When an expression is separable into three equal factors the expression is called a *perfect cube*, and each of the factors is called its *cube root.*

Ex. $27a^3x^6y^9 = 3ax^2y^3 \cdot 3ax^2y^3 \cdot 3ax^2y^3$.

$\therefore 3ax^2y^3$ is the cube root of $27a^3x^6y^9$.

100. Factors of Monomials. Since monomials are formed by simply indicated multiplications, the factors of a monomial are recognized by direct inspection.

Thus, the factors of $7a^2x^3$ are 7, a, a, x, x, x.

101. Factors of Polynomials. When binomials or trinomials are multiplied together, simplifications are often made in the product obtained by the addition of similar terms.

Hence, given the simplified result, the problem of determining the quantities multiplied (or factors) is made more difficult, and different cases must be carefully discriminated.

CASE I.

102. A Polynomial all of whose Terms contain a Common Factor.

Ex. 1. Factor $3x^2 + 6x$.

Each term contains the factor $3x$.

Divide $3x^2 + 6x$ by $3x$, and the quotient is $x + 2$.

The factors are the divisor and quotient.

$\therefore 3x^2 + 6x = 3x(x + 2)$, *Factors.*

Ex. 2. Factor $12x^2y^3 - 16xy^4 + 8x^2y^5$.

$$12x^2y^3 - 16xy^4 + 8x^2y^5 = 4xy^3(3x - 4y + 2xy^2).$$

Hence, in general,

Divide all the terms of the polynomial by the common factor;
The factors will be the divisor and quotient.

EXERCISE 26.

Factor—

1. $2x^3 + 5x^2$.
2. $x^3 - 2x$.
3. $x^2 + x$.
4. $3a^2 - a$.
5. $7a + 14a^3$.
6. $3a^3x^2 - 15a^2x^3$.
7. $18x^5 - 27x^4y$.
8. $x^2 - x^3 - x^4$.
9. $a^2x - 2a^2x^2$.
10. $3a^2 - 6ax + 9x^2$.
11. $2x + 4x^2 - 6x^3$.
12. $10a^3b^2 - 35a^2b^3$.
13. $a^4b^3c - a^3b^3c^3 + 2a^2b^4c^2$.
14. $2x^ny^4 - 8x^{2n}y^3 + 6x^{3n}y^2$.
15. $a^mb^3c^{2n} + 11a^mb^3c^{2n+1}$.

CASE II.

108. A Trinomial that is a Perfect Square.

By Arts. 85 and 86 a trinomial is a perfect square when its first and last terms are perfect squares and positive, and the middle term is twice the product of the square roots of the end terms. The sign of the middle term determines whether the square root of the trinomial is a sum or a difference.

Ex. 1. Factor $16x^2 - 24xy + 9y^2$.

This trinomial satisfies the conditions for being the square of $4x - 3y$.

$$\therefore 16x^2 - 24xy + 9y^2 = (4x - 3y)(4x - 3y), \textit{ Factors.}$$

Ex. 2. Factor $a^4 + 4a^2b + 4b^2$.

$$a^4 + 4a^2b + 4b^2 = (a^2 + 2b)^2, \textit{ Factors.}$$

Hence, in general, to factor a trinomial that is a perfect square,

Take the square roots of the first and last terms, and connect these by the sign of the middle term;

Take the result as a factor twice.

EXERCISE 27.

Factor—

1. $4x^2 + 4xy + y^2$.
2. $16a^2 - 24ay + 9y^2$.
3. $25x^2 - 10x + 1$.
4. $x^2 - 20xy + 100y^2$.
5.* $49c + 28bc^2 + 4b^2c^3$.
6. $a^2b^2 - 6abc + 9c^2$.
7. $xy^2 + 2xy + x$.
8. $2m^2n - 4mn + 2n$.
9. $a^5 + 2a^4 + a^3$.
10. $4x^3 + 44x^2y^2 + 121xy^4$.
11. $81a^3b + 126a^2b^2 + 49ab^3$.
12. $8a^2y - 40axy + 50x^2y$.
13. $2x^4 - 8x^3 + 8x^2$.
14. $30x^2y + 3x^4 + 75y^2$.
15. $a^3x + ax^3 - 2a^2x^2$.
16. $x^{2n} + 2x^ny + y^2$.
17. $(a - b)^2 - 2c(a - b) + c^2$.
18. $9(x + y)^2 + 12z(x + y) + 4z^2$.
19. $16(2a - 3)^2 - 16ab + 24b + b^2$.
20. $25(x - y)^2 - 120xy(x - y) + 144x^2y^2$.
21. $a^2 + b^2 + c^2 + 2ab + 2ac + 2bc$.

CASE III.

104. The Difference of Two Perfect Squares.

From Art. 87, $(a + b)(a - b) = a^2 - b^2$,

hence, $a^2 - b^2 = (a + b)(a - b)$.

But any algebraic quantities may be used instead of a and b. Hence,

Ex. 1. Factor $x^2 - 16y^2$.

$$x^2 - 16y^2 = (x + 4y)(x - 4y), \textit{ Factors.}$$

In general, to factor the difference of two squares,

Take the square root of each square;

The factors will be the sum of these roots and their difference.

Ex. 2. $x^4 - y^4 = (x^2 + y^2)(x^2 - y^2)$

$$= (x^2 + y^2)(x + y)(x - y), \textit{ Factors.}$$

---

* Apply Case I. first.

105. Special Cases under Case III. A. We may also factor by this case the difference of two squares when one or both of the given squares is a compound expression.

Ex. 1. Factor $(a+2b)^2-4x^2$.

$$(a+2b)^2-4x^2=[(a+2b)+2x]\,[(a+2b)-2x]$$
$$=(a+2b+2x)\,(a+2b-2x), \textit{ Factors.}$$

Ex. 2. Factor $(3x+4y)^2-(2x+3y)^2$.

$$(3x+4y)^2-(2x+3y)^2=[(3x+4y)+(2x+3y)]\,[(3x+4y)-(2x+3y)]$$
$$=(3x+4y+2x+3y)\,(3x+4y-2x-3y)$$
$$=(5x+7y)\,(x+y), \textit{ Factors.}$$

EXERCISE 28.

Factor—

1. x^2-9.
2. $25-16a^2$.
3. $4a^2-49b^2$.
4. $3x^2-12y^2$.
5. $100-81m^2$.
6. $m-64mn^2$.
7. $49x^4-1$.
8. x^2y^3-36y.
9.* x^5-x.
10. x^6-25.
11. $3x^3-75xy^6$.
12. $242-2x^4$.
13. $9x^3-xy^4$.
14. a^4-81b^4.
15. a^6-4x^4.
16. $2a^3b^5-98ab$.
17. $144-x^{12}$.
18. x^8-169y^{10}.
19.* x^8-y^8.
20.* $16ax^4-81a$.
21. $225x^{2n}-y^2$.
22. x^6-y^6.
23. $x^{4m}-y^{2n}z^6$.
24. $(x+y)^2-1$.
25. $x^2-(y+1)^2$.
26. $(x-y)^2-9$.
27. $4(x-y)^2-25$.
28. $1-36(x+2y)^2$.
29. $(x+2y)^2-(3x+1)^2$.
30. $25(2a-b)^2-(a-3b)^2$.
31.* $x^{12}y^9-yz^{16}$.
32. $81x^{12}-16y^4$.
33. $x^5-144xy^2z^6$.
34. $(a-b)^2-4(c+1)^2$.
35. $1-100(x^2-x-1)^2$.

---

* May be resolved into four factors.

106. B. **Grouping of Terms.** It may be possible to group or rearrange the terms of a polynomial expression so as to produce the difference of two squares.

Ex. 1. Factor $x^2 - 4xy + 4y^2 - 9z^2$.

$$\begin{aligned} x^2 - 4xy + 4y^2 - 9z^2 &= (x^2 - 4xy + 4y^2) - 9z^2 \\ &= (x - 2y)^2 - 9z^2 \\ &= [(x - 2y) + 3z]\,[(x - 2y) - 3z] \\ &= (x - 2y + 3z)\,(x - 2y - 3z), \textit{ Factors.} \end{aligned}$$

Ex. 2. Factor $2xy - x^2 + a^2 - y^2$.

$$\begin{aligned} 2xy - x^2 + a^2 - y^2 &= a^2 - (x^2 - 2xy + y^2) \\ &= a^2 - (x - y)^2 \\ &= [a + (x - y)]\,[a - (x - y)] \\ &= (a + x - y)\,(a - x + y), \textit{ Factors.} \end{aligned}$$

Ex. 3. Factor $a^2 - x^2 - y^2 + b^2 + 2ab + 2xy$.

$$\begin{aligned} a^2 - x^2 - y^2 + b^2 + 2ab + 2xy &= (a^2 + 2ab + b^2) - (x^2 - 2xy + y^2) \\ &= (a + b)^2 - (x - y)^2 \\ &= (a + b + x - y)\,(a + b - x + y), \\ &\qquad\qquad \textit{Factors.} \end{aligned}$$

EXERCISE 29.

Factor—

1. $x^2 - 16(x - 2y)^2$.
2. $9(a - b)^2 - 25$.
3. $a^2 - 2ab + b^2 - 1$.
4. $9x^2 + 12xy + 4y^2 - z^2$.
5. $x^2 + a^2 - y^2 - 2ax$.
6. $a^2 + y^2 - x^2 + 2ay$.
7. $a^4 - x^4 - 2x^2y - y^2$.
8. $x^2 - y^2 - 1 - 2y$.
9. $1 + 2xy - x^2 - y^2$.
10. $c^2 - a^2 - b^2 + 2ab$.
11. $a^2 + b^2 - c^2 - 2ab$.
12. $4 - x^2 - 4y^2 - 4xy$.
13. $2a + b^2 - a^2 - 1$.
14. $2ab + a^2b^2 + 1 - x^2$.
15. $2z^2 - 4z - 2z^4 + 2$.
16. $9x^2 + y^2 - 25z^2 - 6xy$.

17. $20yz + x^2 - 4y^2 - 25z^2$.

18. $45x^2 - 20x^4 - 5y^2 - 20x^2y$.

19. $a^2 + 2ab + b^2 - c^2 - 2cd - d^2$.

20. $x^2 + 4y^2 - 9z^2 - 1 - 4xy - 6z$.

21. $9a^2 - 25x^2 + 4b^2 - 1 - 10x - 12ab$.

22. $a^2 - 9b^2x^2 - 1 + 6bx - 10ab + 25b^2$.

23. $16x^3 - 16x^2y - 16x - 48xz - 36xz^2 + 4xy^2$.

107. C. **The Addition and Subtraction of Some Quantity** will sometimes transform an expression into a difference of two perfect squares.

Ex. 1. $a^4 + a^2b^2 + b^4 = a^4 + 2a^2b^2 + b^4 - a^2b^2$

$= (a^2 + b^2)^2 - a^2b^2$

$= (a^2 + b^2 + ab)(a^2 + b^2 - ab)$, *Factors.*

Ex. 2. $x^4 - 7x^2y^2 + y^4$.

Add and subtract $9x^2y^2$.

$x^4 - 7x^2y^2 + y^4 = x^4 + 2x^2y^2 + y^4 - 9x^2y^2$

$= (x^2 + y^2)^2 - 9x^2y^2$

$= (x^2 + y^2 + 3xy)(x^2 + y^2 - 3xy)$, *Factors.*

EXERCISE 30.

Factor—

1. $c^4 + c^2x^2 + x^4$.
2. $x^4 + x^2 + 1$.
3. $4x^4 - 13x^2 + 1$.
4. $4a^4 - 21a^2b^2 + 9b^4$.
5. $9x^4 + 3x^2y^2 + 4y^4$.
6. $49c^4 - 11c^2d^2 + 25d^4$.
7. $49x^4 + 34x^2z^2 + 25z^4$.
8. $16x^4 - 9x^2 + 1$.
9. $100x^4 - 61x^2 + 9$.
10. $100x^4 + 11x^2 + 9$.
11. $225a^4b^4 - 4a^2b^2 + 4$.
12. $32a^4 + 2b^4 - 56a^2b^2$.
13. $a^4 + 4b^4$.
14. $1 + 64x^4$.
15. $x^4y^4 + 324$.

EXERCISE 31.

SPECIAL REVIEW.

Factor—

1. $1 - 4x + 4x^2$.
2. $12xy^2 - 3x^3$.
3. $9 + x^2 - 6x - y^2$.
4. $x - 2x^3 + x^5$.
5. $a^2 - 2ab + b^2 - 1$.
6. $x^4 - 14x^2 + 1$.
7. $20x^4y - 45y^3$.
8. $a^2 - b^2 - c^2 + 2bc$.
9. $32x^5 - 2x^3 + 2x$.
10. $2ax^4 - 2a^5$.
11. $49 - 140n^2 + 100n^4$.
12. $4x^3 + x^4 - 4x^2 - 1$.
13. $81 - 18x^2 + x^4$.
14. $48x^4y - 3y^5$.
15. $3x^2 - 3 - 3a^4 + 6a^2$.
16. $(m^2 + n^2)^2 - 4m^2n^2$.
17. $4x^2y^2 - (x^2 + y^2 - 1)^2$.
18. $4a^2x^2 + 25b^2 - 20abx$.
19. $2xy^4 - 2xy^2 - 2x + 4xy$.
20. $1 - 6ab - 9b^2 + 12a^2b - 4a^4 + 9a^2b^2$.
21. $a^2x^2 - b^2 - y^2 + 1 - 2ax + 2by$.
22. $9a^2b^2 - 25 - 16a^2x^2 + 4x^2 - 12abx - 40ax$.
23. $(x^2 + z^2 - 9)^2 - 4x^2z^2$.
24. $49x^4 + 66x^2y^4 + 25y^8$.

CASE IV.

108. A Trinomial of the Form $x^2 + bx + c$.

It was found in Art. 90 that on multiplying two binomials like $x + 3$ and $x - 5$, the product, $x^2 - 2x - 15$, was formed by taking the algebraic sum of $+3$ and -5 to obtain the coefficient of x,—viz. -2,—and taking their product, -15, to form the last term of the result. Hence, in undoing this work to find the factors of $x^2 - 2x - 15$, the essential part of the process is to find two numbers which, added together, will give -2, and multiplied together will give -15.

Ex. 1. Factor $x^2 + 11x + 30$.

The pairs of numbers whose product is 30 are, 30 and 1, 15 and 2, 10 and 3, 6 and 5. Of these, that pair whose sum is also 11 is 6 and 5.

Hence, $x^2 + 11x + 30 = (x + 6)(x + 5)$, *Factors.*

Ex. 2. Factor $x^2 - 8x + 7$.

It is necessary to find two numbers whose product is $+7$, and sum is -8. When the last term is positive, as in this example, the two required numbers must be both positive or both negative, and since their sum is negative, they must be both negative.

$$\therefore x^2 - 8x + 7 = (x - 7)(x - 1), \textit{ Factors.}$$

Ex. 3. Factor $x^2 - x - 30$.

It is necessary to find two numbers whose product is -30, and sum is -1.

Since the sign of the last term is minus, the two numbers must be one positive, the other negative; and since their sum is -1, the greater number must be negative.

$$x^2 - x - 30 = (x - 6)(x + 5), \textit{ Factors.}$$

Ex. 4. Factor $x^2 + 3xy - 10y^2$.

Since $5y$, $-2y$, added give $3y$, multiplied give $-10y^2$,

$$x^2 + 3xy - 10y^2 = (x + 5y)(x - 2y), \textit{ Factors.}$$

Hence, in general, to factor a trinomial of the form $x^2 + bx + c$,

Find two numbers which, multiplied together, produce the third term of the trinomial, and added together give the coefficient of the second term;

x (or whatever takes the place of x), plus the one number, and x plus the other number, are the factors required.

EXERCISE 32.

Factor—

1. $x^2 + 5x + 6$.
2. $x^2 - x - 6$.
3. $x^2 + x - 6$.
4. $x^2 + 7x - 44$.
5. $x^2 - 11x + 30$.
6. $x^2 + x - 30$.
7. $x^2 + 6xy - 16y^2$.
8. $x^2 - 6xy - 16y^2$.
9. $x^2 + 8x + 16$.
10. $x^2 + 5x - 36$.

11. $x^2 - 5x - 36$.
12. $x^4 - 5x^2 - 36$.
13. $x^2 + 3x - 28$.
14. $x^2 - 2x - 48$.
15. $x^2 - 8x - 48$.
16. $x^2 + 16x + 48$.
17. $x^2 + 19x + 48$.
18. $x^2 + 13x - 48$.
19. $x^2 - 22x - 48$.
20. $x^2 - 49x + 48$.
21. $x^2 - 4x - 96$.
22. $x^2y^2 - 23xy + 132$.
23. $x^2 - 5ax - 24a^2$.
24. $x^4 - 9x^2 + 8$.
25. $2a - 14ax - 60ax^2$.
26. $2x^3 - 22x^2 - 120x$.
27. $25x^2y^2 + 5x^3y - 30xy^3$.
28. $56ax^2y + 96axy + 8ax^3y$.
29. $2x^6 - 10x^4 - 28x^2$.
30. $x^5 + x^3 - 20x$.
31. $x^5 - 25x^3 + 144x$.
32. $3x^8 - 51x^4 + 48$.
33. $x^{2n} - x^n - 56$.
34. $a^2b^2 - 11abc^2 - 26c^4$.
35. $3axy^2 - 9ax^2y - 30ax^3$.
36. $5x^5 + 30x^3y^2 - 35xy^4$.
37. $x^2 + (a+b)x + ab$.
38. $x^2 + (2a - 3b)x - 6ab$.
39. $x^2 - (a + 2b^2)x + 2ab^2$.
40. $x^2 + (a + 2b + c)x + (a+b)(b+c)$.
41. $x^2 + (a+b)x + (a-c)(b+c)$.
42. $(x-y)^2 - 3(x-y) - 18$.
43. $2(x^2+2x)^2 - 14(x^2+2x) - 16$.

CASE V.

109. Trinomial of the Form $ax^2 + bx + c$.

From Art. 91 it is evident that the essential part of the process of factoring a trinomial of the form $ax^2 + bx + c$ lies in determining two factors of the first term and two factors of the last term, such that the algebraic sum of the cross products of these factors equals the middle term of the trinomial.

Ex. Factor $10x^2 + 13x - 3$.

The possible factors of the first term are $10x$, x; and $5x$, $2x$. The possible factors of the third term are -3, 1; and 3, -1. In order to determine which of these pairs taken together have the sum of their cross products equal to $+13x$, it is convenient to arrange the pairs thus:

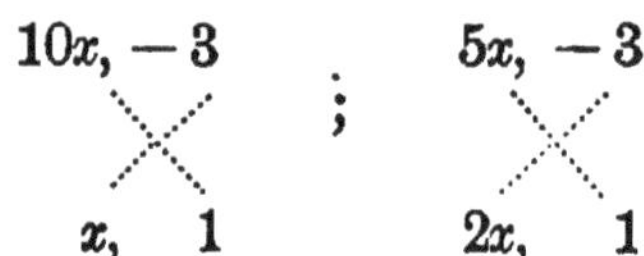

Variations of these may be made mentally by transferring the minus sign from 3 to 1; and also by causing the 3 and the 1 to change places.

It is found that the sum of the cross products of

$$\begin{matrix} 5x, & -1 \\ 2x, & 3 \end{matrix} \quad \text{is } +13x.$$

Hence, $\quad 10x^2 + 13x - 3 = (5x - 1)(2x + 3)$, *Factors.*

Hence, in general, to factor a trinomial of the form $ax^2 + bx + c$,

Separate the first term into two such factors, and the third term into two such factors, that the sum of their cross products equals the middle term of the trinomial;

As arranged for cross multiplication, the upper pair taken together and the lower pair taken together form the two factors.

EXERCISE 33.

Factor—

1. $2x^2 + 3x + 1.$
2. $3x^2 - 14x + 8.$
3. $2x^2 + 5x + 2.$
4. $3x^2 + 10x + 3.$
5. $6x^2 - 7x - 5.$
6. $2x^2 + 5x - 3.$
7. $6x^3 + 20x^2 - 16x$
8. $3x^2 - 4x - 4.$
9. $8x^2 + 2x - 15.$
10. $2x^2 + x - 10.$
11. $12x^2 - 5x - 2.$
12. $4x^2 + 11x - 3.$

13. $5x^2 + 24x - 5$.
14. $9x^3 - 15x^2 - 6x$.
15. $6x^2y - 2xy - 4y$.
16. $16x^2 - 6xy - 27y^2$.
17. $12x^2 + xy - 63y^2$.
18. $42x^2 + 13x - 42$.
19. $32a^2 + 4ab - 45b^2$.
20. $4x^4 - 13x^2 + 9$.
21. $9x^4 - 148x^2 + 64$.
22. $12x^2 - 7xz - 12z^2$.
23. $24x^3 + 104x^2y^2 - 18xy^4$.
24. $25a^4 + 9a^2b^2 - 16b^4$.
25. $16x^4 - 10x^2y^2 - 9y^4$.
26. $3x^{2n} - 8x^ny - 3y^2$.
27. $25a^4 - 41a^2b^2 + 16b^4$.
28. $36x^4 - 97x^2y^2 + 36y^4$.
29. $20 - 9x - 20x^2$.
30. $5 + 32xy - 21x^2y^2$.
31. $(a + b)^2 + 5(a + b) - 24$.
32. $3(x - y)^2 + 7(x - y)z - 6z^2$.
33. $3(x^2 + 2x)^2 - 5(x^2 + 2x) - 12$.
34. $4x(x^2 + 3x)^2 - 8x(x^2 + 3x) - 32x$.
35. $2(x + 1)^2 - 5(x^2 - 1) - 3(x - 1)^2$.

CASE VI.

110. Sum or Difference of Two Cubes.

From Art. 94, $\dfrac{a^3 + b^3}{a + b} = a^2 - ab + b^2$.

Hence, $a^3 + b^3 = (a + b)(a^2 - ab + b^2)$ . . . . (1)

In like manner, $a^3 - b^3 = (a - b)(a^2 + ab + b^2)$ . . . . (2)

But any algebraic expressions may be used instead of a and b in (1) and (2).

Ex. 1. Factor $27x^3 - 8y^3$.

$$27x^3 - 8y^3 = (3x)^3 - (2y)^3.$$

Use $3x$ for a and $2y$ for b in (2) above.

$$27x^3 - 8y^3 = (3x - 2y)(9x^2 + 6xy + 4y^2), \textit{ Factors.}$$

Ex. 2. Factor $a^6 + 8b^9$.

$$a^6 + 8b^9 = (a^2)^3 + (2b^3)^3$$
$$= (a^2 + 2b^3)(a^4 - 2a^2b^3 + 4b^6).$$

Ex. 3. Factor $(a+b)^3-x^3$.

$$(a+b)^3-x^3=[(a+b)-x][(a+b)^2+(a+b)x+x^2].$$

Hence, in general, to factor the sum or difference of two cubes,

Obtain the values of **a** *and* **b** *in the given example, and substitute these values in either the formula* (1) *or* (2).

111. Sum or Difference of any Two like Odd Powers.

Since the *difference* of *two like odd* powers is always divisible by the difference of their roots (see Art. 95), the factors of a^n-b^n, when n is odd, are the divisor, $a-b$, and the quotient.

Ex. 1. $a^5-b^5=(a-b)(a^4+a^3b+a^2b^2+ab^3+b^4)$.

Since the *sum* of *two like odd* powers is divisible by the sum of the roots (see Art. 95), the factors of a^n+b^n, when n is odd, are the divisor, $a+b$, and the quotient.

Ex. 2. $x^5+32y^5=x^5+(2y)^5$.

$$=(x+2y)[x^4-x^3(2y)+x^2(2y)^2-x(2y)^3+(2y)^4]$$
$$=(x+2y)[x^4-2x^3y+4x^2y^2-8xy^3+16y^4], \textit{ Factors.}$$

112. Sum or Difference of any Two Even Powers.

The *difference* of two *even* powers is factored to best advantage by Case III.

Ex. 1.
$$\begin{aligned}x^8-y^8&=(x^4+y^4)(x^4-y^4)\\&=(x^4+y^4)(x^2+y^2)(x^2-y^2)\\&=(x^4+y^4)(x^2+y^2)(x+y)(x-y).\end{aligned}$$

The *sum* of two *even* powers cannot in general be factored by elementary methods unless the expression may be regarded as the sum or difference of two cubes (Art. 110), or other like odd powers.

Ex. 2.
$$\begin{aligned}a^6+b^6&=(a^2)^3+(b^2)^3\\&=(a^2+b^2)(a^4-a^2b^2+b^4), \textit{ Factors.}\end{aligned}$$

But a^2+b^2, a^4+b^4, a^8+b^8, cannot be factored by any elementary method, and are therefore prime expressions.

EXERCISE 34.

Factor—

1. $m^3 - n^3$.
2. $c^3 + 8d^3$.
3. $27 - x^3$.
4. $a^3 + 8b^3c^3$.
5. $x^3 - 125$.
6. $64y^3 - 27$.
7. $a^3b^3 + 1$.
8. $1 - 1000x^3$.
9. $27x^4 + a^3x$.
10. $512x^3 - y^6$.
11. $a + 343a^4$.
12.* $a^6 - x^6$.
13. $x^{12} - y^6$.
14. $a^6 - 64n^{12}$.
15. $250x - 2x^7$.
16. $8x^6 + y^3$.
17. $(a+b)^3 + 1$.
18. $125 + (2b - a)^3$.
19. $8 - (c+d)^3$.
20. $(x-y)^3 - 27x^3$.
21. $16x^4y^6 - 54xz^3$.

---

22. $x^5 + y^5$.
23. $x^7 - y^7$.
24.† $a^6 + m^6$.
25. $x^{12} + y^{12}$.
26. $a^7 - 128b^7$.
27. $a^{11} + x^{11}$.
28. $a^9 + b^9$.
29. $32x^5 - 1$.
30. $a^{11} - b^{11}$.
31. $243 - x^5$.
32. $64 - (a-b)^3$.
33. $8(x-2y)^3 + 1$
34. $a^{10} - b^{10}$.
35. $a^{10} + b^{10}$.
36. $32x^5 - a^{10}$.

CASE VII.

118. A Polynomial whose Terms may be grouped so as to be Divisible by a Binomial Divisor.

Ex. 1. $ax - ay - bx + by = (ax - ay) - (bx - by)$
$= a(x-y) - b(x-y)$
$= (a-b)(x-y)$, *Factors.*

Ex. 2. $1 + 15a^4 - 5a - 3a^3 = 1 - 3a^3 - 5a + 15a^4$
$= (1 - 3a^3) - 5a(1 - 3a^3)$
$= (1 - 3a^3)(1 - 5a)$, *Factors.*

Ex. 3. $x^3 + y^3 + x + y = (x+y)(x^2 - xy + y^2) + (x+y)$
$= (x+y)(x^2 - xy + y^2 + 1)$, *Factors.*

Ex. 4. $a^3 + 3a^2 - 4 = a^3 + 2a^2 + a^2 - 4.$
$= a^2(a+2) + (a+2)(a-2)$
$= (a+2)(a^2 + a - 2)$
$= (a+2)(a+2)(a-1)$, *Factors.*

---

* Use Case III first. † Sum of two cubes.

EXERCISE 35.

Factor—

1. $ax + ay + bx + by.$
2. $x^2 - ax + cx - ac.$
3. $5xy - 10y - 3x + 6.$
4. $3am - 4mn - 6ay + 8ny.$
5. $a^2x + 3ax + acx + 3cx.$
6. $3a^2y + 3aby - 5any - 5bny.$
7. $x^4 + x^3 + 2x^2 + 2x.$
8. $2x^4 - 2x^3 - 2a^2x^2 + 2a^2x.$
9. $y^3 + y^2 + y + 1.$
10. $ax^2 - 2a^2x - x + 2a.$
11. $x^2 + 3y - 3x - xy.$
12. $z^3 - z^2 - z + 1.$
13. $ab - by - a + y.$
14. $x^5 - x^4 - 4x + 4.$
15. $a^2x^2 - b^2x^2 - a^2y^2 + b^2y^2.$
16. $x(x + 4)^2 + 4(x + 4).$
17. $a^2(a + 3) - 3(a + 3).$
18. $2(x^2 - y^2) - (x - y).$
19. $4x(x - 1)^2 + x - 1.$
20. $x^3 - 1 + 2(x^2 - 1).$
21. $4a^2 - a^2x^2 + x^2 - 4.$
22. $4ax^3 + 8ax - 8a - 4ax^2.$
23. $a(3a - x)^2 - 6ax^2 + 2x^3.$
24. $x^3 - 8 - 7(x - 2).$
25. $4(x^3 + 27) - 31x - 93.$
26. $(2x + 1)^3 - (2x + 1)(3x + 4).$
27. $(2x - 3)^3 + 2x^2 - 9x + 9.$
28. $x^3 - 7x - 6.$
29. $x^3 - 3x^2 - 10x + 24.$
30. $x^3 - 8x^2 + 17x - 10.$
31. $6x^3 - 23x^2 + 16x - 3.$

114. General Principles in Factoring. In order that the application of factoring may be as effective as possible, it is important to reduce each expression factored to its prime factors. Hence it is important to use the different methods of factoring in such a way as to give prime factors as a result most readily.

Hence, in factoring any given expression, it is useful to—

1. Observe, first of all, whether all the terms of the expression have a common factor (Case I.); if so, remove it.

2. Determine which other case in Factoring can be used next to the best advantage.

3. If the expression comes under no case directly, try to discover its factors by rearranging its terms, or by adding and subtracting the same quantity to the given expression, or by separating one term into two terms.

4. Continue the process of factoring till each factor can be resolved no further.

Ex. 1. Factor $x^4 - 9x^2$.

This expression as it stands might be factored as the difference of two squares (Case III.), but it is best to apply Case I. first.

$$x^4 - 9x^2 = x^2(x^2 - 9)$$
$$= x^2(x+3)(x-3), \textit{ Factors.}$$

Ex. 2. Factor $a^6 - b^6$.

This expression might be factored by dividing it by $a + b$ or by $a - b$, and taking the divisor and quotient as the factors, but it is factored to best advantage by the use of Case III., and afterward Case VI.

$$a^6 - b^6 = (a^3 + b^3)(a^3 - b^3)$$
$$= (a+b)(a^2 - ab + b^2)(a-b)(a^2 + ab + b^2), \textit{ Factors.}$$

Ex. 3. Factor $x^3 - 6x^2 + 6x - 5$.

$$x^3 - 6x^2 + 6x - 5 = x^3 - 6x^2 + 5x + x - 5$$
$$= x(x-1)(x-5) + (x-5)$$
$$= (x-5)[x(x-1) + 1]$$
$$= (x-5)(x^2 - x + 1), \textit{ Factors.}$$

Ex. 4. Factor $6x^3y - 8x^2y - 2xy^3 - 4x^2y^2 + 2xy$.

$$6x^3y - 8x^2y - 2xy^3 - 4x^2y^2 + 2xy$$
$$= 2xy(3x^2 - 4x - y^2 - 2xy + 1)$$
$$= 2xy(4x^2 - 4x + 1 - x^2 - 2xy - y^2)$$
$$= 2xy[(2x-1)^2 - (x+y)^2]$$
$$= 2xy(3x + y - 1)(x - y - 1), \textit{ Factors.}$$

EXERCISE 36.

Factor—

1. $3x^3 - 3x$.
2. $2x^3 - 8x^2y + 8xy^2$.
3. $x^2 - 11x + 30$.
4. $4x^2 + 5xy - 6y^2$.
5. $12a^2 - 2ab - 30b^2$.
6. $x^4 - 1 - y^2 + 2y$.
7. $40a^3 - 5$.
8. $16x^2 - 40xy + 25y^2$.
9. $x^2 + 3ax - 3a - x$.
10. $3x^7 - 3x$.
11. $4a^4 - 5a^2 + 1$.
12. $2x^8 - 32$.
13. $x^2 + 4x - 45$.
14. $4x^2 + 2a - a^2 - 1$.
15. $5ax^9 - 5a$.
16. $18x^3 - 3x^2 - 36x$.
17. $x^4 + 3x^2z^2 + 4z^4$.
18. $a^2x^2 - 9x^2 - a^2 + 9$.
19. $110 - x - x^2$.
20. $3x^2 + 13xy - 30y^2$.
21. $7a - 7a^3b^4$.
22. $6x^2 + 14x + 8$.
23. $x^4 - (x - 2)^2$.
24. $3a + 3a^4$.
25. $a^3 - a^2 + 2a - 2$.
26. $6x^3 - 2x - 4x^2$.
27. $1 - 23z^2 + z^4$.
28. $128 - 2y^3$.
29. $1 - a^2 - b^2 - 2ab$.
30. $21a^2 - 17a - 30$.
31. $x^{12} + y^{12}$.
32. $8x^3 + 729z^9$.
33. $405x^4y^4 - 45x^4$.
34. $a^5 - 4a^3 + 5a^2 - 20$.
35. $(c + d)^3 - 1$.
36. $(x - y)^2 + 2(x - y)$.
37. $24x^2 + 5xy - 36y^2$.
38. $x^3 - 2x^2y - 4xy^2 + 8y^3$.
39. $(a^2 - 6)^2 - a^2$.
40. $z^4 + z^2 + 1$.
41. $(a^2 - b^2 - c^2)^2 - 4b^2c^2$.
42. $21x^2 - 40xy - 21y^2$.
43. $32 + n^5$.
44. $5x^7 + 5xy^6$.
45. $m^7 + n^7$.
46. $2ax^3 + \frac{1}{4}ay^3$.
47. $1 + x - x^4 - x^5$.
48. $x^2 - 9 - 7(x - 3)^2$.
49. $4a^4 - 37a^2 + 9$.
50. $x^6 - 64$.
51. $x^3 - 27 - 7(x - 3)$.
52. $32x^5y - yz^{10}$.
53. $(x^2 + y^2)^4 - 16x^4y^4$.
54. $x^4 + x^2y^2 - y^2z^2 - z^4$.
55. $ax^4 - ax - x^3y + y$.
56. $4(a^2 - b^2) - 3(a + b)^2$.

57. $a^{12}-1$.

58. $4a^2-9b^2+4a-6b$.

59. $4a^2-9b^2-1-6b$.

60. $36x^3+18x^2-40x$.

61. $(x^2-1)^2+(2x+3)(x-1)^2$.

62. $a^2-b^4-a^2x^3+b^4x^3$.

63. $3x^2-27+ax^2-9a$.

64. $18a^2b+8b-27a^2c-12c$.

65. $3x^3-3x+4x^4-4x^2$.

66. $1-4a^2b^2c^2-9x^2y^2z^2+12abcxyz$.

67. $a^2bcx-amnpx+m^2npy-abcmy$.

68. $4x+4an+x^2-4a^2-n^2+4$.

69. $2(x^3-8)+7x^2-17x+6$.

70. $a^4-4b^4+a^2+2b^2$.

71. $4x^3-19x+15$.

72. $3x^3+7x^2-4$.

---

73. $49x^2-70x+25$.

74. $y^2+4x-1-4x^2$.

75. $49x^4-22x^2y^2+9y^4$.

76. $5x^7y-5xy^4$.

77. x^4+x^3-x-1.

78. $21x^2+2x-55$.

79. $18x^2+52xy-6y^2$.

80. $(x+1)^3-x^6$.

81. $(1-2x)^2-x^4$.

82. $ax^2-cx+ax-c$.

83. $45x^2+8xy-21y^2$.

84. $ax^2+5ax-84a$.

85. $15x^3-5x^2+33x-11$.

86. $x^3y-10x^2y^2z^2+25xy^3z^4$.

87. x^4-79x^2+1.

88. a^2-9+9b^2-6ab.

89. $x^6-4x^4-16x^2+64$.

90. $(x^2+3)^3-64x^6$.

91. $x^4-49y^2+9-6x^2$.

92. $60x^2+119x-60$.

93. $x^4y^4-4x^2+4-y^2-4x^2y^2+4xy$.

94. $a^2nx-bcm^2yz+acmxz-abmny$.

95. $5(x^3+27)-11x^2-46x-39$.

CHAPTER IX.

HIGHEST COMMON FACTOR AND LOWEST COMMON MULTIPLE.

115. Value of Highest Common Factor and Lowest Common Multiple. In the use of factors it is frequently important, in order to do required work most effectively and with least labor, to be able to find the factor of highest degree common to a number of given expressions, or to determine the expression of lowest degree which will contain exactly a number of given expressions.

116. A Common Factor of two or more algebraic expressions is an expression which divides each of the given expressions without a remainder.

The **Highest Common Factor** of two or more algebraic expressions is the product of all their prime common factors.

This product will evidently be the factor highest in degree that will divide each of the original expressions without a remainder.

Ex. 1. The H. C. F. of $4x^2$, $12x^3$, $16x^5y$ is $4x^2$.

Ex. 2. The H. C. F, of $6x^2(x-y)^2$, $15x(x^2-y^2)$ is $3x(x-y)$.

CASE I.

When the Highest Common Factor may be found directly by Inspection.

117. H. C. F. of Monomials.

Ex. Find H. C. F. of $60a^2x^2$, $45ax^3$, $90a^2x^5y$.

By arithmetic the H. C. F. of the coefficients, 60, 45, 90, is 15.

a is common to all of the given expressions, and its least exponent in any of them is 1.

x is common to all the expressions, and its least exponent in any of them is 2.

$$\therefore 15ax^2 \text{ is the H. C. F.}$$

In general,

Take the highest common factor of the coefficients;

Annex the letters common to all of the expressions, giving to each letter the least exponent which it has in any expression.

118. H. C. F. of Polynomials directly Factorable.

Ex. 1. Find the H. C. F. of $x^4 - 3x^3$, $x^3 - 9x$, $x^2 - 6x + 9$.

$$x^4 - 3x^3 = x^3(x - 3)$$
$$x^3 - 9x = x(x + 3)(x - 3)$$
$$x^2 - 6x + 9 = (x - 3)^2.$$
$$\therefore \text{H. C. F.} = x - 3.$$

Ex. 2. Find the H. C. F. of $6x^2y - 12xy^2 + 6y^3$ and $3x^2y^2 + 9xy^3 - 12y^4$.

$$6x^2y - 12xy^2 + 6y^3 = 6y(x - y)^2$$
$$3x^2y^2 + 9xy^3 - 12y^4 = 3y^2(x^2 + 3xy - 4y^2) = 3y^2(x + 4y)(x - y)$$
$$\therefore \text{H. C. F.} = 3y(x - y).$$

Ex. 3. Find H. C. F. of

$$12a^3b^5,\ 8a^2(a - b)^3,\ 16a^2b^5(a + b)^2,\ 4a^3 - 4a^2b.$$
$$\text{H. C. F.} = 4a^2.$$

In general,

Separate each expression into its prime factors;

Multiply together the factors common to all the expressions, taking each common factor the least number of times it occurs in any one expression.

EXERCISE 37.

Find the H. C. F. of—

1. $4a^2b$, $6ab^2$.
2. $5x^3y$, $15x^2y^2$.
3. abc^2, $3a^2bc^3$.
4. $24a^2x^3$, $56a^3x^2$.
5. $14m^2n$, $42am^3$.
6. $24xy$, $48ax^2$, $36x$.
7. $34a^5x^3$, $51ax^5$.
8. a^3x^2y, $a^2x^3y^4z$.
9. $a(a+b)$, a^2-b^2.
10. $(x-y)^2$, x^2-y^2.
11. x^2-3x, x^2-9.
12. $4x^2+6x$, $6x^2+9x$.
13. a^3-x^3, a^2-x^2.
14. x^2+x, x^3+1.
15. $xy-y$, x^3-x.
16. $4a^3+2a^2$, $4a^3-a$.
17. x^2+x, x^2-1, x^2-x-2.
18. x^2+x-12, x^2-x-6, x^2-6x+9.
19. $4a^3x-4ax^3$, $8a^2x^3-8ax^4$, $4a^2x^2(a-x)^2$.
20. $2x^3-2x$, $3x^4-3x$, $4x(x-1)^2$.
21. $6x^2+5xy-4y^2$, $4x^2+4xy-3y^2$.
22. $3x^3-5x^2-2x$, $4x^3-5x^2-6x$, x^3-4x.
23. x^4-81, x^4+8x^2-9, $2x^4+17x^2-9$.
24. $b-a^2b$, $3b-a^2b-2a^4b$, $b^2-a^4b^2$.
25. $1-a^3$, $1-a^6$, $3a+3a^2+3a^3$, $1+a^2+a^4$.
26. $xy+x-y-1$, $14x^2+10x-24$, $3(x^2-1)^2$.

CASE II.

Highest Common Factor Determined Indirectly by Method of Long Division.

119. For **Polynomials that cannot be readily factored** the H. C. F. is found by the same general method that is used in arithmetic to determine the G. C. D. of large numbers. By successive divisions, using the remainder of each division for the next divisor, successive pairs of smaller and smaller numbers are found which have the same H. C. F. as the original numbers, till at last the H. C. F. is determined. The essential parts of the process will be recalled by aid of the following example:

Find the G. C. D. of 182, 299.

```
182)299(1
    182
    117)182(1
        117
         65)117(1
             65
             52)65(1
                52
                13)52(4
                   52
```

∴ G. C. D. of 182, 299 is 13.

120. Principles I. and II. for simplifying the Process of finding H. C. F. by the Division Method.

Let A and B stand for any two algebraic expressions. Then—

I. The H. C. F. of A and B is the same as the H. C. F. of A and mB, or A and $\frac{B}{m}$, provided m is not a factor of A.

That is, one of two algebraic expressions may be multiplied or divided by a quantity which is not a factor of the other expression without changing the H. C. F. of the expressions.

Ex. The H. C. F. of $3x$, $6ax$,
is the same as H. C. F. of $3x$, $12ax$,
and of $3x$, $6x$,
the H. C. F. in all instances being $3x$.

II. The H. C. F. of a pair of expressions, A, B, is the same as H. C. F. of the pair A, $B-mA$.

For any quantity which will divide both A and B will evidently divide $B-mA$.

Conversely, any expression which will divide both A, and $B-mA$ will also divide B.

For any quantity which divides A will divide mA, and, since it divides $B-mA$, must also divide B.

Hence, the H. C. F. of one of these pairs of expressions is the H. C. F. of the other also.

As applied in the method of finding the H. C. F. by the long division method, this principle amounts to this, that the H. C. F. of the *divisor* and *dividend* is the same as the H. C. F. of the simpler pair of quantities, the *divisor*, and *dividend minus quotient* $\times$ *divisor;* that is, of the *divisor* and *remainder*.

Principle I. enables us to use other simplifications in the process of the work.

121. Examples Illustrating the Use of Principles I. and II.

Ex. 1. Find H. C. F. of $4x^2+3x-10$ and $4x^3+7x^2-3x-15$.

Divide the second expression by the first,

$$\begin{array}{r|l|l}
4x^2+3x-10 & 4x^3+7x^2-3x-15 & x+1 \\
 & \underline{4x^3+3x^2-10x} & \\
 & 4x^2+7x-15 & \\
 & \underline{4x^2+3x-10} & \\
 & 4x-5 &
\end{array}$$

By Principle II., Art. 120, the H. C. F. of the two original expressions is the same as that of the simpler pair, $4x^2+3x-10$, $4x-5$.

Proceeding with these,

$$\begin{array}{r|l|l} 4x-5 & 4x^2+3x-10 & x+2 \\ & \underline{4x^2-5x} & \\ & \quad\; 8x-10 & \\ & \quad\; \underline{8x-10} & \end{array}$$

Since $4x-5$ divides the other expression, $4x^2+3x-10$, exactly, it is the H. C. F. of the second pair, and hence of the original pair of expressions.

$$\therefore 4x-5 = \text{H. C. F.}$$

Ex. 2. Find H. C. F. of x^3+4x^2+5x+2 and $3x^3+15x^2+12x$.

The second of these expressions is divisible by $3x$, which is not a factor of the first expression; hence, by Principle I., $3x$ may be removed, and we proceed to find the H. C. F. of

$$x^3+4x^2+5x+2 \text{ and } x^2+5x+4.$$

$$\begin{array}{r|l|l} x^2+5x+4 & x^3+4x^2+5x+2 & x-1 \\ & \underline{x^3+5x^2+4x} & \\ & \quad -x^2+x+2 & \\ & \quad \underline{-x^2-5x-4} & \\ & \qquad\quad 6x+6 & \end{array}$$

We have now to find the H. C. F. of x^2+5x+4 and $6x+6$. But by Prin. I., Art. 120, the factor 6 may be dropped from $6x+6$.

$$\begin{array}{r|l|l} x+1 & x^2+5x+4 & x+4 \\ & \underline{x^2+x} & \\ & \quad 4x+4 & \\ & \quad \underline{4x+4} & \end{array}$$

$$\therefore x+1 = \text{H. C. F.}$$

Ex. 3. Find H. C. F. of $4x^3-4x^2-5x+3$ and $10x^2-19x+6$.

To render the first expression divisible by the second, by Principle I. we may multiply the first expression by 5, which is not a factor of the second expression,

$$
\begin{array}{rlll}
 & 4x^3 - 4x^2 - 5x + 3 & & \\
 & 5 & & \\
10x^2 - 19x + 6 & 20x^3 - 20x^2 - 25x + 15 & 2x & \\
 & 20x^3 - 38x^2 + 12x & & \\
 & \quad 18x^2 - 37x + 15 & & \\
 & \quad 5 & & \\
 & \quad 90x^2 - 185x + 75 & 9 & \\
 & \quad 90x^2 - 171x + 54 & & \\
 & \quad -7 \,\lfloor\, -14x + 21 & & \\
 & \qquad\quad 2x - 3 & 10x^2 - 19x + 6 & 5x - 2 \\
 & & 10x^2 - 15x & \\
 & & \quad - 4x + 6 & \\
 & & \quad - 4x + 6 &
\end{array}
$$

$\therefore 2x - 3 =$ H. C. F.

122. Arrangement of Work. The following will be found a more compact and orderly method of arranging the work of finding the H. C. F. of two expressions (see Ex. 3, Art. 121):

$$
\begin{array}{r|r|l}
10x^2 - 19x + 6 & 4x^3 - 4x^2 - 5x + 3 & \\
 & 5 \qquad\qquad\qquad\quad & \\
 & 20x^3 - 20x^2 - 25x + 15 & 2x \\
 & 20x^3 - 38x^2 + 12x \quad\;\; & \\
 & 18x^2 - 37x + 15 & \\
 & 5 \qquad\qquad\quad & \\
 & 90x^2 - 185x + 75 & 9 \\
10x^2 - 15x & 90x^2 - 171x + 54 & \\
- 4x + 6 & -7 \,\lfloor\, -14x + 21 & \\
- 4x + 6 & \text{H. C. F.} = 2x - 3 & 5x - 2
\end{array}
$$

123. Removal of Simple Factors. It is important for the student to remember that if either one or both of the given polynomials whose H. C. F. is sought have simple factors, these simple factors are to be removed at the outset, and their H. C. F. reserved to be multiplied into the H. C. F. of the remaining polynomial factors as found by the division method.

Ex. Find the H. C. F. of $6x^4 - 30x^3 + 78x^2 - 54x$ and $2x^5 - 4x^4 + 8x^3 - 6x^2$.

$$6x^4 - 30x^3 + 78x^2 - 54x = 6x(x^3 - 5x^2 + 13x - 9)$$
$$2x^5 - 4x^4 + 8x^3 - 6x^2 = 2x^2(x^3 - 2x^2 + 4x - 3).$$

The H. C. F. of $6x$ and $2x^2$ is $2x$.

By the division method let the student determine the H. C. F. of $x^3 - 5x^2 + 13x - 9$ and $x^3 - 2x^2 + 4x - 3$.

This will be found to be $x - 1$.

Combining these results, the H. C. F. of the two original expressions is

$$2x(x - 1).$$

124. The General Process of finding the H. C. F. of two expressions by the division method may now be stated as follows:

Arrange the given expressions according to the descending powers of the same letters;

Remove simple factors of the given expressions, reserving their H. C. F. as a factor of the entire H. C. F.;

Use the expression of lower degree for divisor, or, if both are of the same degree, that whose first term has the smaller coefficient;

Continue each division till the degree of the remainder is lower than the degree of the divisor;

Remove from each remainder each factor that is not a factor of both the given expressions;

If the first term of a dividend is not exactly divisible by the first term of the divisor, multiply the dividend by such a number as will make the term divisible;

Continue the process by using each simplified remainder as a new divisor and the last divisor as a new dividend;

The first divisor to divide its dividend exactly is the H. C. F. of the two original expressions.

EXERCISE 38.

Find the H. C. F. of—

1. $2x^2 - x - 3$ and $4x^3 - 4x^2 - 3x + 5$.
2. $6x^2 - x - 12$ and $6x^3 - 13x^2 - 6x + 18$.
3. $x^3 + x^2 + x - 3$, $x^3 - 3x^2 + 5x - 3$.
4. $3x^3 - 9x^2 + 9x - 3$, $6x^3 - 6x^2 - 6x + 6$.
5. $6x^4 - 5x^3 + 6x^2 + 5x$, $2x^4 - 9x^3 - 9x^2 - 2x$.
6. $3x^3 + x^2 - x + 4$, $3x^3 + 7x^2 + x - 4$.
7. $8x^2 + 2x - 3$, $6x^3 + 5x^2 - 2$.
8. $3x^3 + 7x^2 - 5x + 3$, $2x^3 + 3x^2 - 7x + 6$.
9. $2x^4 + x^3 + 4x - 3$, $3x^4 + 2x^3 - 2x^2 + 3x - 2$.
10. $x^4 - x^3 - x^2 + 7x - 6$, $x^4 + x^3 - 5x^2 + 13x - 6$.
11. $x^4 + 3x^3 + 9x^2 + 12x + 20$, $x^5 + 6x^4 + 6x^3 + 8x^2 + 24x$.
12. $2x^3 - 16x + 6$, $5x^6 + 15x^5 + 5x + 15$.
13. $2x^5 + x^4 + 2x^3 - x^2 - 1$, $5x^4 + 2x^3 + 3x^2 - 2x + 1$.
14. $4x^5 - 10x^4 + 10x^3 - 10x^2 + 6x$, $4x^5 - 14x^4 + 8x^3 + 10x^2 - 6x$.
15. $3x^4 + 2x^3y + 2x^2y^2 + 5xy^3 - 2y^4$, $6x^4 + x^3y + 2x^2y^2 + 2xy^3 - y^4$.
16. $3x^5 + 2x^4 - 8x^3 - 3x^2 + 4x$, $3x^5 - 10x^4 + 14x^3 - 11x^2 + 4x$.
17. $2x^4 - 3x^3 + 2x^2 - 3x + 2$, $3x^4 - 4x^3 + 5x^2 - 6x + 2$.

The H. C. F. of three or more expressions may be obtained by finding that of two of them; then find the H. C. F. of this and another of the quantities; the last H. C. F. thus obtained is the one required.

18. $x^3 - x^2 - x - 2$, $x^3 - 2x^2 + 3x - 6$, $2x^3 - 3x^2 - x - 2$.
19. $2x^3 - 3x^2 - 5x - 12$, $3x^4 - 7x^3 - 2x^2 - 12x$, $x^3 - 9x^2 + 27x - 27$.
20. $2x^4 - 14x^2 + 12x$, $2x^4 + 6x^3 - 32x^2 + 24x$, $6x^4 - 30x^3 + 42x^2 - 18x$.

LOWEST COMMON MULTIPLE.

125. A **Common Multiple** of two or more algebraic expressions is an expression which will contain each of them without a remainder.

The **Lowest Common Multiple** of two or more algebraic expressions is the expression of lowest degree which will contain them all without a remainder.

Ex. 1. The lowest common multiple, or L. C. M., of $3a^3$, $6a^2x$, $4ax^2$ is $12a^3x^2$.

Ex. 2. The L. C. M. of $3x$ and $4y$ is $12xy$.

CASE I.

When the Lowest Common Multiple may be found directly by Inspection.

126. L. C. M. of Monomials.

Take the L. C. M. of the coefficients;

Annex each literal factor that occurs in any of the given expressions, giving the letter the highest exponent which it has in any one expression.

Ex. Find L. C. M. of $4a^2x^2$, $5ax^3$, $10a^2x^5y$.

The L. C. M. of 4, 5, 10 is 20.
The highest exponent of a is 2.
" " " x is 5.
" " " y is 1.
$\therefore 20a^2x^5y =$ L. C. M.

127. L. C. M. of Polynomials readily Factored.

Ex. Find L. C. M. of $x^4 - 3x^3$, $x^3 - 9x$, $x^2 - 6x + 9$.

$$x^4 - 3x^3 = x^3(x-3)$$
$$x^3 - 9x = x(x+3)(x-3)$$
$$x^2 - 6x + 9 = (x-3)^2$$
$$\therefore \text{L. C. M.} = x^3(x+3)(x-3)^2.$$

Hence, in general,

Separate each expression into its prime factors;

Take the product of all the different factors, using each factor the greatest number of times which it occurs in any one expression.

EXERCISE 39.

Find the L. C. M. of—

1. $3a^2b,\ 2ab^2.$
2. $6x^3y,\ 8y^3z.$
3. $12a^2x^2,\ 9a^2y^2.$
4. $16x^2y^3,\ 12x^3y^2.$
5. $2ac,\ 3bc,\ 4ab.$
6. $3a^2b,\ 4ac^2,\ 6b^2c.$
7. $42x^3y^2,\ 28y^3z^2.$
8. $12a^2b,\ 16ab^3,\ 24a^2b^2.$
9. $7a^2,\ 2ab,\ 6b^2,\ 21.$
10. $3x^3,\ 8,\ 6x^2y,\ 12xy^2.$
11. $2x(x+1),\ x^2-1.$
12. $a^2+ab,\ ab+b^2.$
13. $7x^2,\ 2x^2-6x.$
14. $x^3-1,\ x^2-1.$
15. $x^2-y^2,\ x^2-3xy+2y^2.$
16. $3x^3-3x,\ 6x^2-12x+6.$
17. $ax^2(x-y)^2,\ bxy(x^2-y^2).$
18. $x^2-3x-40,\ x^2-9x+8.$
19. $3x^2+2x-8,\ 6x^2+x-12.$
20. $a^2-b^2,\ a^3-b^3,\ a^3+b^3.$
21. $6x^2+6x,\ 2x^3-2x^2,\ 3x^2-3.$
22. $a^2b+ab^2,\ a^2b-ab^2,\ 3a^2-3b^2.$
23. $2x^2+x-1,\ 4x^2-1,\ 2x^2+3x+1.$
24. $3x^3-3,\ 6x^2-12x+6,\ 2x^3+2x^2+2x.$
25. $12x^3-2x^2-140x,\ 18x^2+6x-180,\ 6x^3-39x^2+63x.$
26. $1-x+x^2-x^3,\ 1+x+x^2+x^3,\ 2x-2x^3.$
27. $(x-1)^3,\ 7xy^3(x^2-1)^2,\ 14x^5y(x+1)^3.$
28. $18x^3-12x^2+2x,\ 27x^5-3x^3,\ 18x^3-24x^2+6x.$
29. $36x^4-81x^2,\ 16x^5-48x^4+36x^3,\ 24x^4+72x^3+54x^2.$
30. $x^2-1-2a-a^2,\ x^2-1+a^2+2ax,\ x^2+1-a^2-2x.$
31. $(x-1)(x+3)^2,\ (x+1)^2(x-3),\ (x^2-1)^2,\ x^2-9.$

CASE II.

Lowest Common Multiple Determined Indirectly by the Division Method.

128. If it be required to find the L. C. M. of expressions which cannot be factored readily, we proceed in general as in arithmetic when finding the L. C. M. of two large numbers; that is, we first find the H. C. F. of the two numbers.

Thus, to find the L. C. M. of 182, 299 by the division method, the G. C. D. is found to be 13.

Then, since $182 = 13 \times 14, \quad 299 = 13 \times 23,$

$$\begin{array}{r|l} 13 & 13 \times 14 \;,\; 13 \times 23 \\ \hline & \quad 14 \;\;,\;\; 23 \end{array}$$

$$\therefore \text{L. C. M.} = 13 \times 14 \times 23 = 182 \times 23.$$

In brief, we find the G. C. D. of the two numbers, divide one of the numbers by this G. C. D., and multiply the quotient by the other number.

Similarly, to find the L. C. M. of two algebraic expressions which cannot be readily factored, we first find the H. C. F. of the two expressions by the division method.

Thus, to find the L. C. M. of $4x^2 + 3x - 10$ and $4x^3 + 7x^2 - 3x - 15$, we first find the H. C. F. by the division method; this is $4x - 5$.

Then
$$4x^2 + 3x - 10 = (4x - 5)(x + 2)$$
$$4x^3 + 7x^2 - 3x - 15 = (4x - 5)(x^2 + 3x + 3)$$
$$\therefore \text{L. C. M.} = (4x - 5)(x + 2)(x^2 + 3x + 3)$$
$$= (4x^2 + 3x - 10)(x^2 + 3x + 3).$$

Hence, in general,

Find the H. C. F. of the two given expressions;

Divide one of the expressions by the H. C. F., and multiply the quotient by the other.

EXERCISE 40.

Find the L. C. M. of—

1. x^3-5x-2 and x^3-x+6.
2. $3x^3+x^2-x+4$, $3x^3+7x^2+x-4$.
3. $6x^3-3x^2-9x-3$, $6x^4+9x^3+9x^2+3x$.
4. $6x^3-3x^2-10x+5$, $8x^3-4x^2+20x-10$.
5. $8x^4-20x^3-14x^2+5x+3$, $4x^3-3x-1$.
6. $12x^3-8x^2-27x+18$, $18x^3-27x^2-8x+12$.
7. $3x^3-2x^2-1$, $4x^3-5x+1$.
8. $2x^3+3x^2-x+2$, $3x^3-x^2-9x+10$.

The L. C. M. of three or more expressions may be obtained by finding that of two of them; then finding the L. C. M. of this result and the third expression. The last L. C. M. thus obtained is the one required.

9. x^3-7x+6, x^3+7x^2-36, $x^3-31x+30$.
10. $4x^3-13x+6$, $4x^3-4x^2-5x+3$, $3x^3+7x^2-4$.

Find the H. C. F. and L. C. M. of—

11. $20a^2b^3c^3$, $35a^3b^2d^3$, $14a^3b^3c^2$, $10a^3b^2c^2d^2$.
12. $3a^2b(x^2-1)^2$, $6ab^2(5x^2+3x-2)^2$, $9(3x^2+5x+2)^2$.
13. $x^4+2x^3+x^2-4$, x^4-x^2+4x-4.
14. $4x^3-3x-1$, $2x^3-3x^2+1$, $6x^3-x^2+1$.
15. $3x^3+2x^2-7x+2$, $4x^3-12x+8$, $9x^3+12x^2-11x+2$.

CHAPTER X.

FRACTIONS.

129. Origin and Use of Fractions. It is sometimes necessary to indicate the division of one algebraic expression by another, but apart from this it is often useful to do so. For when a number of indicated quotients is combined in a process, cancellations and other simplifications are possible before making the final reduction, and in this way much labor is saved.

130. A **Fraction** is the quotient of two algebraic expressions indicated in the form $\frac{a}{b}$.

This form of indicating a quotient is preferred, since it enables us readily to discriminate the parts of a fraction from the rest of an expression, and hence to compare the parts of different fractions to the best advantage.

Thus, the fractional part of the expression is more readily perceived in $x^2 + \frac{x-1}{x+2} + 5$, than in $x^2 + (x-1) \div (x+2) + 5$.

131. The **Numerator** is the dividend part of the indicated quotient, or part above the line; the divisor, or part below the line, is called the **Denominator**. The numerator and denominator are called the *Terms* of the fraction.

If $(5x+2) \div 3x^2$ be written as a fraction, we have $\frac{5x+2}{3x^2}$;

that is, the dividing line of a fraction takes the place of a parenthesis, and hence is in effect a vinculum.

132. An **Integral Expression** is one which does not contain a fraction; as, $3x^2 - 2y$.

183. A Mixed Expression is one which is part integral, part fractional.

$$\text{Ex. } 3x^2 + x - 5 + \frac{x+1}{3x^2 - 2}.$$

184. Sign of a Fraction. A fraction has its own sign, which is distinct from the sign of both numerator and denominator. It is written to the left of the dividing line of the fraction.

GENERAL PRINCIPLES.

185. A. *If the numerator and denominator of a fraction be both multiplied or both divided by the same quantity, the value of the fraction is not changed.*

This principle is seen to be true at once, since the terms of a fraction are a dividend and a divisor. It is a useful exercise, however, to derive it from the fundamental laws of algebra (see Art. 33).

$$\frac{a}{b} = a \div b = a \div b \times m \div m$$

$$= a \times m \div b \div m \quad \text{(Comm. Law.)}$$

$$= \frac{a \times m}{b} \div m$$

$$= \frac{a \times m}{b \times m}.$$

Similarly, $$\frac{a}{b} \times m = \frac{am}{b},$$

and $$\frac{a}{b} \div m = \frac{a}{bm}.$$

186. B. **Law of Signs.** By the laws of signs for multiplication and division (see Arts. 52, 64),

$$\frac{a}{b} = \frac{-a}{-b}, \ -\frac{a}{b} = \frac{-a}{b} = \frac{a}{-b}, \ \frac{a}{bc} = -\frac{a}{-b \times c} = \frac{a}{-b \times -c}.$$

Or, in general,

The signs of any **even** *number of factors of the numerator and denominator of a fraction may be changed without changing the sign of the fraction.*

But if the signs of an **odd** *number of factors be changed, the sign of the fraction must be changed.*

TRANSFORMATIONS OF FRACTIONS.

I. To Reduce a Fraction to its Lowest Terms.

137. A fraction is in its *lowest terms* when its numerator and denominator have no common factor.

138. Direct Reduction. When the terms of the fraction are monomials, or polynomials readily factored,

Resolve the numerator and denominator into their prime factors, and cancel the factors common to both.

Ex. 1. Reduce $\dfrac{36a^3x^2}{48a^2x^3y^2}$ to its lowest terms.

Divide both numerator and denominator by $12a^2x^2$ (see Art. 135).

$$\therefore \frac{36a^3x^2}{48a^2x^3y^2} = \frac{3a}{4xy^2}.$$

Ex. 2. $\dfrac{9ab - 12b^2}{12a^2 - 16ab} = \dfrac{3b(3a - 4b)}{4a(3a - 4b)} = \dfrac{3b}{4a}.$

The student should notice particularly that in reducing a fraction to its lowest terms it is allowable to cancel a *factor* which is common to both denominator and numerator, but that it is not allowable to cancel a *term* which is common unless this term be a factor.

Thus, $\dfrac{ab}{ac}$ reduces to $\dfrac{b}{c}$;

but in $\frac{a+x}{a+y}$, a of the numerator will not cancel a of the denominator.

This is a principle very frequently violated by beginners.

139. Finding H. C. F. of Numerator and Denominator by Division Method. When the numerator and denominator of a fraction cannot be factored by inspection,

Find the H. C. F. of the numerator and denominator by the method of Art. 124, and divide both numerator and denominator by their H. C. F.

Ex. Simplify $\frac{6x^3-11x^2+2}{9x^3-22x-8}$.

The H. C. F. of the numerator and denominator is found to be $3x^2-4x-2$.

Dividing both numerator and denominator by this,

$$\frac{6x^3-11x^2+2}{9x^3-22x-8}=\frac{2x-1}{3x+4}, \textit{ Result.}$$

EXERCISE 41.

Reduce to their simplest form—

1. $\frac{8a^3x^4}{12a^2x^5}$.
2. $\frac{12x^4yz^5}{15x^3y^2z^5}$.
3. $\frac{3a^2x}{6a^2-9a^3x}$.
4. $\frac{72x^2y^3z^4}{96xy^5z^3}$.
5. $\frac{2a}{4a^2-2a}$.
6. $\frac{3x-6y}{6ax-12ay}$.
7. $\frac{4x+4y}{4ax+4ay}$.
8. $\frac{x^2-y^2}{(x+y)^2}$.
9. $\frac{8(x^2-1)}{12x-12}$.
10. $\frac{45(x-y)^2}{18(x-y)^3}$.
11. $\frac{a^2b+ab^2}{2a^2b-2ab^2}$.
12. $\frac{6xy}{9x^2y-12xy^2}$.
13. $\frac{2x^2-3xy}{4x^3-9xy^2}$.
14. $\frac{49x^2-64y^2}{14x^3-16x^2y}$.

15. $\dfrac{(x-y)^2(x+y)^3}{(x^2-y^2)^3}$.

16. $\dfrac{2x^2-8y^2}{4x^2-2xy-12y^2}$.

17. $\dfrac{6x^2-xy-2y^2}{6x^2-7xy+2y^2}$.

18. $\dfrac{(a+b)^2-c^2}{a^2-(b+c)^2}$.

19. $\dfrac{1-(a-x)^2}{x^2-(a-1)^2}$.

20. $\dfrac{4-(a+b)^2}{(a-2)^2-b^2}$.

21. $\dfrac{12x^2-2ax-24a^2}{4x^2-2ax-6a^2}$.

22. $\dfrac{x^3-8}{x^2y^2+2xy^2+4y^2}$.

23. $\dfrac{x^4-9x^2}{x^4-x^3-6x^2}$.

24. $\dfrac{x^6-y^6}{x^4+x^2y^2+y^4}$.

25. $\dfrac{x^3-2x-1}{x^3-2x^2+1}$.

26. $\dfrac{18x^4+19x^2y^2-12y^4}{27x^4+6x^2y^2-8y^4}$.

27. $\dfrac{x^3-8x^2+17x-10}{x^4-2x^3-4x^2+11x-6}$.

28. $\dfrac{2ax^3+ax^2-9a}{3x^3-3x-18}$.

29. $\dfrac{3x^3+4x^2-x+6}{2x^3+7x^2+4x-4}$.

30. $\dfrac{2x^5-11x^2-9}{4x^5+11x^4+81}$.

31. $\dfrac{x^2-z^2-4-2xy-4z+y^2}{z^2-x^2-4-2yz-4x+y^2}$.

II. To Reduce an Improper Fraction to an Integral or Mixed Quantity.

140. An **Improper Fraction** is one in which the degree of the numerator equals or exceeds the degree of the denominator.

Since a fraction is an indicated division, to reduce an improper fraction to an integral or mixed expression,

Divide the numerator by the denominator;

If there be a remainder, write it over the denominator, and annex the result to the quotient with the proper sign.

Ex. 1. Reduce $\dfrac{x^3 - y^3}{x + y}$ to an integral or mixed expression.

$$
\begin{array}{rrr|l}
x^3 & & - y^3 & \; x + y \\
x^3 & + x^2y & & \; x^2 - xy + y^2 \\ \hline
& - x^2y & - y^3 & \\
& - x^2y & - xy^2 & \\ \hline
& & xy^2 - y^3 & \\
& & xy^2 + y^3 & \\ \hline
& & - 2y^3. &
\end{array}
$$

$$\therefore \frac{x^3 - y^3}{x + y} = x^2 - xy + y^2 - \frac{2y^3}{x + y}, \textit{ Result.}$$

Ex. 2. Reduce $\dfrac{x^3 + 4x^2 - 5}{x^2 + x + 2}$.

$$
\begin{array}{rrrr|l}
x^3 & + 4x^2 & & - 5 & \; x^2 + x + 2 \\
x^3 & + x^2 & + 2x & & \; x + 3 \\ \hline
& 3x^2 & - 2x & - 5 & \\
& 3x^2 & + 3x & + 6 & \\ \hline
& & - 5x & - 11 &
\end{array}
$$

$$\therefore \frac{x^3 + 4x^2 - 5}{x^2 + x + 2} = x + 3 - \frac{5x + 11}{x^2 + x + 2}, \textit{ Result.}$$

When the remainder is made the numerator of a fraction with the minus sign before it, as in this example, the signs of terms of the remainder must be changed, since the vinculum is in effect a parenthesis (see Art. 48).

EXERCISE 42.

Reduce each to a mixed quantity—

1. $\dfrac{x^2 - 2x + 3}{x}$.
2. $\dfrac{4x^3 + 6x - 5}{2x}$.
3. $\dfrac{10a^3x^3 + 5ax - 7 - a}{5ax}$.
4. $\dfrac{x^3 - 3x^2 + x - 1}{x + 1}$.
5. $\dfrac{x^2 + 3xy - 2y^2 - 1}{x + y}$.
6. $\dfrac{3x^4 - 13x - 28}{x^2 - 3}$.
7. $\dfrac{x^5 - x^3 - x + 2 - a}{x - 1}$.
8. $\dfrac{x^4 + 1}{x^2 + x - 1}$.

9. $\dfrac{x^5}{x^2-x-1}$.

10. $\dfrac{2x^4+7}{x^2+x+1}$.

11. $\dfrac{x^4+x^3-x-1}{x^2+2}$.

12. $\dfrac{9a^3}{3a^2-2b}$.

13. $\dfrac{x^3+x^2-4x+7}{x+3}$.

14. $\dfrac{2a^2}{a+b}$.

15. $\dfrac{x^5-x^4+x^3-2x}{x^3+1}$.

16.* $\dfrac{1}{1+x}$.

17. $\dfrac{1}{1+x-x^2}$.

18. $\dfrac{8}{2+x-x^2}$.

III. To Reduce a Mixed Expression to a Fraction.

141. It is necessary simply to reverse the process of Art. 140 in order to reduce a mixed expression to a fraction. Hence,

Multiply the integral expression by the denominator of the fraction, and add the numerator to the result, changing the signs of the terms of the numerator if the fraction be preceded by the minus sign;

Write the denominator under the result.

Ex. 1. Reduce $a-1+\dfrac{a-2}{a+3}$ to the fractional form.

$$a-1+\frac{a-2}{a+3}=\frac{a^2+2a-3+a-2}{a+3}=\frac{a^2+3a-5}{a+3}, \textit{ Result.}$$

Ex. 2. $x+y-\dfrac{x^2+y^2}{x-y}$

$$=\frac{(x+y)(x-y)-(x^2+y^2)}{x-y}$$

$$=\frac{x^2-y^2-x^2-y^2}{x-y}=\frac{-2y^2}{x-y}$$

$$=\frac{2y^2}{y-x}, \textit{ Result.}$$

* To three integral terms.

EXERCISE 43.

Reduce to a fraction—

1. $a-1+\dfrac{1}{a}$.
2. $x+1+\dfrac{1}{x-1}$.
3. $x^2+x-1-\dfrac{1}{x-1}$.
4. $4x-2-\dfrac{y-2}{2x+1}$.
5. $a-b+\dfrac{2b^2}{a+2b}$.
6. $x-1-\dfrac{x-1}{x^2+x+1}$.
7. $a-x+1-\dfrac{x-1}{a+x}$.
8. $\dfrac{1-a^2}{2a}+a-1$.
9. $\dfrac{2}{a-1}+a+2$.
10. $\dfrac{3a-1}{a-2}+a-1$.
11. $x-a-\dfrac{ay-a^2}{x+a}+y$.
12. $1-\left(x-x^2+\dfrac{1}{1+x}\right)$.
13. $x^2-\left[x-\left(1-\dfrac{2}{x+1}\right)\right]$.
14. $x^3-\left[1+\dfrac{x-1}{x^2+1}\right]$.
15. $x^3-\left\{-x^2-\left[x+1-\dfrac{1}{1-x}\right]\right\}$.
16. $1-\left\{x-\left[2x^2-\dfrac{x^2-x^3}{1+x}-x^3\right]\right\}$.

IV. To Reduce Fractions to Equivalent Fractions of the Lowest Common Denominator.

142. Since by Art. 135 we may multiply the numerator and denominator of a fraction by the same quantity without altering the value of the fraction, we can use the same process as in arithmetic for reducing fractions to their lowest common denominator.

It is supposed at the outset that each fraction has been reduced to its lowest terms.

Find the lowest common multiple of the denominators of the given fractions;

Divide this common multiple by the denominator of each fraction;

Multiply each quotient by the corresponding numerator; the results will form the new numerators;

Write the lowest common denominator under each new numerator.

Ex. 1. Reduce $\frac{2}{3ax}$, $\frac{3}{4a^2x}$, $\frac{5}{6ax^2}$ to equivalent fractions having the lowest common denominator.

The L. C. D. is $12a^2x^2$.

Dividing this by each of the denominators, the quotients are $4ax$, $3x$, $2a$.

Multiplying each of these quotients by the corresponding numerator and setting the results over the common denominator, we obtain

$$\frac{8ax}{12a^2x^2},\ \frac{9x}{12a^2x^2},\ \frac{10a}{12a^2x^2}.$$

Ex. 2. Reduce to their lowest common denominator $\frac{x}{x-y}$, $\frac{x}{x+y}$, $\frac{1}{x^2-y^2}$.

The L. C. D. is $x^2 - y^2$.

Dividing this by each denominator, the quotients are $x + y$, $x - y$, 1.

Multiplying each quotient by the corresponding numerator and setting each result over the common denominator, we obtain

$$\frac{x^2+xy}{x^2-y^2},\ \frac{x^2-xy}{x^2-y^2},\ \frac{1}{x^2-y^2}.$$

EXERCISE 44.

Reduce to equivalent fractions having the lowest common denominator—

1. $\frac{2x}{9}$, $\frac{5x}{6}$.
2. $\frac{12a}{5b}$, $\frac{7}{10}$, $\frac{a}{b}$.
3. $\frac{1}{2ab^2}$, $\frac{2}{a^2b}$, $\frac{1}{ab}$.
4. $\frac{3}{2ac}$, $\frac{2}{4bc}$, $\frac{1}{3ab}$.
5. $\frac{2}{3x}$, 5, $\frac{13}{6x}$.
6. $\frac{2}{3a^2}$, $\frac{3}{4ax}$, $2a$, $\frac{1}{x}$.

7. $\frac{ac}{bd}, \frac{ab}{cd}, \frac{bc}{ad}, \frac{ad}{bc}.$

8. $\frac{1}{x^2-1}, \frac{3}{x+1}.$

9. $\frac{1}{a^2-a}, 2, \frac{3}{a-1}.$

10. $\frac{x}{1+x}, 1, \frac{1}{x}, \frac{1}{x+x^2}.$

11. $\frac{x}{x^2-1}, \frac{1}{x^3-1}.$

12. $\frac{1}{4x^2-9}, \frac{1}{2x+3}, \frac{1}{x}.$

13. $m, \frac{n}{n-m}, \frac{1}{n^2-m^2}.$

14. $\frac{x+1}{(x+3)^2}, 2, \frac{x-3}{x(x+3)}.$

15. $\frac{1+x}{2-2x}, 7, \frac{1-x}{3+3x}.$

16. $\frac{3}{x^3-1}, \frac{4}{x^2+x+1}, 4.$

17. $\frac{3}{a^2b+ab^2}, \frac{4}{a^2b-ab^2}.$

18. $\frac{1}{3x-6}, \frac{5}{2x+4}, \frac{3}{x^2-4}.$

19. $\frac{2}{x-x^3}, \frac{x}{3+3x}, \frac{x}{2-2x}.$

20. $\frac{1}{4-x^2}, \frac{2}{2x+x^2}, \frac{3}{4-2x}.$

21. $\frac{x+1}{x^2+x-6}, 12, \frac{x-1}{x^2+4x+3}.$

22. $\frac{2}{(a+b)^2}, ab, \frac{1}{(a-b)^2}, 4, \frac{ab}{(a^2-b^2)^2}.$

23. $\frac{1}{2x^2+3x-2}, \frac{1}{x^2+3x+2}, \frac{1}{2x^2+x-1}.$

PROCESSES WITH FRACTIONS.

I. Addition and Subtraction of Fractions.

143. By the Distributive Law (Art. 33), inverting the order of the expressions,

$$\frac{a}{c}+\frac{b}{c}=\frac{a+b}{c}.$$

Hence, to add or subtract fractions,

Reduce the fractions to their lowest common denominator;

Add their numerators, changing the signs of the numerator of any fraction preceded by the minus sign;

Set the sum over the common denominator;

Reduce the sum to its lowest terms.

Ex. 1. $\frac{1}{x}-\frac{2}{x+1}+\frac{1}{x+2}$

$$=\frac{(x+1)(x+2)-2x(x+2)+x(x+1)}{x(x+1)(x+2)}$$

$$=\frac{x^2+3x+2-2x^2-4x+x^2+x}{x(x+1)(x+2)}$$

$$=\frac{2}{x(x+1)(x+2)}.$$

Ex. 2. $\frac{a}{a-1}-a+\frac{1}{a^2-a}+\frac{1}{a}.$

$$=\frac{a}{a-1}-\frac{a}{1}+\frac{1}{a^2-a}+\frac{1}{a}$$

$$=\frac{a^2-a^3+a^2+1+a-1}{a(a-1)}$$

$$=\frac{-a^3+2a^2+a}{a(a-1)}=\frac{-a^2+2a+1}{a-1}, \textit{ Result.}$$

EXERCISE 45.

Collect—

1. $\frac{3}{2x}+\frac{2}{x}-\frac{1}{3x}.$
2. $\frac{2}{3a}-\frac{3}{4ax}+\frac{1}{x}.$
3. $\frac{5}{2ac}-\frac{2}{3ab}-\frac{1}{bc}.$
4. $\frac{1}{4x}+1-\frac{2}{x}.$
5. $\frac{3a-b}{2a}+\frac{a-4b}{3b}.$
6. $\frac{a+2b}{2ab}-\frac{6a-1}{6a^2}.$

7. $\dfrac{2a^2x+3}{4ax^2}+1-\dfrac{3a+x}{6x}$.

8. $1+\dfrac{4x-3}{6}-\dfrac{3x+2}{5}$.

9. $x-\dfrac{3x+1}{8}+\dfrac{1-3x}{6}$.

10. $\dfrac{1}{a-b}-\dfrac{1}{a+b}$.

11. $\dfrac{x-1}{x+1}-\dfrac{x+1}{x-1}$.

12. $\dfrac{x+1}{x}-3+\dfrac{7-3x}{3x^2}$.

13. $\dfrac{3a-4b}{2}-\dfrac{2a-b-c}{3}+\dfrac{15a-4c}{12}$.

14. $\dfrac{3x-1}{7}-\dfrac{x-6}{4}+\dfrac{x+2}{28}+2+\dfrac{2x-4}{12}$.

15. $\dfrac{2x^2y-3z}{3x^2y}-\dfrac{xz^2-y^2z}{2xy^2}+\dfrac{y-3xz^2}{6x^2z}-\dfrac{2}{3}$.

16. $\dfrac{a}{a-b}-\dfrac{b}{a+b}$.

17. $\dfrac{1}{x-3}-\dfrac{1}{x-4}$.

18. $\dfrac{x+1}{x-2}+\dfrac{1-x}{x+2}$.

19. $\dfrac{m+1}{(m-1)^2}+\dfrac{2m}{m^2-1}$.

20. $\dfrac{x}{x-1}+1-\dfrac{x}{x+1}$.

21. $\dfrac{3x}{x+2}-\dfrac{2x}{x-2}+\dfrac{10x}{x^2-4}$.

22. $\dfrac{1}{x^2+x}-2+\dfrac{2x^2}{x^2-x}$.

23. $\dfrac{1}{3x-3}-\dfrac{1}{2x+2}+\dfrac{x-5}{6x^2-6}$.

24. $\dfrac{10}{9-a^2}-\dfrac{2}{3+a}-\dfrac{1}{3-a}$.

25. $\dfrac{2}{2x-1}+\dfrac{3}{4x+2}-\dfrac{7x}{4x^2-1}$.

26. $\dfrac{x}{x^2-1}+2-\dfrac{x-1}{x+1}-\dfrac{x-2}{x-1}$.

27. $\dfrac{3}{x+1}-\dfrac{4}{x+2}+\dfrac{2}{x+3}$.

28. $\dfrac{x+2}{2x^2+x-1}-\dfrac{x-3}{4x^2-1}+\dfrac{2x+5}{2x^2+3x+1}$.

29. $\dfrac{b}{a+b}-\dfrac{ab}{(a+b)^2}-\dfrac{ab^2}{(a+b)^3}$.

30. $\dfrac{1}{2x^2-x-1}-\dfrac{1}{2x^2+x-3}-\dfrac{1}{4x^2+8x+3}.$

31. $\dfrac{2xy}{x^2-y^2}+\dfrac{3y}{2x}+\dfrac{3x}{2y}-\dfrac{3x^2-3y^2}{2xy}.$

32. $\dfrac{3x-y}{x+2y}+\dfrac{14xy}{x^2-4y^2}-\dfrac{3x+y}{x-2y}.$

33. $1-\dfrac{2}{x-1}+x-\dfrac{3x-1}{x+1}-\dfrac{2x-5}{2}.$

34. $\dfrac{4b^2}{a^2-b^2}-\dfrac{a-b}{a+b}+2-\dfrac{a+b}{a-b}.$

35. $\dfrac{2}{x+4}-\dfrac{x-3}{x^2-4x+16}-\dfrac{x^2}{x^3+64}.$

36. $\dfrac{2}{x^2-3x+2}+\dfrac{2}{x^2-x-2}-\dfrac{1}{x^2-1}.$

37. $\dfrac{5x}{2(x-3)^2}-\dfrac{7}{3x+9}-\dfrac{26}{4x^2-36}.$

38. $\dfrac{1}{x}-\left\{\dfrac{x}{x+1}-\left[\dfrac{1-x}{x^2-x+1}-\dfrac{1}{x+1}\right]-1\right\}-\dfrac{1}{x^3+1}.$

39. $\dfrac{1}{2}-\left\{\dfrac{x^3-6x-3}{2x^3-2}+\left[\dfrac{1}{x-1}-\dfrac{x}{x^2+x+1}\right]\right\}.$

40. $\dfrac{3}{x-2}-\left[\dfrac{1}{x}+\dfrac{3x^2-4x-1}{x^2-3x+2}-\left(\dfrac{4}{x-1}+3\right)\right].$

144. Changing Signs of Factors. The process of reducing fractions to their lowest comm n denominator is frequently simplified by changing the sign of one or more of the factors of a denominator, at the same time making the necessary change in the sign of the fraction. It is to be remembered from Art. 136 that if the sign of an even number of factors be changed, the sign of the fraction is unchanged; but if the sign of an odd number of factors be changed, the sign of the fraction is changed.

Ex. 1. Simplify $\frac{x^2}{x^2-1}+\frac{x}{x+1}-\frac{x}{1-x}$.

The factors of x^2-1 are $x+1$, $x-1$. Hence, if the sign of the denominator, $1-x$, be changed, it will become $x-1$, and be a factor of x^2-1. But by Art. 136, if the sign of $1-x$ be changed, the sign of the fraction in which it occurs must also be changed. Hence, we have

$$\frac{x^2}{x^2-1}+\frac{x}{x+1}+\frac{x}{x-1}=\frac{x^2+x^2-x+x^2+x}{x^2-1}=\frac{3x^2}{x^2-1}, \text{ Sum.}$$

Where the differences of three letters occur as factors in the various denominators, it is useful to have some standard order for the letters in the factors. It is customary to reduce the factors so that the alphabetical order of the letters be preserved in each factor, except that the last letter be followed by the first.

This is called the *cyclic order.*

Thus, $a-b$, $b-c$, $c-a$ obey the cyclic order.

Ex. 2. Simplify

$$\frac{1}{(a-b)(c-a)}+\frac{1}{(a-b)(c-b)}+\frac{1}{(c-b)(a-c)}.$$

Changing $c-b$ to $b-c$, and $a-c$ to $c-a$ where they occur, we obtain

$$\frac{1}{(a-b)(c-a)}-\frac{1}{(a-b)(b-c)}+\frac{1}{(b-c)(c-a)}$$

$$=\frac{b-c-c+a+a-b}{(a-b)(b-c)(c-a)}$$

$$=\frac{2a-2c}{(a-b)(b-c)(c-a)}=\frac{-2}{(a-b)(b-c)}, \text{ Sum.}$$

EXERCISE 46.

Collect—

1. $\frac{3x}{x^2-1}+\frac{4}{1-x}+\frac{1}{1+x}$.

2. $\frac{2a}{a^2-b^2}+\frac{1}{a+b}+\frac{2}{b-a}$.

3. $\dfrac{3xy}{x^2-4y^2}-\dfrac{y-x}{2y+x}+\dfrac{y+x}{2y-x}$.

4. $\dfrac{1}{b}-\dfrac{1}{a+b}+\dfrac{1}{a-b}+\dfrac{a^2}{b^3-a^2b}$.

5. $\dfrac{1}{x-1}+\dfrac{1}{1+x}+\dfrac{2x}{1-x^2}$.

6. $\dfrac{5}{1+2a}-\dfrac{3a}{1-2a}+\dfrac{4-13a}{4a^2-1}$.

7. $\dfrac{x^2+y^2}{x^2-y^2}-\dfrac{x}{x+y}+\dfrac{y}{y-x}$.

8. $\dfrac{3}{8-8a}+\dfrac{5}{4a+4}-\dfrac{7a}{8a^2-8}$.

9. $\dfrac{3}{x}+\dfrac{2}{x-1}+\dfrac{5x}{1-x^2}-\dfrac{1}{x+1}-\dfrac{3}{x+x^2}$.

10. $\dfrac{1}{(x-2)(3-x)}-\dfrac{1}{10-7x+x^2}-\dfrac{1}{(5-x)(x-3)}$.

11. $\dfrac{2}{(a-3)(b-2)}-\dfrac{3}{(a-2)(2-b)}+\dfrac{4}{(a-2)(3-a)}$
$+\dfrac{5}{(a-3)(2-b)}$.

12. $\dfrac{5a}{6a-18}+\dfrac{5a}{27-3a^2}-\dfrac{a}{4a+12}$.

13. $\dfrac{2b+a}{x+a}-\dfrac{2b-a}{a-x}-\dfrac{4bx-2a^2}{x^2-a^2}$.

14. $\dfrac{x+1}{6x-6}-\dfrac{2x-1}{12x+12}+\dfrac{2}{3-3x^2}-\dfrac{7}{12x}$.

15. $\dfrac{x}{2x-6}+\dfrac{x}{3x+9}-1-\dfrac{1-x}{6x}+\dfrac{15x+3}{18x-2x^3}$.

16.* $\dfrac{x^2-x-6}{x^2+5x+6}-\dfrac{x^2+4x+3}{x^2-4x+3}-\dfrac{15x}{9-x^2}$.

---

* Reduce before adding.

17. $\dfrac{a^2+2ab+b^2}{a^2-b^2}-\dfrac{4a^2-b^2}{2a^2-3ab-2b^2}+\dfrac{a^2-2ab+3b^2}{a^2-3ab+2b^2}.$

18. $\dfrac{1-x^2}{9-x^2}+\dfrac{x^2-9}{3(x+3)^2}-\dfrac{x^2-4x+3}{5(x-3)^2}-\dfrac{2x}{5x^2-45}.$

19. $\dfrac{3x+2}{x^2-5x+6}+\dfrac{x}{8x-x^2-15}-\dfrac{4-x}{7x-x^2-10}.$

20. $\dfrac{a}{(a-b)(a-c)}+\dfrac{b}{(b-c)(b-a)}+\dfrac{c}{(c-a)(c-b)}.$

21. $\dfrac{a^2}{(a-b)(a-c)}+\dfrac{b^2}{(b-c)(b-a)}+\dfrac{c^2}{(c-a)(c-b)}.$

22. $\dfrac{y+z}{(x-y)(x-z)}+\dfrac{z+x}{(y-z)(y-x)}+\dfrac{x+y}{(z-x)(z-y)}.$

23. $\dfrac{yz}{(x-y)(x-z)}+\dfrac{zx}{(y-z)(y-x)}+\dfrac{xy}{(z-x)(z-y)}.$

24. $\dfrac{1+l}{(l-m)(l-n)}+\dfrac{1+m}{(m-n)(m-l)}+\dfrac{1+n}{(n-l)(n-m)}.$

II. Multiplication of Fractions.

145. To find the product of any two fractions, $\dfrac{a}{b}$ and $\dfrac{c}{d}$, we may proceed thus:

$$\frac{a}{b}\times\frac{c}{d}=a\div b\times c\div d=a\times c\div b\div d \quad \text{(Art. 33)}$$

$$=\frac{a\times c}{b}\div d=\frac{a\times c}{b\times d}.$$

Hence, to multiply fractions,

Multiply the numerators together for a new numerator, and multiply the denominators together for a new denominator, canceling factors that are common to the two products.

This reduces the multiplication of fractions to the multiplication of integral expressions, and enables us to use again our knowledge of the latter process.

Ex 1. $\dfrac{2x}{5a^2} \times \dfrac{10a^3y^3}{12b^2x^2} \times \dfrac{4b^2x}{6a^2y}$

$$= \frac{2 \times 10 \times 4a^3b^2x^2y^3}{5 \times 12 \times 6a^4b^2x^2y} = \frac{2y^2}{9a}.$$

Ex. 2. $\dfrac{x+y}{x} \times \dfrac{x^2-y^2}{x^3+xy^2} \times \dfrac{4x^2}{(x+y)^2}$

$$= \frac{x+y}{x} \times \frac{(x+y)(x-y)}{x(x^2+y^2)} \times \frac{4x^2}{(x+y)(x+y)}$$

$$= \frac{4(x-y)}{x^2+y^2}, \textit{ Product.}$$

III. Division of Fractions.

146. To divide any fraction, $\frac{a}{b}$, by any other fraction, $\frac{c}{d}$, we may proceed thus:

Let $x = \frac{a}{b}$, $y = \frac{c}{d}$ $\quad \therefore bx = a, \quad dy = c \quad \therefore \frac{bx}{dy} = \frac{a}{c}$

$\therefore \frac{x}{y} = \frac{ad}{bc} = \frac{a}{b} \times \frac{d}{c}$ but $\frac{x}{y} = \frac{a}{b} \div \frac{c}{d} \quad \therefore \frac{a}{b} \div \frac{c}{d} = \frac{a}{b} \times \frac{d}{c}.$

Hence, to divide one fraction by another,

Invert the divisor and proceed as in multiplication.

This reduces division of fractions to the already-learned process of multiplication of fractions.

Ex. 1. Divide $\dfrac{x^2-4a^2}{ax+2a^2}$ by $\dfrac{x-2a}{2a}$.

$$\frac{x^2-4a^2}{ax+2a^2} \div \frac{x-2a}{2a} = \frac{x^2-4a^2}{ax+2a^2} \times \frac{2a}{x-2a}$$

$$= \frac{(x+2a)(x-2a)}{a(x+2a)} \times \frac{2a}{(x-2a)}.$$

$= 2$, *Quotient.*

Ex. 2. $\dfrac{x^3-1}{x(x+1)} \times \dfrac{(x^2-1)^2}{x^2+x+1} \div \dfrac{(x-1)^3}{(x+1)^2}$

$$= \frac{(x-1)(x^2+x+1)}{x(x+1)} \times \frac{(x^2-1)(x^2-1)}{x^2+x+1} \times \frac{(x+1)^2}{(x-1)^3}$$

$= \dfrac{(x+1)^3}{x}$, *Result.*

EXERCISE 47.

Simplify—

1. $\dfrac{5x^2y}{14a^3c} \times \dfrac{28a^2b^3}{15xy^2}$.

2. $\dfrac{21xy^2}{13z^3} \div \dfrac{28x^3}{39z^4}$.

3. $\dfrac{12b}{25a} \times \dfrac{35ab}{48} \times \dfrac{5}{7b^2}$.

4. $\dfrac{9a^2b}{8c^2x} \times \dfrac{28ax^2}{15b^2c} \div \dfrac{21a^3x}{10bc^3}$.

5. $\dfrac{50x^2z}{49y^n} \times 28xy^{3n} \times \dfrac{-y^n}{40x^3}$.

6. $\dfrac{15x}{2x(2x-1)} \times \dfrac{2x(x+1)}{5x^2}$.

7. $\dfrac{xy-y^2}{2x+2} \times \dfrac{4x+4}{x^2-xy}$.

8. $\dfrac{a^2b^2+3ab}{4a^2-1} \div \dfrac{ab+3}{2a+1}$.

9. $\dfrac{x^2-9}{x^2+x} \div \dfrac{x-3}{x^2-1}$.

10. $\dfrac{(a-1)^3}{a(x+1)^2} \times \dfrac{x+1}{(a-1)^2}$.

11. $\dfrac{4x^2-9}{9x^2-1} \times \dfrac{6x+2}{12x-18}$.

12. $\dfrac{2x^2-x-1}{2x^2+x-1} \times \dfrac{4x^2-1}{x^2-1}$.

13. $\dfrac{a^3y-ax^2y}{a^3x^2+a^2x^2y} \div \dfrac{a^2y-2axy+x^2y}{a^2+ay}$.

14. $\dfrac{a^2-1}{(a+1)^2} \div \dfrac{a^2-a}{(a+1)^2}$.

15. $\dfrac{3x^2+x-2}{4x^2-4x-3} \times \dfrac{6x^2-x-2}{2x^2-x-3}$.

16. $\dfrac{2x^2-x-6}{2x^2+x-1} \div \dfrac{2x^2+x-3}{2x^2+3x-2}$.

17. $\left(x+\dfrac{1}{x-1}\right) \times \dfrac{2x-2}{x^3+1}$.

18. $\left(\dfrac{x}{y}+1\right) \div \left(\dfrac{x^2}{y}+\dfrac{y^2}{x}\right)$.

19. $\dfrac{x^2-y^2}{5xy} \times \dfrac{x-y}{x+y} \times \dfrac{10x^2y^2}{(x-y)^2} \times \dfrac{1}{2xy}$.

20. $\dfrac{3(a-b)^2}{4(a+b)^2} \times \dfrac{7(a^2-b^2)}{9(a-b)^3} \div \dfrac{14ab}{8(a+b)}$.

21. $\dfrac{x^2+2x-3}{x^2+x-12} \times \dfrac{x^2+2x-15}{x^2+2x-3} \div \dfrac{x^3+5x^2}{x^3+4x^2}$.

22. $\dfrac{6x^2y-4xy^2}{45x^2-20y^2} \times \dfrac{30x+20y}{4x^2y^2} \times \dfrac{xy}{x+y}$.

23. $\dfrac{6x^2-5x-4}{2x^2+7x-4} \times \dfrac{6x^2+x-2}{4x^2-4x-3} \times \dfrac{2x^2+5x-12}{9x^2-6x-8}$.

24. $\left(x+1+\dfrac{1}{x}\right)\left(x-1+\dfrac{1}{x}\right) \div \dfrac{x^6-1}{x^2(x^2-1)}$.

25. $\left(\dfrac{a}{b}+1\right)\left(1-\dfrac{b}{a}\right) \times \left(1-\dfrac{a^2+b^2}{a^2-b^2}\right)$.

26. $\dfrac{x^3+y^3}{x^3+x^2y+xy^2} \times \dfrac{x^3-y^3}{x^2-xy+y^2} \times \left(1+\dfrac{y}{x-y}\right)$.

27. $\dfrac{x^2-(a-1)^2}{a^2-(x+1)^2} \times \dfrac{(a+x)^2-1}{1-(a-x)^2} \div \dfrac{a+x-1}{a-x-1}$.

28. $\dfrac{2-b-a}{b-2-a} \times \dfrac{a^2-b^2-4b-4}{a^2+b^2+2ab-4} \times \dfrac{b^2-a^2-4b+4}{b^2-a^2+4a-4}$.

29. $\dfrac{12x^2-xy-20y^2}{12x^2-8xy-15y^2}\times\dfrac{2x^2+5xy-12y^2}{3x^2+5xy-12y^2}\times\dfrac{6x^3+23x^2y+15xy^2}{4x^2+21xy+20y^2}$.

30. $\left(\dfrac{1}{ab}+\dfrac{1}{bc}+\dfrac{1}{ac}\right)\left(\dfrac{a}{b}+\dfrac{b}{c}+\dfrac{c}{a}\right)\div\dfrac{(a+b+c)^2}{a^2b^2c^2}$.

31. $\left(\dfrac{1}{x}-x^2\right)\left(x+\dfrac{x^2}{1-x}\right)\div\left[(x+1)^2-x\right]$.

32. $\left(\dfrac{1}{a^2}+\dfrac{1}{x^2}+\dfrac{2}{ax}-1\right)\div\left[\dfrac{x+a(1-x)}{ax}\times(1+\dfrac{a}{x}+a)\right]$.

33. $\left[\dfrac{m+2n}{m-2n}+\dfrac{m-2n}{m+2n}\right]\div\left[\dfrac{m+2n}{m-2n}-\dfrac{m-2n}{m+2n}\right]$.

IV. Reduction of Complex Fractions.

147. A Complex Fraction is one having a fraction in its numerator or in its denominator, or in both.

In simplifying any complex fraction it is important to write down the entire fraction at each step of the process.

Ex. 1. Simplify $\dfrac{x}{1-\dfrac{x}{y}}$.

$$\frac{x}{1-\dfrac{x}{y}}=\frac{x}{\dfrac{y-x}{y}}=x\times\frac{y}{y-x}=\frac{xy}{y-x},\ \textit{Result.}$$

If the numerator and denominator of the complex fraction each contain fractions, the simplification is often effected most readily by multiplying both the numerator and denominator by the lowest common denominator of the fractions contained in them.

Ex. 2. Simplify $\dfrac{\dfrac{1}{x}+\dfrac{1}{y}+\dfrac{1}{z}}{\dfrac{x}{y}+\dfrac{y}{z}+\dfrac{z}{x}}$.

Multiply both numerator and denominator by xyz, and obtain

$$\frac{yz+xz+xy}{x^2z+xy^2+yz^2}, \text{ \textit{Result.}}$$

Ex. 3. Simplify $\dfrac{1}{x+\dfrac{1}{1-\dfrac{3}{x-2}}}$.

A fraction of this form is called a *continued fraction.* In simplifying a continued fraction begin at the bottom, and reduce by alternate conversions of a mixed quantity into an improper fraction, and divisions of a numerator by a fractional denominator. Thus,

$$\frac{1}{x+\dfrac{1}{1-\dfrac{3}{x-2}}}=\frac{1}{x+\dfrac{1}{\dfrac{x-5}{x-2}}}=\frac{1}{x+\dfrac{x-2}{x-5}}$$

$$=\frac{1}{\dfrac{x^2-4x-2}{x-5}}=\frac{x-5}{x^2-4x-2}, \text{ \textit{Result.}}$$

EXERCISE 48.

Simplify—

1. $\dfrac{\dfrac{4}{x}-x}{1+\dfrac{x}{2}}$.

2. $\dfrac{2-\dfrac{1}{x}}{4-\dfrac{1}{x^2}}$.

3. $\dfrac{x-\dfrac{1}{x}}{1-\dfrac{1}{x}}$.

4. $\dfrac{\dfrac{ab}{c}-2d}{a-\dfrac{2cd}{b}}$.

5. $\dfrac{1-\dfrac{1}{a+1}}{1+\dfrac{1}{a-1}}$.

6. $1+\dfrac{1}{1-\dfrac{1}{a}}$.

7. $\dfrac{2x-\dfrac{1}{4x^2}}{1-\dfrac{1}{2x}}.$

8. $\dfrac{\left(\dfrac{1}{x}+\dfrac{1}{y}\right)^2}{\left(1+\dfrac{x}{y}\right)^2}.$

9. $3-\dfrac{x-1}{2x-\dfrac{5x}{3}}.$

10. $\dfrac{a+\dfrac{1}{a+1}-1}{a+\dfrac{1}{a-1}+1}.$

11. $3-\dfrac{1}{a-\dfrac{a^2}{1+a}}.$

12. $\dfrac{\dfrac{4}{a-1}+\dfrac{a-1}{a}}{\dfrac{1}{a-1}-\dfrac{1}{a}}.$

13. $\dfrac{\dfrac{1}{x}-\dfrac{3}{x^3}-\dfrac{2}{x^2}}{\dfrac{9}{x^2}-1}.$

14. $\dfrac{\dfrac{a}{x^2}+\dfrac{x}{a^2}}{\dfrac{1}{a^2}-\dfrac{1}{ax}+\dfrac{1}{x^2}}.$

15. $1-\dfrac{1}{1+a+\dfrac{2a^2}{1-a}}.$

16. $\dfrac{\dfrac{x}{1+x}+\dfrac{1-x}{x}}{\dfrac{x}{1+x}-\dfrac{1-x}{x}}.$

17. $\dfrac{\dfrac{2x}{y}+1-\dfrac{y}{x}}{\dfrac{2x}{y}+\dfrac{y}{x}-3}.$

18. $\dfrac{\left(\dfrac{1}{a}+\dfrac{1}{b}-\dfrac{1}{c}\right)^2}{\dfrac{c^2(a+b)^2-a^2b^2}{a^2b^2c^2}}.$

19. $\dfrac{\dfrac{x}{a}+\dfrac{a}{x}-2-\dfrac{1}{ax}}{\dfrac{x}{a}-\dfrac{a}{x}-\dfrac{2}{a}+\dfrac{1}{ax}}.$

20. $\dfrac{1-\dfrac{a^3}{8}}{1+\dfrac{a}{2}+\dfrac{a^2}{4}}\div(a-2).$

21. $\dfrac{2(\frac{1}{2}x-\frac{1}{4})}{2x-1}-\frac{1}{2}.$

22. $2a-1-\dfrac{a-1}{2-\dfrac{a}{a-\dfrac{a}{1+a}}}.$

23. $\dfrac{\dfrac{1-(ab-cd)^2}{(ab-1)^2-c^2d^2}}{\dfrac{(cd+1)^2-a^2b^2}{(ab+cd)^2-1}}.$

24. $\dfrac{\dfrac{1}{a}+\dfrac{1}{b+c}}{\dfrac{1}{a}-\dfrac{1}{b+c}}\times\left(1+\dfrac{b^2+c^2-a^2}{2bc}\right).$

25. $\dfrac{1}{x+\dfrac{1}{x+2}}\times\dfrac{1}{x+\dfrac{1}{x-2}}\div\dfrac{x-\dfrac{4}{x}}{x^2+\dfrac{1}{x^2}-2}.$

26. $\dfrac{1+x^3}{1-\dfrac{x}{1+\dfrac{x}{1-x}}}-\dfrac{1-x^3}{1+\dfrac{x}{1-\dfrac{x}{1+x}}}.$

27. $\dfrac{x^3y-y^4}{xy^2+x^2y}\div\left\{\dfrac{x^4+x^3y+x^2y^2}{(x^2-y^2)^2}\div\dfrac{1}{\left(1+\dfrac{y}{x}\right)^2}\right\}.$

28. $\dfrac{\dfrac{1}{a^2}-\left(\dfrac{1}{b}+\dfrac{1}{c}\right)^2}{\dfrac{1}{b^2}-\left(\dfrac{1}{a}+\dfrac{1}{c}\right)^2}\times\dfrac{\dfrac{1}{b^2}-\left(\dfrac{1}{a}-\dfrac{1}{c}\right)^2}{\dfrac{1}{c^2}-\left(\dfrac{1}{a}-\dfrac{1}{b}\right)^2}\times\dfrac{\left(\dfrac{1}{a}+\dfrac{1}{c}-\dfrac{1}{b}\right)^2}{\left(\dfrac{1}{a}-\dfrac{1}{c}\right)^2-\dfrac{1}{b^2}}.$

29. $\dfrac{1-2\dfrac{1-2x}{1+2x}}{1+2\dfrac{1+2x}{1-2x}}+\dfrac{4(\frac{1}{2}-\frac{1}{2}x+x^2)-\frac{2}{3}}{\frac{8}{3}(\frac{1}{2}+x+\frac{1}{2}x^2)-\frac{1}{3}}.$

EXERCISE 49.

REVIEW.

Reduce to their simplest form—

1. $\dfrac{x^3 - 3x + 2}{x^3 + x^2 - 3x - 2}$.

2. $\dfrac{x^3 - 2x^2 + 4x - 3}{x^3 - 5x^2 + 13x - 9}$.

3. $\dfrac{3x - 4}{2x + 3} - \dfrac{3x - 2}{2x + 1}$.

4. $\dfrac{x - 2n}{x + 2n} + \dfrac{2n + x}{2n - x}$.

5. $\dfrac{\frac{3}{2}(\frac{2}{3}x^2 + \frac{1}{3}x - 2)}{\frac{2}{3}(\frac{3}{2}x^2 + \frac{5}{2}x - 1)}$.

6. $\dfrac{1 - \frac{1}{2}[1 - 3(1 - x)]}{1 - \frac{1}{3}[1 - 2(1 - x)]}$.

7. $\dfrac{2(x + \frac{1}{2})}{3(x - \frac{2}{3})} - \dfrac{\frac{1}{3}(x - \frac{1}{2})}{\frac{1}{2}(x + \frac{2}{3})}$.

8. $3 - \dfrac{1}{1 - \dfrac{2x}{3x - \dfrac{3x}{x + 1}}}$.

9. $\dfrac{2 - x}{1 - 2x} - \dfrac{x + 2}{2x + 1} + \dfrac{6x}{4x^2 - 1}$.

10. $\dfrac{\dfrac{x}{y}\left(\dfrac{y^3}{x} + xy - 2y^2\right)}{x^2 + xy - 2y^2}$.

11. $\dfrac{1}{x - 1} + \dfrac{2}{x - 2} - \dfrac{3}{x - 3} + \dfrac{4x - 3}{(x^2 - x)(x - 2)}$.

12. $\dfrac{1}{a - \dfrac{1}{a}} + \dfrac{1}{a + \dfrac{1}{a}} - \dfrac{2}{a - \dfrac{1}{a^3}} + \dfrac{b}{ab + \dfrac{b}{a}}$.

13. $\dfrac{1}{2}\left\{\dfrac{1}{a} - \dfrac{1}{a - b}\right\} + \dfrac{b}{a + b} - \dfrac{b}{2}\left\{\dfrac{2(a - b) - 1}{a^2 - b^2}\right\}$.

14. $\dfrac{1}{9x^2 + 9x + 2} - \dfrac{1}{1 - 9x^2} + \dfrac{1}{4 - 9x^2} - \dfrac{1}{2 - 9x + 9x^2}$

15. $\dfrac{3}{x^2 - 3x + 2} + \dfrac{2}{(x - 1)(3 - x)} + \dfrac{1}{(2 - x)(x - 3)}$.

16. $\dfrac{2x + y}{x + y} - 1 - \dfrac{y}{y - x} + \dfrac{x^2}{y^2 - x^2}$.

17. $\left(\dfrac{a}{x} + \dfrac{x}{a} - 2\right)\left(\dfrac{a}{x} + \dfrac{x}{a} + 2\right) \div \left(\dfrac{a}{x} - \dfrac{x}{a}\right)^2$.

18. $\left\{\dfrac{1 + x}{1 + x^2} - \dfrac{1 + x^2}{1 + x^3}\right\} \div \left\{\dfrac{1 + x^2}{1 + x^3} - \dfrac{1 + x^3}{1 + x^4}\right\}$.

19. $\left(\dfrac{a+\dfrac{1}{x}}{a-\dfrac{x^2}{a}}\right)^2 \times \left(\dfrac{1+\dfrac{a}{x}}{1+\dfrac{1}{ax}}\right)^2 \times \left(\dfrac{1}{a}-\dfrac{x}{a^2}\right)^2.$

20. $\dfrac{x-4-\dfrac{4}{x-4}}{x-4-\dfrac{1}{x-4}} \times \dfrac{x-\dfrac{1}{x-2}-2}{x-\dfrac{4}{x-5}-2}.$

21. $\dfrac{\dfrac{2a+3}{2a^2+a-1}-\dfrac{3a+2}{3a^2+a-2}}{\dfrac{1}{(a+1)^2}-\dfrac{6}{2-7a+6a^2}}.$

22. $\left\{1-\left(\dfrac{x^2+y^2-z^2}{2xy}\right)^2\right\} \div \left\{\left(\dfrac{x^2+z^2-y^2}{2xz}\right)^2-1\right\}.$

23. $\left\{1-\dfrac{\dfrac{9}{x^2}+\dfrac{x^2}{9}}{\dfrac{9}{x^2}-\dfrac{x^2}{9}}\right\} \times \left(\dfrac{3}{x}-\dfrac{x}{3}\right)\left(\dfrac{9}{x^2}+1\right).$

24. $\dfrac{\left(2+\dfrac{a}{b}\right)\left(1+\dfrac{b}{a}\right)}{1+\dfrac{a}{b}+\dfrac{b}{a}} \div \dfrac{3\left(1+\dfrac{a}{b}\right)}{\dfrac{a^3}{b^3}-1}.$

25. $\dfrac{1+8x^3}{1-\dfrac{2x}{1+\dfrac{2x}{1-2x}}} - \dfrac{1-27x^3}{1+\dfrac{3x}{1-\dfrac{3x}{1+3x}}}.$

26. $\dfrac{x^4-5x^2+4}{x^2+1} \times \dfrac{\dfrac{1}{x}-\dfrac{1}{x-2}}{x-\dfrac{1}{x}} \div \dfrac{2+\dfrac{4}{x}}{1+\dfrac{1}{x^2}}.$

27. $\left(\dfrac{x^2}{y^2}-2+\dfrac{y^2}{x^2}\right) \times \left(\dfrac{x^4y^4}{xy+y^2}\right) \times \left(\dfrac{\dfrac{x}{y}-1+\dfrac{y}{x}}{x^3-2x^2y+xy^2}\right).$

28. $\dfrac{1-\dfrac{b}{a}+\dfrac{b^2}{a^2}}{1+\dfrac{b}{a}+\dfrac{b^2}{a^2}} \times \dfrac{\dfrac{a^3}{b^3}-1}{\dfrac{a^3}{b^3}+1} \div \dfrac{\left(\dfrac{1}{a}-\dfrac{1}{b}\right)^2}{\left(\dfrac{1}{a}+\dfrac{1}{b}\right)^2}.$

$$\frac{1+\frac{1}{x^3}}{1-\frac{1}{x^3}} \times \frac{1-\frac{1}{x}}{1+\frac{1}{x}} \div \left\{ \frac{x+\frac{1}{x}}{x-\frac{1}{x}} \times \left(1-\frac{1}{x}\right) \right\}$$

$$\frac{\frac{2}{3}(\frac{1}{2}-3x)-\frac{1}{3}}{\frac{1}{2}(x-1)+\frac{1}{4}} - \frac{\frac{1}{2}(1-x)+\frac{1}{3}(1+x)}{\frac{1}{3}(x+\frac{1}{2})-\frac{2}{3}x} + 5.$$

$$\frac{x-\frac{8}{x^2}}{\frac{x^2}{4}-1} \times \frac{x+\frac{1}{x+1}+3}{\frac{7}{x-1}+x+3} \div \frac{x^2-x+\frac{7}{x+1}+1}{x-1+\frac{3}{x-1}}.$$

$$\frac{a^2-x^2+6a+9}{x^2-a^2-6a-9} \times \frac{a^2-x^2+6x-9}{9-a^2-x^2+2ax} \div \frac{x^2-3x+ax}{a^2-3a-ax}.$$

$$\tfrac{1}{3}\{\tfrac{1}{2}a-\tfrac{1}{2}[2a-\tfrac{2}{5}(3a+2(a-5))]+2+\tfrac{1}{4}a\}.$$

$$3\tfrac{1}{3}-\frac{2a}{3}+\tfrac{2}{3}\{\tfrac{5}{2}a-\tfrac{3}{4}[\tfrac{2}{3}-\tfrac{1}{2}(4a-4(5a-3))]\}.$$

$$\left\{1+\frac{1+\frac{1+x}{1-3x}}{1-3\frac{1+x}{1-3x}}\right\} \div \left\{1-3\frac{1+\frac{1+x}{1-3x}}{1-3\frac{1+x}{1-3x}}\right\}.$$

$$2+\cfrac{1}{2-\cfrac{6}{4+\cfrac{2}{2+\cfrac{3-2a}{a-1}}}}.$$

37. $$\frac{\frac{3}{4}\left[\frac{1+3\frac{1+a}{1-a}}{1-2\frac{1-a}{1+a}}\right]}{2\left[\frac{a+2}{3a-7+4\frac{2}{a+1}}\right]}$$

$$\left\{\left[\frac{\frac{1}{x}}{1-\frac{1}{x}}+\frac{1+\frac{1}{x}}{\frac{1}{x}}\right] \div \left[\frac{\frac{1}{x}-1}{\frac{1}{x}}-\frac{\frac{1}{x}}{1+\frac{1}{x}}\right]\right\} \div \frac{\left(\frac{1}{x}+1\right)^2}{x-\frac{1}{x}}$$

$$\frac{\left(\frac{a+bc}{a-bc}\right)^2+\frac{a+bc}{a-bc}+1}{\left(\frac{a+bc}{a-bc}\right)^2-\frac{a+bc}{a-bc}+1}.$$

$$\frac{\left(\frac{a+1}{a-1}\right)^3-\left(\frac{a+1}{a-1}\right)^2-\frac{a+1}{a-1}+1}{\left(\frac{a+1}{a-1}\right)^3+\left(\frac{a+1}{a-1}\right)^2-\frac{a+1}{a-1}-1}.$$

CHAPTER XI.

FRACTIONAL AND LITERAL EQUATIONS.

148. General Method of Solution. If an equation contain fractions, it is necessary to clear the equation of fractions before transposing terms and solving by the method given in Chapter VI.

Ex. 1. Solve $x + \frac{x+2}{6} = 5 + \frac{x-4}{4}$.

Multiply both members of the equation by 12, the L. C. D. of the fractions.

$$12x + 2(x+2) = 60 + 3(x-4)$$
$$12x + 2x + 4 = 60 + 3x - 12$$
$$12x + 2x - 3x = 60 - 4 - 12$$
$$11x = 44$$
$$x = 4, \textit{ Root}$$

If a fraction be preceded by a minus sign, it is important to remember that the sign of each term of the numerator must be changed on clearing the equation of fractions.

Ex. 2. Solve $\frac{x+1}{2} - \frac{2x-5}{5} = \frac{11x+5}{10} - \frac{x-13}{3}$.

Multiplying by 30,

$$15x + 15 - 12x + 30 = 33x + 15 - 10x + 130$$
$$15x - 12x - 33x + 10x = -15 - 30 + 15 + 130$$
$$-20x = 100$$
$$x = -5, \textit{ Root.}$$

29. $\dfrac{1+\dfrac{1}{x^3}}{1-\dfrac{1}{x^2}} \times \dfrac{1-\dfrac{1}{x}}{1+\dfrac{1}{x}} \div \left\{\dfrac{x+\dfrac{1}{x}}{x-\dfrac{1}{x}} \times \left(1-\dfrac{1}{x}\right)\right\}$

30. $\dfrac{\frac{3}{4}(\frac{1}{2}-3x)-\frac{1}{3}}{\frac{1}{2}(x-1)+\frac{1}{4}} - \dfrac{\frac{1}{2}(1-x)+\frac{1}{3}(1+x)}{\frac{1}{3}(x+\frac{1}{2})-\frac{2}{3}x} + 5.$

31. $\dfrac{x-\dfrac{8}{x^2}}{\dfrac{x^2}{4}-1} \times \dfrac{x+\dfrac{1}{x+1}+3}{\dfrac{7}{x-1}+x+3} \div \dfrac{x^2-x+\dfrac{7}{x+1}+1}{x-1+\dfrac{3}{x-1}}.$

32. $\dfrac{a^2-x^2+6a+9}{x^2-a^2-6a-9} \times \dfrac{a^2-x^2+6x-9}{9-a^2-x^2+2ax} \div \dfrac{x^2-3x+ax}{a^2-3a-ax}.$

33. $\frac{1}{3}\{\frac{1}{2}a-\frac{1}{2}[2a-\frac{2}{5}(3a+2(a-5))]+2+\frac{1}{4}a\}.$

34. $3\frac{1}{3}-\dfrac{2a}{3}+\frac{2}{3}\{\frac{5}{2}a-\frac{3}{4}[\frac{2}{3}-\frac{1}{2}(4a-4(5a-3))]\}.$

35. $\left\{1+\dfrac{1+\dfrac{1+x}{1-3x}}{1-3\dfrac{1+x}{1-3x}}\right\} \div \left\{1-3\dfrac{1+\dfrac{1+x}{1-3x}}{1-3\dfrac{1+x}{1-3x}}\right\}.$

36. $2+\dfrac{1}{2-\dfrac{6}{4+\dfrac{2}{2+\dfrac{3-2a}{a-1}}}}.$

37. $\dfrac{\frac{3}{4}\left[\dfrac{1+3\dfrac{1+a}{1-a}}{1-2\dfrac{1-a}{1+a}}\right]}{2\left[\dfrac{a+2}{3a-7+4\dfrac{2}{a+1}}\right]}$

38. $\left\{\left[\dfrac{\dfrac{1}{x}}{1-\dfrac{1}{x}}+\dfrac{1+\dfrac{1}{x}}{\dfrac{1}{x}}\right] \div \left[\dfrac{\dfrac{1}{x}-1}{\dfrac{1}{x}}-\dfrac{\dfrac{1}{x}}{1+\dfrac{1}{x}}\right]\right\} \div \dfrac{\left(\dfrac{1}{x}+1\right)^2}{x-\dfrac{1}{x}}$

39. $\dfrac{\left(\dfrac{a+bc}{a-bc}\right)^2+\dfrac{a+bc}{a-bc}+1}{\left(\dfrac{a+bc}{a-bc}\right)^2-\dfrac{a+bc}{a-bc}+1}.$

40. $\dfrac{\left(\dfrac{a+1}{a-1}\right)^3-\left(\dfrac{a+1}{a-1}\right)^2-\dfrac{a+1}{a-1}+1}{\left(\dfrac{a+1}{a-1}\right)^3+\left(\dfrac{a+1}{a-1}\right)^2-\dfrac{a+1}{a-1}-1}.$

CHAPTER XI.

FRACTIONAL AND LITERAL EQUATIONS.

148. General Method of Solution. If an equation contain fractions, it is necessary to clear the equation of fractions before transposing terms and solving by the method given in Chapter VI.

Ex. 1. Solve $\quad x+\dfrac{x+2}{6}=5+\dfrac{x-4}{4}.$

Multiply both members of the equation by 12, the L.C.D. [illegible] fractions.

$$12x+2(x+2)=60+3(x-4)$$

$$12x+2x+4=60+3x-12$$

$$12x+2x-3x=60-4-12$$

$$11x=44$$

$$x=4, \textit{Root}$$

If a fraction be preceded by a minus sign, [illegible] that the sign of each term of the numerator [illegible] the equation of fractions.

Ex. 2. Solve $\dfrac{x+1}{2}-\dfrac{2x-5}{5}=$ [illegible]

Multiplying by 30,

$$15x+15-12x \text{ [illegible]}$$

[illegible]

Ex. 3. Solve $\frac{4}{1+x}+\frac{x+1}{1-x}-\frac{x^2-3}{1-x^2}=0.$

Multiplying by the L. C. D., $1-x^2$,

$$4(1-x)+(x+1)^2-x^2+3=0$$
$$4-4x+x^2+2x+1-x^2+3=0$$
$$-2x=-8$$
$$x=4, \textit{ Root.}$$

Hence, in general,

Clear the equation of fractions by multiplying each term by the L. C. D. of all the fractions;

For a fraction preceded by the minus sign the sign of each term of the numerator must be changed when the denominator is removed;

Complete the solution by the methods of Chapter VI.

149. Equations involving Decimal Fractions. If one or more of the coefficients of an equation is a decimal fraction, we may solve the equation by expressing the decimal fractions as common fractions.

If all the coefficients are decimals or whole numbers, it is simpler to solve directly in decimals.

Ex. Solve $.3x-.14x=.012x+.592.$

$$.3x-.14x-.012x=.592$$
$$.148x=.592$$
$$x=4, \textit{ Root.}$$

EXERCISE 50.

Solve—

1. $\frac{3x}{4}+\frac{5x}{6}=\frac{2x}{3}+\frac{11}{2}.$
2. $\frac{x}{3}-\frac{3x}{5}+\frac{7x}{5}=\frac{34}{15}.$
3. $\frac{2x-3}{4}+\frac{x+1}{6}=\frac{5x+2}{12}.$
4. $\frac{3x}{5}+\frac{2x+7}{3}=\frac{4x+5}{15}.$
5. $\frac{x}{2}+\frac{x}{3}+\frac{x}{5}=\frac{x}{4}+\frac{47}{6}.$
6. $\frac{1}{2x}-\frac{3}{x}+\frac{5}{3x}=\frac{3}{4x}-\frac{19}{24}.$

7. $\frac{2x}{3} - \frac{2x+1}{5} = \frac{1}{3}.$

8. $\frac{3x+5}{4} = 1 - \frac{x+4}{6}.$

9. $2x - 8 - \frac{24-2x}{7} = 0.$

10. $\frac{2}{x} = \frac{4}{5}.$

11. $\frac{6}{x} + \frac{2}{3} = 0.$

12. $\frac{3}{x} + 2 = 0.$

13. $\frac{4}{x} - \frac{3}{4} = 0.$

14. $1 + \frac{x}{3} = x.$

15. $\frac{2x}{3} - 2 = x - \frac{3}{2}.$

16. $\frac{3}{4}(x-1) = \frac{1}{3}(x-2).$

17. $\frac{2}{3}(1+x) = \frac{3}{4}(2x+1).$

18. $3(\frac{2}{3}x - \frac{1}{2})(\frac{1}{2}x + \frac{2}{3}) = x^2.$

19. $\frac{x-5}{3} - \frac{2x+3}{6} = 1 - \frac{6-5x}{12}.$

20. $\frac{3-2x}{8} - \frac{x-3}{6} - 1 = \frac{x+4}{3} + \frac{1}{24}.$

21. $\frac{3x-1}{7} - \frac{x+1}{6} - \frac{4x+1}{21} = \frac{3(x-1)}{4} - 3.$

22. $10 - \frac{3x+5}{4} - 3x - 4\frac{1}{6} = 0.$

23. $\frac{2x+5}{5} - \frac{x+1\frac{1}{2}}{10} + x = \frac{5x-10\frac{1}{2}}{20} - \frac{1}{5}.$

24. $\frac{x-1}{2} - \frac{2x+3}{3} + \frac{x}{4} = x - \frac{x}{6}.$

25. $\frac{3-x}{10} - \frac{2}{3}(5+x) + \frac{x-3}{5} - \frac{1}{2}(2x+5) = \frac{7x-4}{6}.$

26. $2(x+\frac{1}{3}) + x\left(1 - \frac{1}{2x}\right) = \frac{6x+5}{12} + \frac{x+5}{4}.$

27. $\frac{x+5}{7} - \frac{x+7}{5} + \frac{x+1}{2} - \frac{2x-5}{10} = \frac{x+22}{70}.$

28. $\frac{2x+3}{5}+\frac{5x+1}{6}-\frac{3x+1}{4}=\frac{1}{3}+\frac{6x-1}{15}+\frac{x+9}{30}.$

29. $\frac{2}{3}(5x+2)-\frac{3}{4}(7x-2)+\frac{1}{2}(3x-2)=x-\frac{x}{2}.$

30. $\frac{y+6}{11}-\frac{2y-18}{3}+\frac{2y+3}{4}=5\frac{1}{3}+\frac{3y+4}{12}.$

31. $\frac{7y+9}{8}-\frac{3y+1}{7}=\frac{1}{6}y-3\frac{1}{4}-\frac{3}{14}(5y+11).$

32. $0.5x-0.4x=0.3.$

33. $1.5x-5=x.$

34. $1.25x+1.9=-1.125x.$

35. $0.6x-1.5=0.2-0.15x.$

36. $2+\frac{x}{1.2}=\frac{x}{3}-\frac{4.5}{9}.$

37. $\frac{1.5x-1.6}{1.2}=\frac{3.5x-2.4}{0.8}.$

38. $\frac{3.2x-3.4}{4.5}=\frac{0.6x+4}{2.5}.$

39. $\frac{.0032x-1}{0.1x}=\frac{1.005}{.0125x}.$

40. $\frac{.0001}{80x}=\frac{1.00005}{200.01}.$

41. $\frac{0.6x+.045}{0.4}-\frac{5x-1.78}{6}=.3825.$

42. $\frac{x-1}{x+1}=\frac{3}{5}.$

43. $\frac{5}{2x-1}=\frac{8}{3x+1}.$

44. $\frac{2x+5}{5x+3}=\frac{2x+1}{5x+2}.$

45. $\frac{2x-5}{2x+2}=\frac{3x-5}{3x-3}.$

46. $\frac{6x-5}{3x-3}=\frac{8x-7}{4x+4}.$

47. $\frac{x}{3}-\frac{x^2-5x}{3x-7}=\frac{2}{3}.$

48. $\frac{5}{1-x}+\frac{6}{1+x}=\frac{7}{1-x^2}.$

49. $\frac{3}{3-x}+\frac{4}{3+x}=\frac{8x+3}{9-x^2}$

50. $\frac{2x+1}{2x-1}-\frac{10}{4x^2-1}=\frac{2x-1}{2x+1}$

51. $\frac{1}{x+1}+\frac{1}{x-1}=\frac{2}{x+2}.$

52. $\dfrac{x-1}{x^3-8}=\dfrac{3}{x-2}+\dfrac{1-3x}{x^2+2x+4}.$

53. $\dfrac{x^2-x+1}{x-1}=2x-\dfrac{x^2+x+1}{x+1}.$

54. $\dfrac{x-3}{2x+6}-\dfrac{\frac{1}{6}x^2+1}{x^2-9}=\dfrac{x+3}{3x-9}.$

55. $\dfrac{3}{x-1}+\dfrac{4}{x+1}-\dfrac{5}{2x-2}=\dfrac{11}{3x+3}+\dfrac{4}{1-x^2}.$

56. $\dfrac{x+1}{2x-3}-\dfrac{x^2+7}{4x^2-9}=\dfrac{2}{2x+3}-\dfrac{x-1}{6-4x}.$

57. $\dfrac{3x^2-5}{3x-6}-\dfrac{7}{6x+12}-2-x=\dfrac{7}{2x^2-8}.$

58. $\dfrac{4}{3x^2+9x+6}+\dfrac{2}{x^2-3x}-\dfrac{3}{x^2-x-6}+\dfrac{1}{x^2-2x-3}=$
$\dfrac{4}{3x^2+6x}.$

59. $\dfrac{x+1}{x^2+x-6}+\dfrac{5}{3x-x^2-2}=\dfrac{x}{x^2+2x-3}.$

60. $\dfrac{6x+6}{2x^2+5x+3}-\dfrac{2x+1}{2x^2-x-1}=\dfrac{2x}{x^2+2x}.$

150. Special Methods. The work of solving an equation may frequently be diminished by using some special method or device adapted to the peculiarities of the given equation.

1st Special Method. *If the denominators of some fractions are monomials, and of some are polynomials,* it is best to make two steps of the process of clearing the equation of fractions, the first step being to remove the monomial denominators and simplify as far as possible before proceeding to the second step, which is to remove the remaining polynomial denominators.

Ex. 1. Solve $\frac{2x+8\frac{1}{2}}{9}-\frac{13x-2}{17x-32}+\frac{x}{3}=\frac{7x}{12}-\frac{x+16}{36}$.

Multiplying by 36, the L. C. D. of the monomial denominators,

$$8x+34-\frac{36(13x-2)}{17x-32}+12x=21x-x-16.$$

Transposing all terms except the fraction to right-hand side,

$$-\frac{36(13x-2)}{17x-32}=-50$$

Dividing by -2, $\frac{18(13x-2)}{17x-32}=25$

$$234x-36=425x-800$$

$$191x=764$$

$$x=4, \textit{ Root.}$$

2d Special Method. Before clearing an equation of fractions it is often best to combine some of the fractions into a single fraction.

Ex. 2. Solve $\frac{x-1}{x-2}-\frac{x-2}{x-3}=\frac{x-3}{x-4}-\frac{x-4}{x-5}$.

In this equation it is best to combine the fractions in the left-hand member, and those in the right-hand member, each into a single fraction, before clearing of fractions. We obtain

$$\frac{-1}{(x-2)(x-3)}=\frac{-1}{(x-4)(x-5)}.$$

Clearing and solving, $x=\frac{7}{2}$.

EXERCISE 51.

Solve—

1. $\frac{3x-1}{6}+\frac{4x}{3x+2}=\frac{x+5}{2}$.

2. $\frac{3-2x}{4}+\frac{x}{6}-\frac{1-6x}{15-7x}=\frac{2-3x}{9}$.

3. $2\frac{1}{2}-\frac{3}{2x+4}=\frac{2x-1}{4}-\frac{x}{2}.$

4. $\frac{5x+13}{12}=\frac{2x+5}{6}+\frac{23}{4x-36}-\frac{5-\frac{1}{4}x}{3}.$

5. $\frac{5}{7-x}-\frac{2\frac{1}{4}x-3}{4}-\frac{x+11}{8}+\frac{11x+5}{16}=0.$

6. $\frac{3x-1}{30}+\frac{4x-7}{15}=\frac{x}{4}-\frac{2x-3}{12x-11}+\frac{7x-15}{60}.$

7. $\frac{6x-7}{11x+5}-\frac{x+1}{15}+\frac{2x-1}{30}=\frac{199}{10}.$

8. $3+\frac{x+4}{7x+11}=\frac{4-3x}{8}+\frac{4x+9}{12}-\frac{4-x}{24}+\frac{5}{4}.$

9. $\frac{3x-2\frac{1}{2}}{9}-\frac{7}{12}+\frac{x-\frac{1}{2}}{\frac{4}{3}x+11}=\frac{2x-1\frac{1}{2}}{3}-\frac{2x+3\frac{1}{3}}{6}.$

10. $\frac{1}{5}\left[\frac{5-4x}{2}-3x-1+10\left(\frac{2\frac{1}{2}x+1}{11x+6}\right)-\frac{3}{2}\right]=1-x.$

11. $\frac{2}{3}\left[\frac{2x}{9}-\frac{1}{2}\left\{\frac{3x-1}{6}+\frac{2x-5}{7x+8}+x\right\}\right]+\frac{19x+3}{54}=0.$

12. $\frac{0.1x-100.01}{1.001x+.0002}=\frac{10}{.05}.$

13. $\frac{1-1.4x}{2+x}=\frac{0.7(x+1)}{1-0.5x}.$

14. $\frac{0.2x-1.6}{0.4}=\frac{2x-0.2}{4x+0.5}+\frac{x}{2}.$

15. $\frac{2x-3}{0.3x-0.4}=\frac{0.4x-0.9}{.06x-.07}.$

16. $\frac{1.2x-1.5}{1.5}+\frac{0.4x+1}{0.2x-0.2}=\frac{0.4x+1}{0.5}.$

17. $\frac{x-0.25}{0.125}=\frac{0.4x-1}{.05}-\frac{5x-20}{3x-0.2}.$

18. $\frac{1}{x-2}-\frac{1}{x-3}=\frac{1}{x-4}-\frac{1}{x-5}.$

19. $\frac{x-1}{x-2}-\frac{x-3}{x-4}=\frac{x-5}{x-6}-\frac{x-7}{x-8}.$

20. $\dfrac{x-7}{x-8}-\dfrac{x-8}{x-9}=\dfrac{x-4}{x-5}-\dfrac{x-5}{x-6}.$

21. $\dfrac{3}{3x-2}-\dfrac{2}{2x-3}=\dfrac{2}{2x+3}-\dfrac{3}{3x+2}.$

22.* $\dfrac{2x+1}{x+1}+\dfrac{2x+9}{x+5}=\dfrac{2x+3}{x+2}+\dfrac{2x+7}{x+4}.$

23. $\dfrac{4x-17}{x-4}+\dfrac{10x-13}{2x-3}-\dfrac{8x-30}{2x-7}-\dfrac{5x-4}{x-1}=0.$

151. Literal Equations are equations in which some or all of the known quantities are denoted by letters; as, $a, b, c \ldots$, or $m, n, p \ldots$

The methods used in solving literal equations are the same as those employed in numerical equations.

Ex. 1. Solve $a(x-a)=b(x-b)$.

$$ax-a^2=bx-b^2$$
$$ax-bx=a^2-b^2$$
$$(a-b)x=a^2-b^2$$
$$x=a+b.$$

Ex. 2. Solve $\dfrac{a-b}{x-c}=\dfrac{a+b}{x+2c}.$

$$(a-b)(x+2c)=(a+b)(x-c)$$
$$ax+2ac-bx-2bc=ax+bx-ac-bc$$
$$-2bx=-3ac+bc$$
$$x=\frac{c(3a-b)}{2b}.$$

EXERCISE 52.

Solve—

1. $3x+2a=x+8a.$
2. $9ax-3b=2ax+4b.$
3. $5ax-c=ax-5c.$
4. $ax+b=bx+2b.$

* Transpose the second and third fractions.

5. $3cx = a - (2b - a + cx).$

6. $5x - 2ax = 3 - b.$

7. $2ax - 3b = cx + 2d.$

8. $(x + a)(x - b) = x^2.$

9. $ab(x + 1) = a^2 + b^2x.$

10. $(x - 1)(x - 2) = (x - a)^2.$

11. $a^2x = (a - b)^2 + b^2x.$

12. $(a - b)x = a^2 - (a + b)x.$

13. $\frac{ax}{b} + \frac{bx}{a} = \frac{a}{b} - \frac{b}{a}.$

14. $\frac{a + x}{a - 2x} = \frac{a - x}{a + 2x}.$

15. $\frac{3}{4}\left(\frac{x}{a} - 1\right) = \frac{2}{3}\left(\frac{x}{a} + 1\right).$

16. $\frac{2a}{3}\left(\frac{x}{a} - a\right) = \frac{3a}{5}\left(\frac{x}{a} + a\right).$

17. $\frac{4x - a}{2x - a} - 1 = \frac{x + a}{x - a}.$

18. $\frac{x}{a} + \frac{x}{b} + \frac{x}{c} = d.$

19. $\frac{ax - b}{ab} + \frac{bx - c}{bc} + \frac{cx - a}{ac} = 0.$

20. $\frac{ax}{3a + b} + \frac{bx}{3a - b} = \frac{3a^2x + b^2}{9a^2 - b^2}.$

21. $\frac{x}{a} - \frac{x}{a - b} = \frac{1}{a + b} - \frac{x}{b}.$

22. $\frac{a^2 - x}{c} - \frac{b^2 - x}{a} - \frac{c^2 - x}{b} = \frac{a^2}{c} - \frac{b^2}{a}.$

23. $\frac{5a^2 - 7x}{3ab} + \frac{ab^2 + 10x}{5ac} = \frac{10c^2 + 3x}{6bc} + \frac{5(a - c)}{3b} + \frac{b^2}{5c}.$

24. $\dfrac{\frac{a + x}{a}}{x - a} = \dfrac{1 - \frac{x}{a}}{\frac{1}{a} - x}.$

25. $\dfrac{\frac{a}{x}}{\frac{b}{c}} = \dfrac{\frac{1}{b}}{\frac{c}{a}}.$

26. $\frac{a + bx}{3b + 2ax} = \frac{a - bx}{b - 2ax}.$

27. $\dfrac{\frac{x - 1}{a}}{\frac{a + 1}{x}} = \dfrac{\frac{x}{a - 1}}{\frac{a}{a + x}}.$

28. $\dfrac{\dfrac{a}{a-1}}{\dfrac{1}{x+1}} = \dfrac{\dfrac{a}{a+1}}{\dfrac{1}{x-1}}.$

30. $\dfrac{\frac{2}{3}a - \frac{1}{2}bx}{\frac{3}{4}a - \frac{2}{3}bx} = \dfrac{\frac{2}{3}b + \frac{3}{2}ax}{\frac{3}{2}b + 2ax}.$

29. $\dfrac{1 + \dfrac{x}{a}}{1 - \dfrac{x}{a}} = \dfrac{1 + \dfrac{x}{b}}{2 - \dfrac{x}{b}}.$

31. $a + \dfrac{x}{1 + \dfrac{1}{a}} = 1 + \dfrac{x}{1 - \dfrac{1}{a^2}}.$

152. Problems involving Simple Equations containing Fractions.

Ex. 1. A has $8\frac{3}{4}$ dollars more than $\frac{2}{3}$ as much as B has, and together they have $56\frac{1}{4}$ dollars. How many has each?

Let $x =$ the number of dollars B has.

Then $\frac{2}{3}x + 8\frac{3}{4} =$ " " A has,

and $\frac{5}{3}x + 8\frac{3}{4} =$ " " both have.

$$\therefore \quad \frac{5x}{3} + \frac{35}{4} = \frac{225}{4}.$$

$$20x + 105 = 675$$

Hence, $x = 28\frac{1}{2}$, the number of dollars B has,

and $\frac{2}{3}x + 8\frac{3}{4} = 27\frac{3}{4}$, " " A has.

Ex. 2. Divide the number 100 into two such parts that the eighth of the larger part exceeds the eleventh of the less part by 3.

Let $x =$ the larger part,

Then $100 - x =$ the less part.

$\dfrac{x}{8} =$ the eighth of the larger.

$\dfrac{100 - x}{11} =$ the eleventh of the less.

$$\therefore \quad \frac{x}{8} - \frac{100 - x}{11} = 3$$

$$11x - 800 + 8x = 264$$

$$19x = 1064$$

$x = 56$, the greater part.

$100 - x = 44$, the less part.

EXERCISE 53.

1. The sum of the third and fourth parts of a number is 14. Find the number.

2. Find that number whose fifth and sixth parts together are $16\frac{1}{2}$.

3. What is that number whose third, fourth, and fifth parts are together 13 less than the number itself?

4. The difference between the seventh and third parts of a number is 5 more than one ninth of the number. Find it.

5. There are two consecutive numbers such that one seventh of the greater exceeds one ninth of the less by one. Find them.

6. There are three consecutive numbers, such that if the first be divided by 6, the next by 7, and the largest by 8, the sum of the three quotients is $\frac{1}{4}$ more than $\frac{1}{7}$ of the sum of the three numbers. Find them.

7. The difference of two numbers is 9, and $\frac{5}{11}$ of the less, increased by 3, is $\frac{3}{7}$ of the greater. Find the numbers.

8. A man left half his property to his wife, one fifth to his children, a twelfth to a friend, and the remainder, $2600, to a hospital. How much property had he?

9. In a certain orchard there are apple, pear, and cherry trees: ten less than one half are apple, twelve more than one third are pear, and four more than an eighth are cherry trees. How many trees are there?

10. Find three consecutive numbers, such that if they are divided by 2, 3, and 4 respectively, the sum of the quotients will be the next higher number.

11. Divide $130 among A, B, and C, so that A receives $\frac{2}{3}$ as much as B, and C, $\frac{5}{8}$ as much as A and B together.

12. The sum of two numbers is 97, and if the greater be divided by the less, the quotient is 5 and the remainder 1. Find the numbers.

HINT. The divisor multiplied by the quotient is equal to the dividend diminished by the remainder.

13. Divide the number 100 into two such parts that the greater part will contain the less 3 times with a remainder of 16.

14. The difference between two numbers is 40, and the less is contained in the greater 3 times with a remainder of 12. Find the numbers.

15. Five years hence a boy will be $\frac{5}{3}$ as old as he was 3 years ago. How old is he now?

16. A's age is $\frac{3}{5}$ of B's age, and in 7 years he will be $\frac{2}{3}$ as old as B. How old is each?

17. Eight years ago a father was $3\frac{1}{3}$ times as old as his son, and 1 year hence he will be $2\frac{1}{3}$ times as old. How old is each now?

18. A tax of $5000 was paid by four men, A, B, C, and D, A paying $\frac{4}{5}$ as much as B, C half as much as A and B together, and D $400 less than A and B together. How much did each pay?

19. A man sold 4 acres more than $\frac{3}{7}$ of his farm, and had 6 acres less than $\frac{3}{5}$ of it left. How many acres had he?

20. Find two consecutive numbers, such that $\frac{3}{4}$ of the less exceeds $\frac{2}{5}$ of the greater by $\frac{1}{3}$ of the greater.

21. If a boy can do a piece of work in 15 days which a man can do in 9 days, how long would it take both working together?

SOLUTION. Let $x =$ number of days both require.

Then $\frac{1}{x} =$ the part they both can do in 1 day.

But $\frac{1}{15} =$ " the boy " "

$\frac{1}{9} =$ " the man " "

And $\frac{1}{15} + \frac{1}{9} =$ " both " "

Hence, $\frac{1}{15} + \frac{1}{9} = \frac{1}{x}$. Or, $x = 5\frac{5}{8}$.

Therefore, they both together require $5\frac{5}{8}$ days.

These are called *common-time* examples.

22. A can accomplish a piece of work in 6 days, and B can do the same in 8 days. How long will it take them together to do the work?

23. A can spade a garden in 3 days, B in 4 days, and C in 6 days. How long will they require working together?

24. A and B can together mow a field in 4 days, but A alone could do it in 12 days. In how many days can B mow it?

25. If A, B, and C can together do a certain amount of work in $5\frac{1}{3}$ days, which B alone could do in 24 days, or C in 16 days, how long would A require?

26. A and B together can dig a certain ditch in $1\frac{5}{7}$ days; A and C in 2 days, but A alone in 3 days. How many days would it take B and C together to dig it?

27. A and B in $5\frac{1}{7}$ days accomplish a piece of work which A and C can do in 6 days or B and C, in $7\frac{1}{5}$ days. If they all work together, how many days will they require to do the same work?

28. Two inflowing pipes can fill a cistern in 27 and 54 minutes respectively, and an outflowing pipe can empty it in 36 minutes. All pipes are open and the cistern is empty; in how many minutes will it be full?

HINT. Since emptying is the opposite of filling, we may consider that a pipe which empties $\frac{1}{36}$ of a cistern in a minute will *fill* $-\frac{1}{36}$ of it each minute.

29. A tank has four pipes attached, two filling and two emptying. The first two can fill it in 40 and 64 minutes respectively, and the other two can empty it in 48 and 72 minutes respectively. If the tank is empty and the pipes all open, in how many minutes will it be full?

30. A man labors 8 days upon a piece of work which he could complete in 4 more days, but he is then joined by a boy, and they finish it in $2\frac{1}{2}$ days. In how many days could the boy do the entire task?

31. At what time between 3 and 4 o'clock are the hands of a watch pointing in opposite directions?

SOLUTION. At 3 o'clock the minute-hand is 15 minute-spaces behind the hour-hand, and finally is 30 spaces in advance: therefore the minute-hand moves over 45 spaces more than the hour-hand.

Let x = the number of spaces the minute-hand moves.

Then $x - 45$ = " " " " " hour-hand "

But the minute-hand moves 12 times as fast as the hour-hand;

hence, $x = 12(x - 45)$. Solving, $x = 49\frac{1}{11}$.

Thus the required time is $49\frac{1}{11}$ min. past 3.

32. When are the hands of a clock pointing in opposite directions between 4 and 5? Between 1 and 2?

33. What is the time when the hands of a clock are together between 6 and 7? Between 10 and 11?

34. At what instants are the hands of a watch at right angles between 4 and 5 o'clock? Between 7 and 8?

35. A courier travels 5 miles an hour for 6 hours, when another follows him at the rate of 7 miles an hour. In how many hours will the second overtake the first?

SOLUTION. Let x = the number of hours the second travels.

Then $x + 6$ = " " " " " first "

$5(x + 6)$ = " " " miles " " "

$7x$ = " " " " " second "

They travel equal distances,

hence, $7x = 5(x + 6)$. Solving, $x = 15$.

Therefore the second courier requires 15 hours.

36. A courier who travels $5\frac{1}{2}$ miles an hour was followed after 8 hours by another, who went $7\frac{1}{8}$ miles an hour. In how many hours will the second overtake the first?

37. A messenger rides for 6 hours at the rate of 13 miles in 2 hours, when he is followed by another at the rate of 8 miles an hour. How many miles will each travel before the first is overtaken?

38. A train running 40 miles an hour left a station 45 minutes before a second train running 45 miles an hour. In how many hours will the second train round the first?

39. An express train which runs 55 miles an hour leaves a station 4 hours after a freight traveling 11 miles an hour. How many miles from the station will the express round the freight?

40. A boy starts on a bicycle $2\frac{1}{2}$ hours after his sister, who rode 8 miles an hour, and overtook her in 5 hours. How fast did he ride?

41. A gentleman has 10 hours at his disposal. He walks out at the rate of $3\frac{1}{2}$ miles an hour and rides back $4\frac{1}{2}$ miles an hour. How far may he go?

42. A and B start out at the same time from P and Q, respectively, 82 miles apart. A walked 7 miles in 2 hours, and B 10 miles in 3 hours. How far and how long did each walk before coming together, if they walked toward each other? If A walked toward Q, and B in the same direction from Q?

43. A hare takes 7 leaps while a dog takes 5, and 5 of the dog's leaps are equal to 8 of the hare's. The hare has a start of 50 of her own leaps. How many leaps will the dog take to catch her?

SOLUTION. Let $x =$ the number of leaps the dog takes.

Then $\frac{7}{5}x =$ " " " hare takes *in same time.*

Also, let $n =$ " " feet in 1 leap of the hare.

Then $\frac{8}{5}n =$ " " " " dog.

Hence $x \times \frac{8n}{5} = \frac{8nx}{5}$ = the number of feet in the whole distance.

And $\left(\frac{7x}{5} + 50\right) \times n = \frac{7nx}{5} + 50n$ = the number of feet in the whole distance.

Therefore $\frac{8nx}{5} = \frac{7nx}{5} + 50n$; or, $x = 250$.

Thus the dog will take 250 leaps.

44. A hare is 50 leaps in advance of a hound, and takes 5 leaps to the hound's 3, but 2 of her leaps are equal to 1 of his. How many leaps must each take before the hare is caught?

45. A greyhound pursues a fox which has a start of 60 leaps, and makes 3 leaps while the greyhound makes 2. Three of the dog's leaps are equivalent to 7 of the fox's. How many leaps does each take before the hound catches the fox?

46. A has a certain sum of money, from which he gives B \$3 and $\frac{1}{5}$ of what remains; he then gives C \$6 and $\frac{1}{4}$ of what remains, and finds that he has given away half his money. How many dollars had A?

47. A colonel in arranging his troops in a solid square found he required 51 men to complete it; but on making each side contain one man less, had 32 men more than the square required. How many men had he?

HINT. Let x = the number of men in each side of his first square.

48. A regiment is arranged in the form of a hollow square 15 men deep, containing 1800 men. How many men on the outer side of the square?

49. An officer arranged his troops in a rectangle, with 3 men more on a side than on an end; if he should form a square each side of which is equal to one more than the width of the rectangle, he would have 44 men left. How many men had he?

50. If a bushel of oats is worth 40 cents and a bushel of corn is worth 55 cents, how many bushels of each grain must a miller use to produce a mixture of 100 bushels worth 48 cents a bushel?

51. A man has \$5050 invested, some at 4%, and some at 5%. How much has he at each rate if the annual income is \$220?

52. Divide the number 54 into 4 parts, such that the first increased by 2, the second diminished by 2, the third multiplied by 2, and the fourth divided by 2, will all produce equal results.

53. Two laborers are hired at \$3 and \$4 a day each; together they did 35 days' work, and each received the same sum. How many days was each employed?

54. Divide 180 into two such parts that if the less be subtracted from $\frac{3}{4}$ of the greater, the remainder is $\frac{1}{4}$ the difference of the two parts.

55. A boy bought some apples at the rate of 5 for 2 cents: he sold $\frac{1}{3}$ of them at the rate of 2 for a cent, and the rest at $\frac{2}{5}$ of a cent apiece; he made 5 cents. How many apples did he have?

56. A lady in reading a book read the first day half the pages and 1 more; the second day half the remainder and 1 more; the third day half the rest of it and 1 page more, and still had 40 pages to read. How many pages were there in the book?

57. Find two numbers whose difference is 12, such that if $\frac{1}{3}$ the less be added to $\frac{1}{5}$ the greater, the sum shall be equal to $\frac{1}{3}$ the greater diminished by $\frac{1}{5}$ the less.

58. A man buys two pieces of cloth, one of which contains 3 yards more than the other. For the less piece he pays at the rate of \$5 for 3 yards; for the other \$1.50 per yard; he sells the whole at the rate of \$9 for 5 yards and gains \$36. How many yards were there in the less piece?

59. A and B together can do a piece of work in $2\frac{2}{9}$ days; A and C in $2\frac{2}{5}$ days; and B and C in $2\frac{8}{11}$ days. How many days will each require to do the work alone?

60. At what instants are the hands of a watch at right angles between 10 and 11 o'clock?

61. A fox is 100 leaps ahead of a dog in pursuit. The fox makes 3 leaps in the same time that the dog makes 2, but 3 of the dog are equal to 5 of the fox. How many leaps will each take before the fox is caught?

62. A does $\frac{4}{7}$ of a piece of work in 10 days, when he receives the aid of B and C. They finish it in 3 days. If B could do the entire task in 30 days, in what time could C do it alone?

63. A man has a hours at his disposal. How far may he ride in a coach which travels b miles an hour, and return home in time, walking c miles an hour?

64. Separate a into two parts such that the greater divided

by the less may give b for a quotient and c for a remainder. Prove your result.

65. The fore wheel of a carriage is a feet in circumference and the hind wheel is b feet. What is the distance passed over when the fore wheel has made c revolutions more than the hind wheel?

153. Problems in Interest. The power of algebra as compared with arithmetic is well illustrated by the algebraic treatment of questions relating to interest.

Let $p =$ the number of dollars in the principal,
$r =$ rate of interest expressed decimally,
$t =$ time expressed in years,
$i =$ the interest on the principal for the given time and rate,
$a =$ amount, *i. e.* the sum of the principal and interest.

154. I. Problems involving Interest, Principal, Rate, Time.

The principal multiplied by the rate, that is, pr, gives the interest for one year, and prt gives the interest for t years.

$$\therefore\ i = prt \qquad (1)$$

If interest, rate, and time be given, to find the principal, in equation (1) i, r, t, represent known quantities, and p the unknown quantity. Solving (1) for p,

$$p = \frac{i}{rt} \qquad (2)$$

In like manner, if interest, principal, and time are given, to find rate, solve (1) with reference to r,

$$r = \frac{i}{pt} \qquad (3)$$

So also if interest, principal, and rate are given,

$$t = \frac{i}{pr} \qquad (4)$$

Hence, in general, if any three of the four quantities, i, p, r, t, are given, the remaining quantity may be found by substituting for the three given quantities in equation (1), and solving the equation for the remaining unknown quantity, or by substitution in one of the formulas (2), (3), (4).

Ex. 1. At what rate will $50 produce $20 interest in 6 years and 8 months?

We have given
$$p = \$50$$
$$i = \$20$$
$$t = 6\tfrac{2}{3} \text{ years},$$
to find r.

Using formula (3),

$$r = \frac{20}{50 \times 6\frac{2}{3}} = \frac{20 \times 3}{50 \times 20} = \frac{3}{50} = .06$$

$$\therefore \quad r = 6\%, \textit{ Rate.}$$

155. II. Problems involving Amount, Principal, Rate, and Time.

Since amount = principal + interest,

$$a = p + prt \quad . \quad . \quad . \quad . \quad . \quad . \quad . \quad . \quad . \quad . \quad (5)$$

Any three of the four quantities, a, p, r, t, being given, the remaining quantity may be found by solving equation (5).

Thus,
$$p = \frac{a}{1 + rt} \quad . \quad . \quad . \quad . \quad . \quad . \quad . \quad . \quad . \quad . \quad (6)$$

$$r = \frac{a - p}{pt} \quad . \quad . \quad . \quad . \quad . \quad . \quad . \quad . \quad . \quad . \quad (7)$$

$$t = \frac{a - p}{pr} \quad . \quad . \quad . \quad . \quad . \quad . \quad . \quad . \quad . \quad . \quad (8)$$

Thus all the problems of arithmetic relating to interest, principal, rate, time, and amount may be solved by means of equations (1) and (5).

Ex. 1. Find the principal that will amount to $335.30 in 8 years and 6 months at 5%.

Here, given $a = 335.30$, $t = 8\frac{1}{2}$, $r = .05$, find p.

Using formula (6), $p = \dfrac{335.30}{1 + 8.5 \times .05} = \dfrac{335.30}{1.425}$

$= \$235.30$, *Principal.*

EXERCISE 54.

1. Find the interest of $650 for 4 years at 6 per cent.

2. Find the time in which $325 will produce $84.50 interest at 5 per cent.

3. Find the rate at which $176 will yield $43.56 interest in 5 years 6 months.

4. What principal at 5 per cent. will produce $102.30 in 7 years and 9 months?

5. Find the time required for $123.45 to amount to $197.52 at 6 per cent.

6. Find the rate at which $75.60 will amount to $91.98 in 5 years.

7. At what rate will $15 amount to $16 in 10 months?

8. What principal will produce a dollar a month at $3\frac{3}{5}$ per cent.?

9. In what time will the interest on a sum of money equal the principal at 4 per cent.? At 6 per cent.? At $5\frac{1}{8}$ per cent.?

10. What principal in 2 years 4 months will amount to $609.30 at $5\frac{1}{2}$ per cent.?

11. In what time will $76.80 amount to $80 at $6\frac{1}{4}$ per cent.?

CHAPTER XII.

SIMULTANEOUS EQUATIONS.

156. Simultaneous Equations are equations in which more than one unknown quantity is used, but in any set of which equations the same symbol for an unknown quantity stands for the same unknown number.

Thus, in the group of three simultaneous equations,

$$x+y+2z=13,$$
$$x-2y+z=0,$$
$$2x+y-z=3,$$

x stands for the same unknown number in all of the three equations, y for another unknown number, and z for still another.

157. Independent Equations are those which cannot be derived one from the other.

Thus, $x+y=10,$
and $2x=20-2y,$

are not independent equations, since by transposing $2y$ in the second equation and dividing it by 2, the second equation may be converted into the first.

But $\left.\begin{array}{r}3x-2y=5\\5x+y=6\end{array}\right\}$ are independent equations.

158. Elimination is the process of combining two equations containing two unknown quantities so as to form a single equation with only one unknown quantity; or, in general, the process of combining several simultaneous equations so as to form equations one less in number and containing one less unknown quantity.

159. Value of Simultaneous Equations. In simultaneous equations we have given the relations of a set of unknown quantities to known quantities in the shape of a group of equations. By the use of the methods of Chapters VI. and XI., and by elimination, we reduce this complex set of relations to more and more simple ones, till at last we arrive at relations so simple that the value of each unknown quantity can be perceived at once.

160. Methods of Elimination. There are three principal methods of elimination:

I. ADDITION AND SUBTRACTION.

II. SUBSTITUTION.

III. COMPARISON.

These methods are presented to best advantage in connection with illustrative examples.

161. I. Elimination by Addition and Subtraction.

Ex. Solve
$$\begin{cases} 12x + 5y = 75 \quad \dots\dots\dots \quad (1) \\ 9x - 4y = 33 \quad \dots\dots\dots \quad (2) \end{cases}$$

Multiply equation (1) by 4, and (2) by 5,

$$48x + 20y = 300 \quad \dots\dots\dots \quad (3)$$
$$45x - 20y = 165 \quad \dots\dots\dots \quad (4)$$

Add equations (3) and (4) $\quad 93x = 465$

Divide by 93, $\quad x = 5$, *Root.*

Substitute for x its value 5, in equation (1),

$$60 + 5y = 75$$
$$\therefore y = 3, \textit{ Root.}$$

Since y was eliminated by adding equations (3) and (4), this process is called elimination by *addition.*

The same example might have been solved by the method of subtraction.

Thus, multiply equation (1) by 3, and (2) by 4,

$$36x + 15y = 225 \quad \ldots\ldots\ldots\ldots (5)$$

$$36x - 16y = 132 \quad \ldots\ldots\ldots\ldots (6)$$

Subtract (6) from (5), $31y = 93,$

$$y = 3,$$

and $x = 5.$

It is important to select in all cases the smallest multipliers that will cause one of the unknown quantities to have the same coefficient in both equations. Thus, in the last solution given above, instead of multiplying equation (1) by 9, and (2) by 12, we divide these multipliers by their common factor, 3, and get the smaller multipliers, 3 and 4.

Hence, in general,

Multiply the given equations by the smallest numbers that will cause one of the unknown quantities to have the same coefficient in both equations;

If the equal coefficients have the same sign, subtract the equations; if they have unlike signs, add the equations.

EXERCISE 55.

Solve by addition and subtraction—

1. $3x - 2y = 1.$
   $x + y = 2.$
2. $2x - 7y = 9.$
   $5x + 3y = 2.$
3. $4x + 3y = 1.$
   $2x - 6y = 3.$
4. $5x - 3y = 1.$
   $3x + 5y = 21.$
5. $x + 5y = -3.$
   $7x + 8y = 6.$
6. $3x - 2y = 4.$
   $5x - 4y = 7.$
7. $2y + x = 0.$
   $4x + 6y = -3.$
8. $9x - 8y = 5.$
   $15x + 12y = 2.$
9. $4x - 6y + 1 = 0.$
   $5x - 7y + 1 = 0.$
10. $8x + 5y = 6.$
    $6y + 2x = 11.$

11. $5x - 3y = 36.$
$7x - 5y = 56.$

12.* $\frac{x}{2} - \frac{y}{3} = 1.$
$\frac{x}{4} - \frac{y}{9} = 1.$

13. $\frac{x}{3} - \frac{y}{5} = 3.$
$\frac{x}{5} + \frac{y}{2} = 8.$

14. $\frac{2x}{3} + \frac{y}{4} = 1.$
$\frac{3x}{2} + \frac{5y}{8} = 2.$

15. $\frac{4x}{5} + \frac{3y}{2} = -7.$
$\frac{3x}{4} + \frac{2y}{5} = \frac{7}{2}.$

16. $\frac{5x}{6} - \frac{8y}{9} = -6.$
$\frac{3x}{4} - \frac{5y}{6} = -6.$

162. II. Elimination by Substitution.

Ex. Solve $5x + 2y = 36$ . . . . (1)
$2x + 3y = 43$ . . . . (2)

From (1), $5x = 36 - 2y$

$\therefore \quad x = \frac{36 - 2y}{5}$ . . . . (3)

In eq. (2) substitute for x its value given in (3),

$$2\left(\frac{36 - 2y}{5}\right) + 3y = 43$$

$$\frac{72 - 4y}{5} + 3y = 43$$

$$72 - 4y + 15y = 215$$

$$11y = 143$$

$$y = 13$$

Substitute for y in (3), $x = \frac{36 - 26}{5} = 2$

Hence, in general,

In one of the given equations obtain the value of one of the unknown quantities in terms of the other unknown quantity;

Substitute this value in the other equation and solve.

* Clear of fractions before eliminating.

EXERCISE 56.

Solve by substitution.

1. $3x - y = 2.$
 $2x + 5y = 7.$

2. $3x - 4y = 1.$
 $4x - 5y = 1.$

3. $2x + 3y = 1.$
 $3x + 4y = 2.$

4. $y - 3x = 9.$
 $2x + 7y = -6.$

5. $3x + 10y = -1.$
 $2x + 7y = -\frac{1}{2}.$

6. $7x + 8y = 19.$
 $5x + 6y = 13\frac{1}{2}.$

7. $4x + 5y = 10.$
 $7x + 3y = 6.$

8. $6x - 5y = 3.$
 $5x - 6y = 8.$

9. $5x - 1 = 4y.$
 $7y + 19 = 3x.$

10.* $\frac{x}{2} + \frac{y}{3} = 10.$
 $\frac{x}{3} + \frac{y}{2} = 10.$

11. $\frac{x}{2} - \frac{3y}{4} = -12.$
 $\frac{2x}{3} + \frac{y}{5} = 8.$

12. $\frac{3x}{5} - \frac{7y}{10} = 2.$
 $\frac{5x}{3} + \frac{4y}{5} = 33.$

13. $\frac{2x}{5} - \frac{3y}{4} = 3\frac{2}{5}.$
 $\frac{3x}{10} - \frac{y}{2} = 2\frac{1}{2}.$

14. $\frac{2x}{7} - \frac{5y}{2} = 4\frac{1}{8}.$
 $\frac{4y}{5} - \frac{5x}{3} = -6\frac{5}{6}.$

15. $7x - \frac{5y}{3} = 33.$
 $5y + \frac{3x}{2} = -9.$

16. $\frac{3x}{7} = \frac{4y}{5} + 23.$
 $\frac{3y}{8} = \frac{2x}{3} - 1.$

* Clear of fractions before eliminating.

168. III. Elimination by Comparison.

Ex. Solve $2x - 3y = 23$ . . . . (1)

$5x + 2y = 29$ . . . . (2)

From (1) $2x = 23 + 3y$ . . . . (3)

From (2) $5x = 29 - 2y$ . . . . (4)

From (3) $x = \frac{23 + 3y}{2}$ . . . . (5)

From (4) $x = \frac{29 - 2y}{5}$ . . . . (6)

Equate the two values of x in (5) and (6),

$$\frac{23 + 3y}{2} = \frac{29 - 2y}{5},$$

Hence,

$$115 + 15y = 58 - 4y$$

$$19y = -57$$

$$y = -3, \textit{ Root.}$$

Substitute for y in (5), $x = \frac{23 - 9}{2} = 7$ *Root.*

Hence, in general,

Select one unknown quantity, and find its value in terms of the other in each of the given equations;

Equate these two values, and solve the resulting equation.

EXERCISE 57.

Solve by comparison—

1. $5x - 3y = 2.$
   $x + 2y = 3.$
2. $y - 2x = 3.$
   $3y + x = 2.$
3. $x - 2y = 6.$
   $5x + 3y = 4.$
4. $2x + 7y + 1 = 0.$
   $5x + 8y - 7 = 0.$
5. $3x + 2y = 5.$
   $4x - 3y = -1\frac{5}{6}.$
6. $3x - 5y = 19.$
   $4x + 7y = -2.$
7. $\frac{1}{2}x + \frac{1}{3}y = 5.$
   $\frac{1}{3}x + \frac{1}{2}y = 5.$
8. $\frac{1}{4}x - \frac{1}{6}y = 1.$
   $\frac{1}{3}x - \frac{1}{4}y = 1.$

9. $\frac{3x}{4}+\frac{4y}{5}=37.$
$\frac{5x}{6}+\frac{6y}{7}=40.$

10. $\frac{x+y}{2}=2\tfrac{1}{2}.$
$\frac{x-y}{3}=3\tfrac{1}{6}.$

11. $\frac{x}{3}+3y=\frac{11}{12}.$
$\frac{x}{4}-2y=-\tfrac{3}{8}.$

12. $\frac{7x}{2}-\frac{y+1}{3}=12.$
$\frac{x+3}{4}-\frac{2y}{5}=-\tfrac{1}{4}.$

13. $\frac{2x+5y}{3}=\frac{x}{4}+5.$
$\frac{2y}{5}+\frac{x+2y}{2}=4\tfrac{1}{6}.$

14. $\frac{x+y}{y-3}=\frac{1}{6}.$
$\frac{2x+5}{y+x}=-9.$

164. Fractional Simultaneous Equations.

Ex. Solve
$$\begin{cases} x-\frac{2x+y}{4}=\frac{5}{4}-\frac{2y+x}{4} \\ 1-\frac{2x-y}{4}=y-\frac{2y-x}{3}. \end{cases}$$

Clear the given equations of fractions,
$$\begin{cases} 4x-2x-\ \ y=5-2y-x \\ 12-6x+3y=12y-8y+4x \end{cases}$$

Transpose terms,
$$\begin{cases} 3x+y=5 \\ -10x-y=-12 \end{cases}$$
$$-7x=-7$$
$$x=1$$
$$y=2.$$

Hence,

Simplify each of the given equations by the methods used (Chapter XI.) for an equation of one unknown quantity;
Solve the resulting equations by one of the three methods of elimination.

EXERCISE 58.

Solve—

1. $\frac{x+1}{3} - \frac{y}{4} = -1.$

$\frac{3x+1}{4} - \frac{y}{3} = 0.$

2. $x - \frac{y-2}{7} = 5.$

$4y - \frac{x+10}{3} = 3.$

3. $\frac{4x+5y}{40} = x - y.$

$\frac{2x-y}{3} + 2y = \frac{1}{2}.$

4. $\frac{3x-1}{5} + \frac{1-3y}{7} = 2.$

$\frac{3y+x}{11} + x = 9.$

5. $\frac{2x-y}{5} + \frac{3x+2y}{11} = 2.$

$\frac{4x+y-1}{4} - \frac{2x}{3} = 1.$

6. $\frac{3x+8y+2}{5x-3y+3} = \frac{13}{5}.$

$\frac{\frac{2x-y}{3} + \frac{x-2y}{4}}{\frac{3x-2y}{12}} = 1.$

7. $0.7x - 0.02y = 2.$

$0.7x + 0.02y = 2.2.$

8. $0.4x - 0.3y = 0.7.$

$0.7x + 0.2y = 0.5.$

9. $2x + 1.5y = 10.$

$0.3x - 0.05y = 0.4.$

10. $0.5x + 4.5y = 2.6.$

$1.3x + 3.1y = 1.6.$

11. $0.8x - 0.7y = .005.$

$2x = 3y.$

12. $3x = 0.4y + 0.1.$

$.06y = 0.5x - .02.$

13. $\frac{1-3y}{x + \frac{2-3x}{3}} = x + 13.$

$\frac{10y+1}{5} = y - \frac{2x+3}{x+3-\frac{1+2x}{2}}.$

14. $\dfrac{x-y}{x+y}=\dfrac{1}{5}.$

$\dfrac{\dfrac{y}{3}-\dfrac{3x}{2}}{11\frac{1}{2}}-\dfrac{\dfrac{5y}{12}-\dfrac{2x}{3}}{1\frac{3}{4}}=2.$

15. $(x-5)(y+3)=(x-1)(y+2).$

$xy+2x=x(y+10)+72y.$

16. $\dfrac{x-2}{5}+\dfrac{x+10}{3}+\dfrac{10-y}{4}=13.$

$\dfrac{2y+6}{3}-\dfrac{4x+y+6}{8}+4=0.$

17. $\dfrac{6y+5}{8}-\dfrac{3x+5\frac{1}{2}}{5x-2y}=\dfrac{9y-4}{12}.$

$\dfrac{2y+3}{4}+\dfrac{x+y}{3x-2y}=\dfrac{4y+7}{8}.$

18. $\dfrac{3x-2}{5}=\dfrac{6x-5}{10}-\dfrac{x+y+6\frac{1}{2}}{6x+y}.$

$\dfrac{3y-2}{12}=\dfrac{2y-5}{8}+\dfrac{3+7x}{10y-3x}.$

165. Literal Equations.

Ex. Solve

$$ax+by=c \quad \ldots\ldots\ldots\ldots (1)$$
$$a'x+b'y=c' \quad \ldots\ldots\ldots\ldots (2)$$

Multiply (1) by a', and (2) by a,

$$aa'x+a'by=a'c \quad \ldots\ldots\ldots\ldots (3)$$
$$aa'x+ab'y=ac' \quad \ldots\ldots\ldots\ldots (4)$$

Subtract (4) from (3), $(a'b-ab')y=a'c-ac'$

$$\therefore y=\frac{a'c-ac'}{a'b-ab'}$$

Again, multiply (1) by b', (2) by b,

$$ab'x+bb'y=b'c \quad \ldots\ldots\ldots\ldots (5)$$
$$a'bx+bb'y=bc' \quad \ldots\ldots\ldots\ldots (6)$$

Subtract (6) from (5), $(ab'-a'b)x=b'c-bc'$

$$\therefore x=\frac{b'c-bc'}{ab'-a'b}.$$

It is to be observed that after finding the value of x, it is best not to find the value of y, as in numerical equations, by substituting the value of x in one of the original equations and reducing, but rather by taking both the original equations and eliminating anew.

EXERCISE 59.

Solve—

1. $3x + 4y = 2a.$
   $5x + 6y = 4a.$

2. $2ax + 3by = 4ab.$
   $5ax + 4by = 3ab.$

3. $ax + by = 1.$
   $a'x + b'y = 1.$

4. $x - y = 2n.$
   $mx - ny = m^2 + n^2.$

5. $2bx + ay = 4b + a.$
   $abx - 2aby = 4b + a.$

6. $ax - by = a^2 + b^2.$
   $bx + ay = 2(a^2 + b^2).$

7. $ax + by = c.$
   $mx + ny = d.$

8. $bx + ay = a + b.$
   $ab(x - y) = a^2 - b^2.$

9. $c^2x - d^2y = c - d.$
   $cd(2dx - cy) = 2d^2 - c^2.$

10. $\frac{x + m}{y - n} = \frac{n}{m}.$
    $x + y = 2n.$

11. $(a + 1)x - by = a + 2.$
    $(a - 1)x + 3by = 9a.$

12. $(a + b)x + cy = 1.$
    $cx + (a + b)y = 1.$

13. $\frac{x}{a + b} + \frac{y}{a - b} = 2.$
    $x - y = 2b.$

14. $(a - b)x = (a - d)y.$
    $x - y = 1.$

15. $\frac{(a - b)x + (a + b)y}{a^2 + b^2} = 1.$
    $ax - 2by = a^2 - 2b^2.$

16. $\frac{x - b}{a + b} + \frac{y - b}{a - b} = -1.$
    $\frac{x + 2a}{a - b} + \frac{y - 2b}{a + b} = \frac{a^2 + b^2}{a^2 - b^2}.$

17. $\frac{x - 1}{b - 1} + \frac{y - a}{b - a} = 1.$
    $\frac{x + 1}{b} + \frac{y - 1}{1 - a} = \frac{1}{b}.$

18. $(x-1)(a+b)=a(y+a+1).$
    $(y+1)(a-b)=b(x-b-1).$

166. Three or More Simultaneous Equations. If three simultaneous equations, containing three unknown quantities, be given, we may take any pair of the given equations and eliminate one of the unknown quantities; then take a different pair of the given equations, and eliminate the same unknown quantity. The result will be two equations with two unknown quantities, which may be solved by the methods already given.

In like manner, if we have n simultaneous equations, containing n unknown quantities, by taking different pairs of the n equations, we may eliminate one of the unknown quantities, leaving $n-1$ equations, with $n-1$ unknown quantities, and so on.

Ex. Solve
$$\begin{cases} 3x+4y-5z=32 \quad \ldots\ldots\ldots (1) \\ 4x-5y+3z=18 \quad \ldots\ldots\ldots (2) \\ 5x-3y-4z=2 \quad \ldots\ldots\ldots (3) \end{cases}$$

If we choose to eliminate z first, multiply (1) by 3, and (2) by 5,

$$9x+12y-15z=96 \quad \ldots\ldots\ldots (4)$$

$$20x-25y+15z=90 \quad \ldots\ldots\ldots (5)$$

Add (4) and (5), $$29x-13y=186 \quad \ldots\ldots\ldots (6)$$

Also multiply (2) by 4, (3) by 3,

$$16x-20y+12z=72 \quad \ldots\ldots\ldots (7)$$

$$15x-9y-12z=6 \quad \ldots\ldots\ldots (8)$$

Add (7) and (8), $$31x-29y=78 \quad \ldots\ldots\ldots (9)$$

We now have the pair of simultaneous equations,

$$\begin{cases} 29x-13y=186 \\ 31x-29y=78 \end{cases}$$

Solving these, obtain $$x=10$$
$$y=8$$

Substitute for x and y in equation (1),

$$30+32-5z=32,$$
$$z=6.$$

EXERCISE 60.

Solve—

1. $x + y + z = 6.$
   $3x + 2y + z = 10.$
   $3x + y + 3z = 14.$

2. $3x - y - 2z = 11.$
   $4x - 2y + z = -2.$
   $6x - y + 3z = -3.$

3. $5x - 6y + 2z = 5.$
   $8x + 4y - 5z = 5.$
   $9x + 5y - 6z = 5.$

4. $3x + 2y - z = 9.$
   $7x + 5y + 2z = 3.$
   $2x - 7y + 5z = 0.$

5. $2x + 3y = 7.$
   $3y + 4z = 9.$
   $5x + 6z = 15.$

6. $2x + 4y + 3z = 6.$
   $6y - 3x + 2z = 7.$
   $3x - 8y - 7z = 6.$

7. $x + 3y + 3z = 1.$
   $3x - 5z = 1.$
   $9y + 10z + 3x = 1.$

8. $u + v - w = 4.$
   $u + v - x = 1.$
   $v + w + x = 8.$
   $u - w + x = 5.$

9. $\frac{1}{3}x + \frac{1}{6}y + \frac{1}{6}z = 2.$
   $\frac{1}{2}x + \frac{1}{3}y + \frac{1}{3}z = 9.$
   $\frac{1}{6}x + \frac{1}{2}y + \frac{1}{3}z = 3.$

10. $2x + 2y - z = 2a.$
    $3x - y - z = 4b.$
    $5x + 3y - 3z = 2(a + b).$

11. $\frac{2x}{3} + \frac{3y}{4} - \frac{4z}{5} = 18.$
    $\frac{5x}{6} - \frac{5y}{8} + \frac{3z}{4} = -5.$
    $\frac{3x}{2} - \frac{7y}{5} + \frac{3z}{10} = -41.$

12. $x + y + 2z = 2(a + b).$
    $x + z + 2y = 2(a + c).$
    $y + z + 2x = 2(b + c).$

13. $x + y - z = 3 - a - b.$
    $x + z - y = 3a - b - 1.$
    $y + z - x = 3b - a - 1.$

14. $3x + 2y = \frac{13}{6}a.$
    $6z - 2x = \frac{5}{8}b.$
    $5y - 13z + x = 0.$

15. $-x + y + z + v = a.$
    $x - y + z + v = b.$
    $x + y - z + v = c.$
    $x + y + z - v = d.$

167. Use of $\frac{1}{x}$ and $\frac{1}{y}$ as Unknown Quantities.

Some equations which would otherwise be difficult of solution are readily solved by regarding $\frac{1}{x}$ and $\frac{1}{y}$ as the unknown quantities and eliminating them by the usual methods.

Ex. 1 Solve
$$\begin{cases} \frac{5}{x}+\frac{13}{y}=49 \quad \ldots\ldots\ldots (1) \\ \frac{7}{x}+\frac{3}{y}=23 \quad \ldots\ldots\ldots (2) \end{cases}$$

Multiply (1) by 7, and (2) by 5,
$$\begin{cases} \frac{35}{x}+\frac{91}{y}=343 \quad \ldots\ldots\ldots (3) \\ \frac{35}{x}+\frac{15}{y}=115 \quad \ldots\ldots\ldots (4) \end{cases}$$

Subtract (4) from (3),
$$\frac{76}{y}=228$$
$$\therefore \quad \frac{1}{y}=3$$
$$y=\tfrac{1}{3}, \textit{ Root,}$$

Substitute the value of y in (2), hence,
$$x=\tfrac{1}{2}, \textit{ Root.}$$

Ex. 2. Solve
$$\begin{cases} \frac{3}{2x}+\frac{5}{3y}=11 \quad \ldots\ldots\ldots (1) \\ \frac{2}{x}-\frac{1}{4y}=\frac{29}{4} \quad \ldots\ldots\ldots (2) \end{cases}$$

When x and y in the denominators have coefficients, as in this example, it is usually best first to remove these coefficients by multiplying each equation by the L. C. M. of the coefficients of x and y in the denominators of that equation. Hence,

Multiply (1) by 6, and (2) by 4,

$$\begin{cases} \frac{9}{x} + \frac{10}{y} = 66 & \ldots\ldots\ldots\ldots (3) \\ \frac{8}{x} - \frac{1}{y} = 29 & \ldots\ldots\ldots\ldots (4) \end{cases}$$

Multiply (4) by 10, $\frac{80}{x} - \frac{10}{y} = 290 \ldots\ldots\ldots\ldots (5)$

Add (3) and (5), $\frac{89}{x} = 356$

$\therefore x = \frac{1}{4},$

from (4), $y = \frac{1}{3}.$

EXERCISE 61.

Solve—

1. $\frac{1}{x} + \frac{2}{y} = 1.$
   $\frac{2}{x} + \frac{5}{y} = 1.$

2. $\frac{4}{x} + \frac{7}{y} = 3.$
   $\frac{3}{x} + \frac{5}{y} = 2.$

3. $\frac{1}{x} - \frac{2}{y} = 7.$
   $\frac{3}{x} + \frac{4}{y} = 1.$

4. $\frac{5}{x} + \frac{6}{y} = 2.$
   $\frac{7}{x} + \frac{8}{y} = 2.$

5. $\frac{4}{x} - \frac{3}{y} = 2.$
   $\frac{10}{x} - \frac{9}{y} = 3.$

6. $\frac{1}{2x} + \frac{1}{3y} = 1.$
   $\frac{2}{3x} + \frac{3}{2y} = -5.$

7. $\frac{3}{4x} - \frac{5}{3y} = 11\frac{1}{2}.$
   $\frac{5}{8x} - \frac{3}{2y} = 10\frac{1}{4}.$

8. $\frac{1}{x} + \frac{1}{y} = \frac{1}{n}.$
   $\frac{1}{x} - \frac{1}{y} = n.$

9. $\frac{a}{x}+\frac{b}{y}=\frac{a-b}{a}.$

$\frac{b}{x}+\frac{a}{y}=\frac{b-a}{a}.$

10. $\frac{m}{x}+\frac{1}{y}=m^2+n.$

$\frac{1}{x}+\frac{n}{y}=m+n^2.$

11. $\frac{a}{bx}+\frac{b}{ay}=2.$

$\frac{b}{x}+\frac{a}{y}=\frac{a^3+b^3}{ab}.$

12. $\frac{a+b}{x}+\frac{a-b}{y}=2a.$

$\frac{a}{x}+\frac{b}{y}=a+b.$

13. $5y-3x=7xy.$

$15x+60y=16xy.$

14. $\frac{1}{x}+\frac{1}{y}+\frac{1}{z}=2.$

$\frac{2}{x}-\frac{1}{y}+\frac{1}{z}=7.$

$\frac{3}{x}+\frac{2}{y}+\frac{5}{z}=14.$

15. $\frac{3}{x}-\frac{1}{y}=3\frac{1}{2}.$

$\frac{5}{y}+\frac{3}{z}=-7.$

$\frac{2}{x}-\frac{1}{z}=0.$

16. $\frac{2}{3x}-\frac{1}{2y}+\frac{4}{5z}=3\frac{1}{30}.$

$\frac{3}{x}+\frac{4}{y}-\frac{3}{2z}=12.$

$\frac{5}{2x}-\frac{3}{4y}-\frac{1}{3z}=1\frac{5}{12}.$

17. $\frac{1}{x}-\frac{1}{y}-\frac{1}{z}=\frac{1}{a}.$

$\frac{1}{y}-\frac{1}{z}-\frac{1}{x}=\frac{1}{b}.$

$\frac{1}{z}-\frac{1}{x}-\frac{1}{y}=\frac{1}{c}.$

18. $\frac{a}{x}+\frac{b}{y}-\frac{c}{z}=l.$

$\frac{a}{x}+\frac{c}{z}-\frac{b}{y}=m.$

$\frac{b}{y}+\frac{c}{z}-\frac{a}{x}=n.$

19. $5yz+6xz-3xy=8xyz.$

$4yz-9xz+xy=19xyz.$

$yz-12xz-2xy=9xyz.$

CHAPTER XIII.

PROBLEMS INVOLVING TWO OR MORE UNKNOWN QUANTITIES.

168. In the solution of problems we are sometimes obliged to employ more than one letter to represent the unknown quantities. We must always obtain from the conditions of the problem as many independent equations as there are letters thus involved.

Ex. 1. Find two numbers such that the greater exceeds twice the less by 1, and twice the greater added to three times the less equals 23.

Let $x =$ the greater number,

and $y =$ the less.

Then, $x - 2y = 1$

and $2x + 3y = 23$ } from the conditions of the problem.

Solving these equations by any of the common methods of elimination, one obtains $x = 7$; $y = 3$.

Hence 7 and 3 are the numbers required.

Ex. 2. There is a fraction such that if 2 be added to both numerator and denominator, it becomes $\frac{1}{2}$; but if 7 be added to both numerator and denominator, it reduces to $\frac{2}{3}$. Find it.

Let $\frac{x}{y}$ represent the fraction.

Then, $$\frac{x+2}{y+2} = \frac{1}{2},$$

and $$\frac{x+7}{y+7} = \frac{2}{3}.$$

Clearing these equations, and collecting like terms,

$$2x - y = -2$$
$$3x - 2y = -7$$

The solution shows $x = 3$ and $y = 8$.

Therefore $\frac{3}{8}$ is the required fraction.

Ex. 3. Two-fifths of A's age is 3 years less than $\frac{3}{5}$ of B's age; but $\frac{5}{9}$ of A's age equals B's age 10 years ago.

Let $x =$ the No. of years in A's age.

$y =$ the No. of years in B's age.

and $y - 10 =$ B's age ten years ago.

Then, $\frac{2}{5}x = \frac{3}{5}y - 3,$

$\frac{5}{9}x = y - 10.$

Clearing and solving, $x = 45$ and $y = 35.$

Thus, A is 45, and B, 35 years of age.

EXERCISE 62.

1. Find two numbers whose sum is 23 and whose difference is 5.

2. Twice the difference of two numbers is 6, and $\frac{1}{8}$ their sum is $3\frac{1}{2}$. What are the numbers?

3. Find two numbers such that twice the greater exceeds 5 times the less by 6; but the sum of the greater and twice the less is 12.

4. If 1 be added to the numerator of a certain fraction, its value becomes $\frac{1}{3}$; but if 1 be subtracted from its denominator, its value is $\frac{1}{4}$. Find the fraction.

5. Two pounds of flour and five pounds of sugar cost 31 cents, and five pounds of flour and three pounds of sugar cost 30 cents. Find the value of a pound of each.

6. What is that fraction whose numerator is 3 less than the denominator, but 5 times the numerator, less 1, is equal to 3 times the denominator?

7. A number consists of two digits whose sum is 13, and if 4 is subtracted from double the number, the order of the digits is reversed. Find the number.

8. A man hired 4 men and 3 boys for a day for \$18; and, for another day, at the same rate, 3 men and 4 boys for \$17. How much did he pay each man and each boy a day?

9. There is a fraction such that if 4 be added to its

numerator it will become $\frac{4}{5}$, and if 3 be subtracted from the denominator it will become $\frac{2}{3}$. What is the fraction?

10. In an orchard of 100 trees there are 5 more apple trees than $\frac{2}{3}$ of the number of pear trees. How many are there of each?

11. A farmer sells to one man 4 sheep and 9 calves for \$79, and to another, at the same rate, 7 sheep and 5 calves for \$63. Required the price of a sheep and of a calf.

12. Three times A's money is \$60 more than 4 times B's; and $\frac{1}{3}$ of A's is \$20 less than $\frac{1}{2}$ of B's. How much has each?

13. One-seventh of A's age is 2 years less than half of B's; and 3 times B's age is equal to what A's age was 1 year ago. Find their ages.

14. Find two numbers such that 3 times the difference of their halves is 4 more than 4 times the difference of their thirds, and $\frac{1}{7}$ of the larger is equal to $\frac{1}{5}$ of the smaller.

15. In 8 hours A walks 8 miles more than B does in 7 hours; and in 12 hours B walks 3 miles more than A does in 10 hours. How many miles does each walk in one hour?

16. A party of boys purchased a boat, and upon payment discovered that if they had numbered 3 more, they would have paid a dollar apiece less; but if they had been 2 less, they would have paid a dollar apiece more. How many boys were there, and what did the boat cost?

HINT. Let $x =$ the number of boys, and y the number of dollars each paid. Then xy represents the number of dollars the boat actually cost.

17. If a rectangle were 3 inches longer and an inch narrower it would contain 7 square inches more than it now does; but if it were 2 inches shorter and 2 inches wider its area would remain unchanged. What are its dimensions?

18. Two persons, A and B, can perform a piece of work in 16 days. They work together for four days, when B is left alone and completes the task in 36 days. In what time could each do it separately?

19. If A gives B \$10, A will have half as much as B; but

if B gives A \$30. B will have $\frac{4}{5}$ as much as A. How much has each?

20. A train maintained a uniform rate for a certain distance. If this rate had been 8 miles more each hour, the time occupied would have been two hours less; but if the rate had been 10 miles an hour less, the time would have been 4 hours more. Required the distance.

21. A certain fraction becomes $\frac{4}{9}$ if $1\frac{3}{5}$ be added to both numerator and denominator; but $\frac{1}{5}$ if $2\frac{1}{4}$ be subtracted from numerator and denominator. What is the fraction?

22. If the greater of two numbers is divided by the less, the quotient is 3 and the remainder 3, but if 3 times the greater be divided by 4 times the less, the quotient is 2 and the remainder 20. Required the numbers.

23. Find two fractions, with numerators 11 and 7 respectively, such that their sum is $3\frac{11}{12}$, but when their denominators are interchanged, their sum becomes $3\frac{7}{12}$.

24. If $\frac{2}{3}$ be added to the numerator of a certain fraction, its value is increased by $\frac{2}{21}$; but if $2\frac{1}{3}$ be taken from its denominator, the fraction becomes $\frac{6}{7}$. Find the fraction.

25. The sum of the digits of a certain number of two figures is 5, and if 3 times the units' digit is added to the number, the order of the digits will be reversed. What is the number?

26. There are two numbers consisting of the same two digits; the difference of the digits is 1 and the sum of the numbers is 121. What are the numbers?

27. The sum of the digits of a number of two figures is $\frac{1}{4}$ of the number, but if 18 be added to the number, the order of the digits is reversed. What is the number?

28. Twice the units' digit of a certain number is 2 greater than the tens' digit; and the number is 4 more than 6 times the sum of its digits. Find the number.

29. Of a number consisting of 3 digits the tens' digit is 5, and the units' digit 1 less than twice the hundreds' digit,

and the number will be increased by 99 if the order of the figures be reversed. What is the number?

30. What is that number of three figures, the first and last of which are alike, the tens' digit is 1 more than twice the sum of the other two, and if the number is divided by the sum of its digits, the quotient is 21 and the remainder 4?

31. One woman buys 4 yards of silk and 7 of satin, and another at the same rate buys 5 yards of silk and $5\frac{4}{5}$ of satin, each paying \$17.70. What is the price of a yard of each?

32. Find three numbers such that if each be added to half the sum of the other two, the three sums thus resulting will be 38, 40 and 42.

33. One cask contains 18 gallons of wine and 12 gallons of water; another, 4 gallons of wine and 12 of water. How many gallons must be taken from each cask so that when mixed there may be 21 gallons, half wine and half water?

34. A gentleman gave a sum of money to his four sons; giving to the eldest half the sum of the shares of the other three; to the second, one third the sum of the other three shares; and to the third, one fourth the sum of the other three shares. The share of the eldest exceeded that of the youngest by \$70. What was the whole sum, and what did the eldest receive?

35. A's money with twice B's and $2\frac{1}{2}$ times C's is \$340; $\frac{2}{3}$ of A's and B's together is equal to \$4 less than $\frac{4}{5}$ of C's, and A and C have $\frac{12}{13}$ of B's and C's together. How much has each?

36. A merchant found that he had \$37 in silver dollars, half-dollars and quarters; altogether he had 84 pieces. He also noticed that $\frac{1}{4}$ of the dollars, $\frac{1}{5}$ of the halves, and $\frac{1}{6}$ of the quarters would have been worth only \$7.50. How many of each did he have?

37. There are two fractions having the same denominator, whose sum is $\frac{4}{5}$. If 1 be added to each numerator, the sum of the fractions becomes $1\frac{1}{2}$, but if 1 be subtracted from each numerator, their difference is $\frac{1}{2}$. Find the fractions.

HINT. Let $\frac{x}{z}$ be the larger, and $\frac{y}{z}$ the less fraction.

38. A gives to B and C as much as each of them has; afterward, B gives to A and C as much as each of them now has; and then C gives to A and B as much as each of them has, when, finally, they each have $16. How much had each at first?

39. Three men, A, B, and C, have each a certain sum, such that if A should give B $40, he would have 3 times as much as A had left; if B should give C $10, he would have $2\frac{1}{2}$ times as much as B had left; and if C should give A $30, he would have $1\frac{5}{8}$ times as much as C had left. How much had each?

40. To distribute $50 to some men, women, and children, I gave $2 to each man, $2.50 to each woman, and $1.50 to each child, with $4 left over; if I had given $2.50 to each man and woman, and $1.20 to each child, I would have had $3 left; but when I gave $1.75 to each man, and $2.25 to each woman and child, nothing remained. How many of each were there?

41. A boy spent $3 for oranges, some at 30 cents a dozen, and some at 40 cents; he sold them at 3 cents apiece, and gained 24 cents. How many dozen of each kind did he buy?

42. A rectangle has the same area as another 4 yards longer and 2 yards narrower, and the same as a third 3 yards shorter and $2\frac{2}{3}$ yards wider. What are its dimensions?

43. If a rectangle were made 3 feet shorter and $1\frac{1}{2}$ feet wider, or if it were 7 feet shorter and $5\frac{1}{4}$ feet wider, its area would remain unchanged. What are its dimensions?

44. The fore wheel of a carriage makes 6 more revolutions than the hind wheel in going 180 yards, but if the circumference of the fore wheel were increased by $\frac{1}{5}$ of itself, and that of the hind wheel decreased by $\frac{1}{6}$ of itself, the fore wheel would make 6 revolutions less than the hind wheel in the same distance. Find the circumference of each.

45. A boy rows 18 miles down a stream and back in 12 hours: he finds that he can row 3 miles with the stream while he rows 1 mile against it. Find his rate and that of the stream.

SOLUTION. Let x = the number of miles the boy can row an hour in still water.

And y = the number of miles the stream flows an hour.

Then $x+y$ = his rate down stream,

and $x-y$ = " up "

$\frac{18}{x+y}$ = number of hours required to row down.

$\frac{18}{x-y}$ = " " " " up.

Hence,

$$\frac{18}{x+y}+\frac{18}{x-y}=12, \quad \text{or} \quad \frac{3}{x+y}+\frac{3}{x-y}=2 \; . \; . \; . \; . \; . \; (1)$$

He rows down three times as fast as he does up; hence, he requires only $\frac{1}{3}$ as much time.

$$\therefore \frac{18}{x+y}=\frac{1}{3}\left(\frac{18}{x-y}\right), \quad \text{or} \quad \frac{3}{x+y}=\frac{1}{x-y} \; . \; . \; . \; . \; (2)$$

Solving (1) and (2) by method of exercise 61,

$x+y=6$, and $x-y=2$. Hence $x=4$, and $y=2$.

Thus he rows 4 and the stream flows 2 miles an hour.

46. A boatman rows 20 miles down a river and back in 8 hours: he can row 5 miles down the river while he rows 3 miles up the river. Find his rate down stream.

47. A man rows for 4 hours down a stream which runs at the rate of 3 miles an hour: in returning it takes $14\frac{1}{2}$ hours to reach a point 3 miles below his place of starting. Find the distance he rowed down the stream and his rate in still water.

48. A man rows down a stream 20 miles in $2\frac{2}{3}$ hours, and rows back only $\frac{3}{5}$ as fast. At what rate does the water flow?

49. Two bins contain a mixture of corn and oats, the one twice as much corn as oats, and the other 3 times as much oats as corn. How much must be taken from each bin to fill a third bin holding 40 bushels, half to be oats and half corn?

50. A and B are walking along 2 roads which cross each other: when A is at the point of crossing B has 560 yards yet to walk before reaching it; in 4 minutes they are equally distant from this point, and again in 24 minutes more they are equally distant from it. How fast does each walk?

51. A sets out from P toward Q, and 3 hours later B starts from Q toward P, traveling 2 miles an hour faster than A. When they meet B has walked 3 miles more than A. If A had walked a mile an hour faster, they would each have walked an hour less than they did, and B, 7 miles less than A. How fast and how far did each walk?

52. A, B, and C were engaged to mow a field. The first day A worked 3 hours, B, 2 hours, and C, 4, and together they mowed an acre; the second day A worked 6 hours, B, 13, and C, 2, and they mowed 2 acres; the third day they worked 12, 8, and 8 hours respectively, and mowed 3 acres. How many hours would each alone have required to mow an acre?

53. A and B run a mile race. In the first heat A gives B a start of 60 yards and wins by 50 yards. In the second, A gives B a start of 25 seconds and is beaten by 3 seconds. Find the rate of running of each.

SOLUTION. Let $x =$ the number of yards A runs in 1 sec.

$y =$ " " " B " "

1st { B runs 1650 yards while A runs 1760.

$$\therefore \frac{1650}{y} = \frac{1760}{x}.$$

2d { B runs $1760 - 25y$ yards while A runs $1760 - 3x$.

$$\therefore \frac{1760 - 25y}{y} = \frac{1760 - 3x}{x}.$$

or,
$$\frac{1760}{y} - \frac{1760}{x} = 22. \quad \text{Solving, } x = 5\tfrac{1}{2};\ y = 5.$$

That is, A runs $5\frac{1}{2}$, and B, 5 yards, a second.

54. In a mile race A gives B a start of 40 yards and wins by 70 yards. In another, he gives B a start of 32 seconds, and is beaten by $5\frac{1}{3}$ seconds. Find the rate of each.

55. A and B run a mile. In the first race A receives 48 yards the start, and is beaten by 1 second; in the next race, A receives 12 seconds start, and wins by 11 yards. How many minutes does each require to run a mile?

56. In a thousand-yard race A gives B a start of 5 seconds, and wins by $31\frac{1}{4}$ yards. But if he gives B a start of 100 yards, B wins by 6 seconds. How long would it take each to run a mile?

57. A in a given time accomplishes $\frac{2}{3}$ as much as B in the same time, and they together engage to reap a field in 15 days. Before the work is completed they are obliged to call in C, who could have done the entire task in 36 days. But if A had begun 4 days earlier and B a day later, and if they had called in C two days later than they did, they could have finished the field in the promised time. How many days did C help, and how many days would it have taken A alone to reap the field?

58. A gives B and C each half as much as each already has; then B gives A and C each half as much as each now has; afterward, C gives A and B half as much as each has, when they have $27 apiece. How much had each at first?

EXERCISE 63.

REVIEW.

1. Tell the degree of each term of—

$$5x^3 - 4x^2y^2 - 11x - xy + xy^3 - x^4y + 3x^2 - 7y + 11.$$

2. Find the H. C. F. and L. C. M. of—

$$4x^3 - 13x + 6 \text{ and } 4x^3 - 2x^2 - 9.$$

Simplify—

3. $$\frac{4x-5}{45} - \frac{4+x}{30} + \frac{2}{3} - \frac{x-5}{18}.$$

4. $\dfrac{1-5x}{6x^2-6}+\dfrac{3x+5}{4x+4}+\dfrac{2x-3}{3-3x}$.

5. $\dfrac{2}{(x-1)^3}+\dfrac{1}{(1-x)^2}-\dfrac{2}{1-x}-\dfrac{1}{x}$.

6. $\dfrac{2x^3-x^2-x-3}{2x^3-5x^2+x+3}$.

7. $\dfrac{\frac{3}{5}\left(\frac{2}{3}x^2-\frac{11}{6}x-\frac{5}{2}\right)}{\frac{1}{3}\left(\frac{4}{3}x^2-\frac{2}{3}x-2\right)}$.

8. $\dfrac{\left(x+\frac{1}{x}\right)^2-2\left(1+\frac{1}{x^2}\right)}{\left(x-\frac{1}{x}\right)^2}$.

9. $1-\dfrac{x}{a-\dfrac{a}{1-\dfrac{a}{a+x}}}$.

Solve—

10. $1\frac{5}{6}+x=\dfrac{x-1}{2}+\dfrac{1}{3}-\dfrac{x-3}{5}$.

11. $3+\dfrac{x}{3}-\dfrac{1}{2}\left(4-\dfrac{x}{2}\right)=\dfrac{1}{3}\left\{\dfrac{x}{3}-\left(2x-\dfrac{x+1}{2}\right)\right\}$.

12. $\dfrac{3x-2}{6}-\dfrac{5x+14}{7x+15}-\dfrac{3-2x}{4}=x-1\frac{1}{4}$.

13. $\dfrac{x}{x-2}-\dfrac{x+1}{x-1}=\dfrac{x-8}{x-6}-\dfrac{x-9}{x-7}$.

14. $\dfrac{3x}{2}-\dfrac{y}{3}=\dfrac{4}{9}$.
$\dfrac{5x}{4}+\dfrac{3y}{2}=3\frac{1}{3}$.

15. $2y-x=4xy$.
$\dfrac{4}{y}-\dfrac{3}{x}=9$.

16. $\dfrac{x-2}{5}-\dfrac{10-x}{3}=\dfrac{y-10}{4}$.
$\dfrac{2y+4}{3}-\dfrac{2x+y}{8}=\dfrac{x+13}{4}$.

17. $(a-b)x+(a+b)y=a+b$.
$(x-y)(a^2-b^2)=a^2+b^2$.

CHAPTER XIV.

INEQUALITIES.

169. AN **Inequality** is a statement in symbols that one algebraic expression represents a greater or less number than another.

Ex. $x + y < a^2 + b^2$.

It is to be remembered that any positive number is greater than any negative number, and that of two negative numbers the smaller is the greater.

Thus,
$$2 > -5$$
$$-2 > -3$$

The **First Member** of an inequality is the expression on the left of the sign of inequality; the **Second Member** is the expression on the right of this sign.

170. Two inequalities are said to be of the **same kind**, or to subsist in the same sense, when the greater member occupies the same relative position in each inequality; that is, is the left-hand member in each, or the right-hand member. Hence, in inequalities of the same kind the signs of inequality point in the same direction,

Thus,
$$x > 2x - 3$$
$$\frac{2x+1}{3} > 5x - 4$$
are of the same kind;

but
$$a < b$$
$$2a > b - \frac{a}{2}$$
are of opposite kinds.

171. Properties of Inequalities. The following primary properties of inequalities are recognized as true:

(1) **Adding and Subtracting Quantities.** *An inequality will be unchanged in kind if the same quantity be added to or subtracted from each member.* Hence,

(2) **Terms Transposed.** *A term may be transposed from one member of an inequality to the other, provided its sign be changed.*

(3) **Signs Changed.** *The signs of all the terms of an inequality may be changed, provided the sign of the inequality be reversed.*

(4) **Positive Multiplier.** *An inequality will be unchanged in kind if all its terms be multiplied or divided by the same positive number.*

(5) **Raised to a Power.** *An inequality will be unchanged in kind if both members be positive and both be raised to the same power.*

(6) **Inequalities Combined.** *If the corresponding members of two inequalities of the same kind be added, the resulting inequality will be of the same kind; but if the members of an inequality be subtracted from the corresponding members of another inequality of the same kind, the resulting inequality will not always be of the same kind.*

172. Application of Primary Principles. By use of these primary properties complicated relations of inequality may be reduced to simple relations, giving more or less definite results of value.

Ex. 1. Given that x is an integer, determine its value from the inequalities.

$$\begin{cases} 4x - 7 < 2x + 3 \\ 3x + 1 > 13 - x \end{cases}$$

Transposing terms, $$\begin{cases} 2x < 10 \\ 4x > 12 \end{cases}$$

Dividing by coefficient of x in each inequality,

$$\begin{cases} x < 5 \\ x > 3 \end{cases}$$

$$\therefore x = 4, \textit{ Result.}$$

Ex. 2. Prove that the sum of the squares of any two unequal quantities is greater than twice their product.

Let a be the greater of the two quantities, and b the less. Then,

$$a - b > 0$$
$$\therefore\ (a-b)^2 > 0$$
$$\therefore\ a^2 - 2ab + b^2 > 0$$
$$a^2 + b^2 > 2ab.$$

Ex. 3. Prove $(a+b)(b+c)(a+c) > 8abc$.

The left-hand member when expanded becomes

$$a(b^2+c^2) + b(a^2+c^2) + c(a^2+b^2) + 2abc.$$

But from Ex. 2, $a(b^2+c^2) > a(2bc) \ldots\ldots (1)$

$b(a^2+c^2) > b(2ac) \ldots\ldots (2)$

$c(a^2+b^2) > c(2ab) \ldots\ldots (3)$

Also, $2abc = 2abc \ldots\ldots (4)$

Adding (1), (2), (3), (4),

$$(a+b)(b+c)(a+c) > 8abc.$$

EXERCISE 64.

Reduce—

1. $(x+1)^2 < x^2 + 3$.
2. $(3-x)^2 > (x-4)^2$.
3. $7ax + b > 3ax + 5b$.
4. $\frac{4x-3}{3} > \frac{x}{2} + \frac{3x+8}{21}$.
5. $\frac{10-4x}{3} < x - \frac{11}{8} - 7$.
6. $\frac{x+1}{x-2} > \frac{x+3}{x-4}$.
7. $\frac{a+x}{a-x} > \frac{b+x}{2b-x}$.
8. $\frac{4(x+3)}{9} < \frac{8x+37}{18} - \frac{7x-29}{5x-12}$.

Find limits of x—

9. $3x + 1 > 2x + 7$.
   $2x - 1 < x + 6$.
10. $3(x-4) + 2 > 4(x-3)$.
    $2(x+1) < 4(x-1) + 3$.

11. What number is that whose fifth plus its sixth is greater than 6, while its third minus its eighth is less than 4?

12. A certain integer decreased by $\frac{2}{5}$ of itself is greater than $\frac{1}{6}$ of the number, increased by $5\frac{1}{2}$; but if $\frac{1}{2}$ of itself be added to the number, the sum is less than 20. Find the number.

If the letters employed in each are positive and unequal, prove:

13. $3a^2 + b^2 > 2a(a + b)$.

14. $a^3 - b^3 > 3a^2b - 3ab^2$.

15. $a^3 + b^3 > a^2b + ab^2$.

16. $\frac{a}{b} + \frac{b}{a} > 2$.

17. $a + b > 2\sqrt{ab}$.

18. $a^2 + b^2 + c^2 > ab + ac + bc$.

19. $6abc < a(b^2 + ab + c^2) + c(b^2 + bc + a^2)$.

20. $ab(a + b) + ac(a + c) + bc(b + c) < 2(a^3 + b^3 + c^3)$.

21. $a^3 + b^3 + c^3 > 3abc$.

CHAPTER XV.

INVOLUTION AND EVOLUTION.

INVOLUTION.

173. Involution is the operation of raising an expression to any required power.

Since a power is the product of equal factors, involution is a species of multiplication. In this multiplication the fact that the quantities multiplied are equal leads to important abbreviations of the work.

POWERS OF MONOMIALS.

174. Law of Exponents or Index Law.

Since $a^3 = a \times a \times a$,

$$(a^3)^4 = (a \times a \times a)(a \times a \times a)(a \times a \times a)(a \times a \times a)$$
$$= a^{3 \times 4} = a^{12}.$$

In general, in raising a^n to the mth power, we have the factor a taken $m \times n$ times, or

$$(a^n)^m = a^{mn} \quad . \; . \; . \; . \; . \; . \; . \; . \; . \; . \quad \text{I.}$$

This law enables us to abbreviate the process of finding the power of a factor affected by an exponent into a mere multiplication of exponents.

Also, $(ab)^n = ab \times ab \times ab \ldots\ldots$ to n factors

$= (a \times a \times a \ldots\ldots$ to n factors$)(b \times b \times b \ldots\ldots$ to n factors$)$

by the Commutative Law for Multiplication.

$$\therefore (ab)^n = a^n b^n \quad . \; . \; . \; . \; . \; . \; . \; . \; . \quad \text{II.}$$

This law enables us to reduce the process of finding the

power of a product to the simpler process of finding the power of each factor of the product.

175. Law of Signs. It is evident from the law of signs in multiplication that—

(1) *An even power of a quantity (whether plus or minus) is always positive.*

Exs. $(-3)^2 = 9, \qquad (-ab)^4 = a^4b^4.$

(2) *An odd power of a quantity has the same sign as the original quantity.*

Exs. $(-a)^7 = -a^7, \qquad (+a)^5 = a^5.$

176. Involution of Monomials in General. Hence, to raise a monomial to a required power,

Raise the coefficient to the required power;
Multiply the exponent of each literal factor by the index of the required power;
Prefix the proper sign to the result.

Ex. 1. Find the cube of $3x^2y$.

$$(3x^2y)^3 = 27x^6y^3.$$

Ex. 2. $(-2ab^3)^5 = -32a^5b^{15}$.

177. Powers of Fractions. By a method similar to that used in Art. 174, it can be shown that

$$\left(\frac{a^n}{b^m}\right)^p = \frac{a^{pn}}{b^{pm}}.$$

Hence, to raise a fraction to a required power,

Raise both numerator and denominator to the required power, and prefix the proper sign to the resulting fraction.

$$\text{Ex. } \left(-\frac{2a^3x}{5b^2y^3}\right)^4 = \frac{16a^{12}x^4}{625b^8y^{12}}.$$

EXERCISE 65.

Write the square of—

1. $7a^2b$.
2. $-5xy^3$.
3. $\frac{2}{3}x^2y^5$.
4. $-\frac{3x}{5z^3}$.
5. $-\frac{6ab}{11}$.
6. $-13x^ny$.
7. $\frac{x^ny^m}{10z}$.
8. $-4y^{n+1}$.
9. $-\frac{1}{8cd}$.

Write the cube of—

10. $3xy$.
11. $-2x^3$.
12. $\frac{1}{2}x^3y^2$.
13. $-5x^ny$.
14. $-\frac{3x}{4y^3}$.
15. $-\frac{5c^2d^5}{7x^{2m}}$.

Write the value of—

16. $(7ab^2c^3)^3$.
17. $(11a^5b^4)^2$.
18. $(\frac{3}{2}x^2y)^4$.
19. $\left(\frac{2x^2y^3}{x^nz}\right)^5$.
20. $(-\frac{1}{3}a^3x^n)^4$.
21. $(-2x^3)^7$.
22. $(-\frac{1}{2}m^2)^6$.
23. $(3\frac{1}{2}x^{2n})^3$.
24.* $(a^2-2b)^2$.
25. $(a+\frac{3}{4})^2$.
26. $(-x^2-\frac{3}{2}y)^2$.
27. $\left(\frac{x}{y}-\frac{y}{2x}\right)^2$.
28. $(\frac{3}{5}x^2+\frac{5}{6}yz)^2$.
29. $(x^n-7y^{2m})^2$.
30. $(a-2b+3)^2$.
31. $(1-a+a^2-a^3)^2$.
32. $(1-2x+\frac{1}{2}x^2)^2$.
33. $(\frac{3}{2}x^2-\frac{2}{3}xy+\frac{1}{4}y^2)^2$.

POWERS OF BINOMIALS.

178. General Process. In obtaining a required power of a binomial, economies are possible still greater than those used in the involution of a monomial.

It is sufficient in taking up the subject for the first time to

* For this and the succeeding examples in this exercise see Art. 181.

obtain several powers of a binomial by actual multiplication, and, by comparing them, obtain a general method for writing out the power of any binomial. A formal proof of the method obtained is given later.

$$(a+b)^2 = a^2 + 2ab + b^2.$$
$$(a+b)^3 = a^3 + 3a^2b + 3ab^2 + b^3.$$
$$(a+b)^4 = a^4 + 4a^3b + 6a^2b^2 + 4ab^3 + b^4.$$
$$(a+b)^5 = a^5 + 5a^4b + 10a^3b^2 + 10a^2b^3 + 5ab^4 + b^5.$$

If b is negative, the terms containing odd powers of b will be negative; that is, the second, fourth, sixth and all even terms will be negative.

Comparing the results obtained, it is perceived that

I. The **Number of Terms** equals the exponent of the power of the binomial, plus one.

II. **Exponents.** The exponent of a in the first term equals the index of the required power, and diminishes by 1 in each succeeding term. The exponent of b in the second term is 1, and increases by 1 in each succeeding term.

III. **Coefficients.** The coefficient of the first term is 1; of the second term it is the index of the required power. In each succeeding term the coefficient is found by *multiplying the coefficient of the preceding term by the exponent of a in that term, and dividing by the exponent of b increased by* 1.

IV. **Signs of Terms.** If the binomial is a difference, the signs of the even terms are minus; otherwise the signs of all the terms are plus.

Ex. $(a+b)^7 = a^7 + 7a^6b + 21a^5b^2 + 35a^4b^3 + 35a^3b^4 + 21a^2b^5 + 7ab^6 + b^7.$

To form the coefficient of the third term we have

$$\frac{7 \times 6}{2} = 21.$$

The other coefficients are determined similarly. It is to be observed that the coefficients of the latter half of the expansion are the same as those of the first half in reverse order.

179. Binomials with Complex Terms. If the terms of the given binomial have coefficients or exponents other than unity, it is usually best to separate the process of writing out the required power into two steps.

Ex. 1. Obtain the cube of $2x + 5y^2$.

Since $(a+b)^3 = a^3 + 3a^2b + 3ab^2 + b^3$,

substituting $2x$ for a, and $5y^2$ for b,

$$(2x+5y^2)^3 = (2x)^3 + 3(2x)^2(5y^2) + 3(2x)(5y^2)^2 + (5y^2)^3$$
$$= 8x^3 + 60x^2y^2 + 150xy^4 + 125y^6.$$

Ex. 2. $$(2x^3 - \tfrac{1}{4}y^2)^4 = (2x^3)^4 - 4(2x^3)^3(\tfrac{1}{4}y^2) + 6(2x^3)^2(\tfrac{1}{4}y^2)^2$$
$$-4(2x^3)(\tfrac{1}{4}y^2)^3 + (\tfrac{1}{4}y^2)^4$$
$$= 16x^{12} - 8x^9y^2 + \tfrac{3}{2}x^6y^4 - \tfrac{1}{8}x^3y^6 + \tfrac{1}{256}y^8.$$

180. Application to Polynomials. By properly grouping its terms a polynomial may be put into the form of a binomial, and any power of the polynomial obtained by use of the above method for involution of a binomial.

Ex. $$(x+2y+3z)^3 = [(x+2y)+3z]^3$$
$$= (x+2y)^3 + 3(x+2y)^2(3z) + 3(x+2y)(3z)^2$$
$$+ (3z)^3$$
$$= x^3 + 6x^2y + 12xy^2 + 8y^3 + 9x^2z + 36xyz$$
$$+ 36y^2z + 27xz^2 + 54yz^2 + 27z^3.$$

181. Cases of Involution Previously Considered. In Arts. 85, 86, 89 important special cases of involution have been considered, and should be here recalled:

1. The square of the sum or difference of two quantities;
2. The square of any polynomial.

EXERCISE 66.

Expand—

1. $(a-b)^3$.
2. $(x+1)^3$.
3. $(1-x)^4$.
4. $(a-2)^3$.
5. $(2+x^2)^4$.
6. $(a-2b)^5$.
7. $(x+3)^5$.
8. $(a^2-2b)^3$.
9. $(2c-d^2)^5$.
10. $(a-3b^2)^4$
11. $(7-3x^2)$.^{3}
12. $(xy^2+2)^4$.
13. $\left(2+\frac{x}{a}\right)^3$.
14. $(3-\frac{1}{2}c^2)^5$.
15. $\left(1-\frac{c^2}{2b}\right)^4$.
16. $\left(\frac{c}{2x}+1\right)^5$.
17. $(x^2+x-1)^3$.
18. $(x^2-3x-1)^3$.
19. $(a^2+ac+c^2)^4$.
20. $(x-y+z)^3$.
21. $(2x^2-x+3)^3$.
22. $(1+x-x^2)^4$.

EVOLUTION.

182. The **Root** of a quantity is that quantity which taken as a factor a given number of times will produce the given quantity.

183. Evolution is the process of finding a required root of a quantity.

Hence evolution is the inverse of involution.

184. The **Radical**, or **Root Sign**, it will be recalled, is $\sqrt{\ }$. The number indicating the required root is written as a small figure over the radical sign and is called the index of the root. For the square root the index number is omitted.

Thus, $\sqrt{a}$, $\sqrt[3]{a}$, $\sqrt[n]{a}$, indicate the square root, the cube root, and the nth root of a respectively.

EVOLUTION OF MONOMIALS.

185. Index Law. Since $(a^m)^n = a^{mn}$ (Art. 174) it follows that

$$\sqrt[n]{a^{mn}} = a^m \quad . \; . \; . \; . \; . \; . \; . \; . \; . \; . \; . \; \text{I.}$$

where m and n are positive integers.

This reduces the process of finding the root of a quantity affected by an exponent to a division of exponents.

Also, $\sqrt[n]{ab} = \sqrt[n]{a}\sqrt[n]{b}$ . . . . II.

For let $\sqrt[n]{a} = x$, $\sqrt[n]{b} = y$;

$\therefore x^n = a, \ldots (1) \qquad y^n = b \ldots (2)$

But $x^n y^n = (xy)^n$ (by Art. 174)

Substitute for x^n and y^n fro.. (1) and (2),

$ab = (\sqrt[n]{a}\sqrt[n]{b})^n$ . . . . (3)

Extract the nth root of each member of (3),

$$\sqrt[n]{ab} = \sqrt[n]{a}\sqrt[n]{b}.$$

This reduces the process of finding the nth root of a product to the simpler process of finding the root of each factor.

186. Law of Signs. From the law of signs for multiplication it follows that—

(1) *Any even root of a positive quantity may be either positive or negative.*

Ex. $\sqrt{9} = +3$, or -3.

It is convenient for the present to consider only positive roots of even powers.

(2) *No negative quantity can have an even root.*

Ex. The square root of -4 is neither $+2$ nor -2, since neither of these multiplied by itself will give -4.

(3) *The odd root of a quantity has the same sign as the quantity itself.*

Ex. $\sqrt[3]{-27} = -3$.

187. Entire Process. Hence, to extract a required root of any monomial,

Extract the required root of the coefficient;

Divide the exponent of each letter by the index of the required root;

Prefix the proper sign to the result.

EXERCISE 67.

Write the square root of—

1. $9x^2y^4$.
2. $25a^6$.
3. $144y^{2n}$.
4. $16x^2y^2$.
5. $\dfrac{36a^8}{49x^{10}}$.
6. $\frac{4}{9}x^2y^{12}$.
7. $\dfrac{121a^6x^{2m}}{81y^{2n+2}}$.

Write the cube root of—

8. $27x^6$.
9. $125y^3z^9$.
10. $-\frac{1}{8}a^3b^6$.
11. $\dfrac{27a^{12}x^9}{64y^{3n}}$.
12. $-\dfrac{216x^{27}}{343y^9}$.
13. $\dfrac{8x^{15}y^3}{z^{3n+3}}$.
14. $-\dfrac{1000}{x^{6n-9}}$.

Write the value of—

15. $\sqrt[3]{-512x^6}$.
16. $\sqrt[4]{16y^{12}}$.
17. $\sqrt[5]{-32x^{10}y^5}$.
18. $\sqrt[6]{64a^{30}x^{6n}}$.
19. $\sqrt[4]{\frac{1}{81}x^8y^{20}}$.
20. $\sqrt[5]{\frac{32}{243}x^{25}}$.
21. $\sqrt[4]{\dfrac{625x^{20}}{y^{4n+8}}}$.
22. $\sqrt[5]{-\frac{1}{32}x^{30n}y^5}$.
23. $\sqrt{25x^2+20x+4}$.
24. $\sqrt{9x^4-42x^2+49}$.
25. $\sqrt{a^2b^2-16abc+64c^2}$.
26. $\sqrt{1+18x^2y^2+81x^4y^4}$.

SQUARE ROOT.

188. Square Root of Polynomials. Our object is to discover such a relation between the terms of a binomial (or in general of a polynomial) and the terms of its square (as, for instance, between $a+b$ and its square, $a^2+2ab+b^2$), that we can state this relation in the inverse form as a general method for readily determining the square root of any polynomial which is a square.

$$\begin{array}{r|l} a^2 + 2ab + b^2 & a + b \\ a^2 & \\ \hline 2a + b \;|\; 2ab + b^2 & \\ 2ab + b^2 & \end{array}$$

The first term of the root, a, is the square root of the first term, a^2, of the square expression. The second term of the root, b, occurs in the second term, $2ab$, of the square expression, and may be obtained from it by dividing by twice the first term, or $2a$ (called the trial divisor). If we take $2a$ and add b to it (giving $2a + b$, called the complete divisor), and multiply the sum by b, we get $2ab + b^2$, which is the rest of the square expression after a^2 has been subtracted. This last step, therefore, furnishes a test of the accuracy of the work.

Ex. Extract the square root of $16x^2 - 24xy + 9y^2$.

$$\begin{array}{r|l} 16x^2 - 24xy + 9y^2 & 4x - 3y \\ 16x^2 & \\ \hline 8x - 3y \;|\; -24xy + 9y^2 & \\ -24xy + 9y^2 & \end{array}$$

Taking the square root of the first term, $16x^2$, we obtain $4x$, which is placed to the right of the given expression as the first term of the root. Subtract the square of $4x$ from the given polynomial.

Taking twice the first term of the root, $8x$, as a trial divisor, and dividing it into the first term of the remainder, we obtain the second term of the root, $-3y$. This is annexed to the first term of the root and also to the trial divisor to make the complete divisor, $8x - 3y$.

189. Square Root to Three or More Terms. In squaring a trinomial, $a + b + c$, we may regard $a + b$ as a single quantity, and denote it by a symbol, as p, and obtain the square in the form $p^2 + 2pc + c^2$.

Evidently we may reverse this process, and extract a square root to three terms, by regarding two terms of the root when found as a single quantity. So a fourth term of a root, or any number of terms, may be found by regarding in each case the root already found as a single quantity.

Ex. Extract the square root of $x^4 - 6x^3 + 19x^2 - 30x + 25$.

$$\begin{array}{r|l|l}
 & x^4 - 6x^3 + 19x^2 - 30x + 25 & \underline{x^2 - 3x + 5} \\
 & x^4 & \\
2x^2 - 3x & -6x^3 + 19x^2 & \\
 & \underline{-6x^3 + 9x^2} & \\
2x^2 - 6x + 5 & +10x^2 - 30x + 25 & \\
 & \underline{+10x^2 - 30x + 25} &
\end{array}$$

The first two terms of the root, $x^2 - 3x$, are found as in the example in Art. 188.

To continue the process, we consider the root already found, $x^2 - 3x$, as a single quantity, and multiply it by 2 to make it a trial divisor.

Dividing the first term of the remainder, $10x^2$, by the first term of the trial divisor, $+ 2x^2$, we obtain the next term of the root, $+ 5$.

The process is then continued as before.

Hence, in general, to extract the square root of a polynomial,

Arrange the terms according to the powers of some letter;

Extract the square root of the first term, set down the result as the first term of the root, and subtract its square from the given polynomial;

Take twice the root already found as a trial divisor, and divide it into the first term of the remainder;

Set down the quotient as the next term of the root, and also annex it to the trial divisor to form a complete divisor;

Multiply the complete divisor by the last term of the root, and subtract the product from the first remainder;

Continue the process till all terms of the root are found.

EXERCISE 68.

Find the square root of—

1. $x^4 - 4x^3 + 6x^2 - 4x + 1$.
2. $1 - 2a - a^2 + 2a^3 + a^4$.
3. $9x^4 - 12x^3 + 10x^2 - 4x + 1$.
4. $25 + 30x + 19x^2 + 6x^3 + x^4$.

5. $n^6 - 4n^5 + 4n^4 + 6n^3 - 12n^2 + 9.$

6. $4x^6 + 12x^5 + x^4 - 24x^3 - 14x^2 + 12x + 9.$

7. $1 + 16m^6 - 40m^4 + 10m - 8m^3 + 25m^2.$

8. $46n^2 + 25n^4 + 4n^6 + 25 - 44n^3 - 40n - 12n^5.$

9. $9x^6 + 9y^6 + 24x^5y + 24xy^5 - 8x^4y^2 - 8x^2y^4 - 50x^3y^3.$

10. $m^2 + 9 + x^2 + 6m + 6x + 2mx.$

11. $1 + 5x^2 + 2x^4 + x^6 - 4x^5 + 2x^3 + 2x.$

12. $28x^2 - 47x^4 + 49x^6 - 42x^5 - 4x^3 + 16x + 4.$

13. $\frac{1}{4}x^2 - 5x + 25.$

14. $\frac{9}{4}x^2 - 5xy + \frac{25}{9}y^2.$

15. $\dfrac{x^2}{4y^2} - \dfrac{2x}{y} + 4.$

16. $x^4 + 2x^3 - x + \frac{1}{4}.$

17. $\frac{1}{4}a^4 - \frac{1}{3}a^3 + \frac{55}{9}a^2 - 4a + 36.$

18. $\dfrac{a^2}{x^2} + \dfrac{6a}{x} + 11 + \dfrac{x^2}{a^2} + \dfrac{6x}{a}.$

19. $\frac{4}{9}x^4 - \frac{4}{5}x^3 + \frac{277}{75}x^2 - 3x + \frac{25}{4}.$

20. $\frac{9}{25}x^4 - \frac{2}{5}x^3 + \frac{29}{18}x^2 - \frac{5}{6}x + \frac{25}{16}.$

21. $1 + a - \frac{5}{12}a^2 - \frac{7}{3}a^3 - \frac{8}{9}a^4 + \frac{2}{3}a^5 + a^6.$

22. $\dfrac{c^4}{4} + cx + \dfrac{x^2}{c^2} + \dfrac{1}{4x^2} + \dfrac{c^2}{2x} + \dfrac{1}{c}.$

23. $\dfrac{a^2}{x^2} - ax + \dfrac{x^2}{a^2} - 2 + \dfrac{a^3}{x} + \dfrac{a^4}{4}.$

Find to three terms the square root of—

24. $1 + 4x.$

25. $1 - 2a.$

26. $x^2 - 6.$

27. $a^2 + 4b.$

28. $9a^2 - 4x.$

29. $4a^2 - 6ab.$

30. $x^2 - 1.$

31. $x^2 + 3.$

32. $a^2 + 3ab - 2b^2.$

190. Square Root of Arithmetical Numbers. The same general method as that used in Art. 188 can be used to extract the square root of arithmetical numbers. The details of the process, however, are somewhat different, owing to the fact that all the numbers which compose a given square number are given united as a single number.

Thus, $(43)^2 = (40 + 3)^2 = 1600 + 240 + 9 = 1849$.

Hence, given 1849 to extract its square root, the square of the first number, 1600, is not presented explicitly as it would be in an algebraic expression, but must be determined indirectly.

The first step is to mark off the figures of the given number whose root is to be extracted into periods of two figures each, beginning at the decimal point, and then to determine the largest square number represented in the first period of figures at the left as a trial number. If the first figure of the root be in the tens' place, and therefore followed by one zero (as 4 in 40 above), its square will be followed by two zeros, as in 1600. If the first figure of the root had been in the hundreds' place, and therefore followed by two zeros, its square would have been followed by four zeros; that is, there are two additional zeros in the square for each additional zero following the first figure of the root. Hence comes the significance of separating the given square number into periods of two figures each, and extracting the approximate square root of the left-hand period of figures.

We will illustrate by an example, using the algebraic formula $(a + b)^2 = a^2 + 2ab + b^2$, to show the essential identity of the arithmetical and algebraic processes.

Ex. Extract the square root of 1849.

$$\begin{array}{rr|l}
 & \widehat{18}\widehat{49}. & 40 + 3 \\
a^2 = & 1600 & \\
\text{Trial divisor, } 2a = 80 & 249 & \\
b = \ \ 3 & 249 & \\
\text{Complete divisor, } 2a + b = 83 & &
\end{array}$$

This work may be put in the following abbreviated form:

$$\begin{array}{r|l}
 & \widehat{18}\widehat{49}. \ | \ 43 \\
 & 16 \\
83 & 249 \\
 & 249
\end{array}$$

191. Square Root of Decimal Numbers. If it be required to extract the square root of a decimal number, as 28.09, we may proceed thus:

$$\sqrt{28.09} = \sqrt{\frac{2809}{100}} = \frac{\sqrt{2809}}{\sqrt{100}} = \frac{53}{10} = 5.3.$$

It is better, however, to put this work into a different form by marking off the given number into periods of two figures each, beginning at the decimal point and marking both to the right and left. If necessary annex a zero to complete the last period of figures to the right; in such cases, however, the root cannot be exactly extracted.

Ex. Extract the square root of 18.550249.

```
      18.55 02 49 | 4.307, Root
      16
  83 | 255
     | 249
8607 | 60249
     | 60249
```

192. Square Root of Common Fractions. If the denominator of the fraction whose square root is to be extracted is a perfect square, extract the root of the numerator and denominator separately and divide the one result by the other.

$$\text{Ex.}\quad \sqrt{\frac{289}{324}} = \frac{\sqrt{289}}{\sqrt{324}} = \frac{17}{18}.$$

If the denominator is not a perfect square, reduce the fraction to a decimal and extract the root of the decimal.

$$\text{Ex.}\quad \sqrt{\tfrac{2}{3}} = \sqrt{0.66666666+}$$

```
        0.66 66 66 66+ | 0.8164+
          64
  161 | 266
      | 161
 1626 | 10566
      |  9756
16324 | 81066
      | 65296
```

Hence, in general, to extract the square root of an arithmetical number,

Separate the number into periods of two figures each, beginning at the decimal point;

Find the greatest square in the left-hand period, and set down its root as the first figure of the required root;

Square this figure, subtract the result from the left-hand period, and to the remainder bring down the next period;

Double the root already found for a trial divisor, divide it into the remainder (omitting last figure of the remainder), and annex the quotient obtained to the root and also to the trial divisor.

Multiply the complete divisor by the figure of the root last found, and subtract the result from the remainder;

Proceed in like manner till all the periods of figures have been used.

EXERCISE 69.

Find the square root of—

1. 7225.
2. 2601.
3. 8464.
4. 105625.
5. 182329.
6. 337561.
7. 567009.
8. 11573604.
9. 36144144.
10. 8114.4064.
11. 199.204996.
12. 10.30731025.
13. 254046.2409.
14. .0291419041.
15. 1513689.763041.

Find to four decimal places the square root of—

16. 7.
17. 11.
18. 12.5.
19. $3\frac{1}{8}$.
20. $2\frac{1}{9}$.
21. 0.9.
22. $6\frac{7}{8}$.
23. $1\frac{5}{8}$.
24. $\frac{7}{60}$.
25. .049.
26. 1.0064.
27. $36\frac{1}{11}$.

Compute to three decimal places the value of—

28. $\sqrt{2+\sqrt{3}}$.
29. $\sqrt{\sqrt{5}-1}$.
30. $\sqrt{\sqrt{13}-\sqrt{3}}$.
31. $\sqrt{3\sqrt{3}+\sqrt{5}}$.
32. $\sqrt{2\sqrt{7}+3\sqrt{2}}$.
33. $\sqrt{3\sqrt{6}-2\sqrt{7}}$.
34. $\sqrt{\frac{\sqrt{5}-\sqrt{2}}{4}}$.
35. $\sqrt{\frac{5(\sqrt{3}-\sqrt{2})}{2}}$.
36. $\sqrt{\frac{7\sqrt{3}+2\sqrt{5}}{4}}$.

CUBE ROOT.

193. Cube Root of Polynomials. Our object is to determine such a relation between the terms of a binomial, or, in general, of a polynomial, and the terms of its cube (as between $a+b$, and its cube, $a^3+3a^2b+3ab^2+b^3$), that we may be able to state this relation in the inverse form as a general method for determining the cube root of any polynomial which is a perfect cube.

$$
\begin{array}{l|l}
 & a^3+3a^2b+3ab^2+b^3 \;\underline{|\,a+b} \\
 & \underline{a^3} \\
3a^2 & +3a^2b+3ab^2+b^3 \\
\underline{\quad +3ab+b^2} & \\
3a^2+3ab+b^2 & \underline{+3a^2b+3ab^2+b^3}
\end{array}
$$

The first term of the root, a, is the cube root of the first term, a^3, of the cube expression. The second term of the root, b, occurs in the second term of the cube expression, $3a^2b$, and may be obtained from it by dividing it by $3a^2$; that is, by three times the square of the first term of the root (called the trial divisor). If we take the trial divisor, and add to it three times the product of the first term of the root by the second term, $3ab$, and also the square of the second term of the root, b^2, we get $3a^2+3ab+b^2$ (called the complete divisor); this multiplied by the second term of the root gives $3a^2b+3ab^2+b^3$, the rest of the cube expression after a^3 has been subtracted.

This last step, therefore, furnishes a test of the accuracy of the work.

194. Three or More Terms in the Root. In cubing a trinomial, $a+b+c$, we may regard $a+b$ as a single quantity, and denote it by p, and obtain the cube in the form $p^3+3p^2c+3pc^2+c^3$. Evidently we may reverse this process, and extract a cube root to three terms, by regarding two terms of the root when found as a single quantity. So a fourth term or any number of terms of a root may be found by regarding, in each case, the root already found as a single quantity.

We will now extract the cube root of a polynomial expression indicating at each step the trial divisor and complete divisor.

Ex. Extract the cube root of $8x^6 + 36x^5 - 6x^4 - 153x^3 + 15x^2 + 225x - 125$.

$$8x^6 + 36x^5 - 6x^4 - 153x^3 + 15x^2 + 225x - 125 \,|\, \underline{2x^2 + 3x - 5}$$
$$8x^6$$

Trial Divisor $= 3(2x^2)^2 = 12x^4$ | $+ 36x^5 - 6x^4 - 153x^3$

$3(2x^2)(3x) = \quad + 18x^3$

$(3x)^2 = \quad + 9x^2$

Complete Divisor $= 12x^4 + 18x^3 + 9x^2$ | $36x^5 + 54x^4 + 27x^3$

Trial Divisor $= 3(2x^2 + 3x)^2 = 12x^4 + 36x^3 + 27x^2$ | $- 60x^4 - 180x^3 + 15x^2 + 225x - 125$

$3(2x^2 + 3x)(-5) = \quad - 30x^2 - 45x$

$(-5)^2 = \quad + 25$

Complete Divisor $= 12x^4 + 36x^3 - 3x^2 - 45x + 25$ | $- 60x^4 - 180x^3 + 15x^2 + 225x - 125$

Hence, in general, to extract the cube root of a polynomial,

Arrange the terms according to the powers of some letter;

Extract the cube root of the first term and set down the result as the first term of the root, and subtract its cube from the given polynomial;

Take three times the square of the root already found as a trial divisor; divide the first term of the remainder by it; set down the quotient as the next term of the root;

Complete the trial divisor by adding to it three times the product of the first and second terms of the root, and the square of the second term;

Multiply the complete divisor by the term of the root last found, and subtract the product from the remainder;

Continue in like manner till all the terms of the root are found.

EXERCISE 70.

Find the cube root of—

1. $a^3+6a^2x+12ax^2+8x^3$.
2. $27-27a+9a^2-a^3$.
3. $1-12x+48x^2-64x^3$.
4. $a^6-3a^5-3a^4+11a^3+6a^2-12a-8$.
5. $x^6-3x^5+6x^4-7x^3+6x^2-3x+1$.
6. $1-9x+21x^2+9x^3-42x^4-36x^5-8x^6$.
7. $12x^4-36x+64x^6-6x^2-8+117x^3-144x^5$.
8. $95a^3+72a^4-72a^2+15a^5+15a+a^6-1$.
9. $114x^4-171x^2-27-135x+8x^6-60x^5+55x^3$.
10. $8+27n^6-36n-81n^5+90n^2-135n^3+135n^4$.

11. $\dfrac{x^3}{8}-\dfrac{x^2}{4}+\dfrac{x}{6}-\dfrac{1}{27}$. 12. $\dfrac{x^6}{8y^3}-\dfrac{x^3}{2y}+\dfrac{2y}{3}-\dfrac{8y^3}{27x^3}$.

13. $x^3-3x^2+6x-7+\dfrac{6}{x}-\dfrac{3}{x^2}+\dfrac{1}{x^3}$.

14. $1+\dfrac{3}{a}-\dfrac{6}{a^2}-\dfrac{17}{a^3}+\dfrac{18}{a^4}+\dfrac{27}{a^5}-\dfrac{27}{a^6}$.

15. $x^6+\dfrac{6x^5}{y}+\dfrac{15x^4}{2y^2}-\dfrac{45x^2}{4y^4}+\dfrac{27x}{2y^5}-\dfrac{27}{8y^6}-\dfrac{10x^3}{y^3}$.

195. Cube Root of Arithmetical Numbers. The same general method as that used in Art. 194 can be used to extract the cube root of arithmetical numbers. As in square root, the process is slightly different from the algebraic one, owing to the fact that all the numbers which compose a given cube are given united or fused into a single number.

Thus, $(42)^3=(40+2)^3=40^3+3\times 40^2\times 2+3\times 40\times 2^2+2^3$
$=64000+9600+480+8$
$=74088.$

Hence, given 74088, to extract its cube root, the cube of the first number, or 64000 is not given explicitly, as it would be in an algebraic expression, but must be determined indirectly. This is done by marking off the given number into periods or groups of three figures each, beginning at the decimal point, and then determining the largest cube represented in the first period of figures, and taking its cube root as a trial number for the first figure of the root. The reason for marking off the given number into periods of three figures each may be briefly stated thus: If the first figure of the root be in the tens' place and therefore followed by one zero (as 40 above), its cube will be followed by three zeros (as 64000). If the first figure of the cube root be in the hundreds' place, and therefore followed by two zeros, its cube would be followed by six zeros. For every additional zero in the root there are three additional zeros in the cube. Hence arises the significance of separating the given number into periods of three figures each, and extracting the approximate cube root of the left-hand period.

We will now illustrate the general process of extracting an arithmetical cube root, using the algebraic formula $(a+b)^3 = a^3 + 3a^2b + 3ab^2 + b^3$, to show the essential identity of the arithmetical and algebraic processes.

Ex. Extract cube root of 74088.

			$\overset{\frown}{74}\overset{\frown}{088}$	40 + 2
	$a^3 = 40^3 =$		64000	
Trial Divisor,	$3a^2 = 3 \times 40^2 =$	4800	10088	
	$3ab = 3 \times 40 \times 2 =$	240		
	$b^2 = 2^2 =$	4		
Complete Divisor,	$3a^2 + 3ab + b^2 =$	5044	10088	

Abbreviated form of work—

		74088	42
		64	
$3 \times 40^2 =$	4800	10088	
$3 \times 40 \times 2 =$	240		
$2^2 =$	4		
	5044	10088.	

196. Cube Root of Decimal Numbers and Fractions. For a reason similar to that given in Art. 191 for square root of decimal numbers, in extracting the cube root of decimal numbers we mark off the decimal numbers into periods of three figures each, beginning at the decimal point, and sup-

14

plying a sufficient number of zeros when the right-hand period is incomplete.

Ex. 1. Extract the cube root of 130.323843.

```
                                         130.323843 | 5.07
                                         125
Trial Divisor = 3 × (500)² = 750000 | 5323843
                  3 × 500 × 7 =  10500 |
                           7² =     49 |
       Complete Divisor = 760549 | 5323843
```

Ex. 2. Extract the cube root of $\frac{5}{12}$ to 4 decimal places.

$$\tfrac{5}{12} = 0.416666666666+.$$

```
                            0.416666666 + | 0.7469+
                              343
     3 × (70)² = 14700 | 73666
   3 × (70 × 4) =   840 |
            4² =    16 |
                 15556 | 62224
    3 × (740)² = 1642800 | 11442666
 3 × (740 × 6) =   13320 |
            6² =      36 |
                 1656156 |  9936936
  3 × (7460)² = 166954800 | 1505730666
                          | 1502593200
                              3137466
```

The first three figures of the root are found directly. The last figure is then found by division of the remainder, using three times the square of the root already found as a divisor. The number of figures of the root that may thus be found by division is two less than the number of figures already found.

Hence, in general, to extract the cube root of an arithmetical number,

Separate the number into periods of three figures each, beginning at the decimal point;

Find the greatest cube in the left-hand period, and set down its cube root as the first figure of the required root;

Cube this figure, and subtract the result from the left-hand period, and annex the next period of figures to the remainder;

Take three times the square of the root already found with zero annexed, as a trial divisor; divide the remainder by it, and set down the quotient as the next figure of the root;

Complete the trial divisor by adding to it three times the product of the first figure of the root with zero annexed, multiplied by the last figure, and the square of the last figure;

Multiply this complete divisor by the figure of the root last found, and subtract the result from the remainder;

Proceed in like manner till all the periods have been used.

EXERCISE 71.

Find the cube root of—

1. 3375.
2. 753571.
3. 1906624.
4. 43614208.
5. 32891033664.
6. 520688691.125.
7. 344324.701729.
8. .000127263527.
9. 0.991026973.

Find to three decimal places the cube root of—

10. 75.
11. 6.
12. 5.6.
13. $3\frac{3}{8}$.
14. $7\frac{5}{11}$.
15. $19\frac{7}{9}$.
16. $\frac{1}{16}$.
17. $\frac{1}{101}$.
18. $1\frac{1}{15}$.
19. $8\frac{1}{25}$.

Compute the value of—

20. $\sqrt{5+2\sqrt[3]{5}}$.
21. $\sqrt[3]{3\sqrt{10}-2\sqrt[3]{10}}$.
22. $\sqrt{3\sqrt[3]{0.8}-2\sqrt{1.935}}$.

197. Higher Roots Obtained by Successive Extractions. By the law of exponents the square of the square of any quantity gives the fourth power of the quantity. Hence, reversing the process, the *fourth root* of a quantity is the *square root* of the *square root* of the quantity. Similarly, the *sixth root* of a quantity is the *square root* of the *cube root* of the quantity. The *eighth, ninth, tenth* . . . . . roots of a quantity may be found by similar methods.

Ex. Extract the fourth root of

$$81a^4 + 108a^3 + 54a^2 + 12a + 1.$$

Obtain first the square root of the given expression, which is $9a^2 + 6a + 1$. Extracting the square root of this, we obtain $3a + 1$, the fourth root of the original expression.

EXERCISE 72.

Find the fourth root of—

1. 130321. 2. 3418801. 3. 90. 4. 0.8.

5. $1 - 12ab + 54a^2b^2 - 108a^3b^3 + 81a^4b^4.$

6. $x^4 - 2x^3 + \frac{3}{2}x^2 - \frac{1}{2}x + \frac{1}{16}.$

7. $\dfrac{16x^4}{y^4} - \dfrac{16x^2}{y^2} + 6 - \dfrac{y^2}{x^2} + \dfrac{y^4}{16x^4}.$

8. $64x^2 - 56x^4 + 16x^5 + x^8 + 16 - 32x^3 + 16x^6 - 8x^7 + 64x.$

Find the sixth root of—

9. 7529536. 10. 1544804416. 11. 15.

12. $x^6 + 1215x^2 + 729 - 1458x + 135x^4 - 540x^3 - 18x^5.$

13. $64a^6 - 192a^4 + \dfrac{60}{a^2} - \dfrac{12}{a^4} + \dfrac{1}{a^6} - 160 + 240a^2.$

14. $4096x^{12} - 3072x^{10} + 960x^8 - 160x^6 + 15x^4 - \frac{3}{4}x^2 + \frac{1}{64}.$

CHAPTER XVI.

EXPONENTS.

198. Positive Integral Exponents. Using a^3 as a brief symbol for $a \times a \times a$, and a^m as a brief symbol for $a \times a \times a \times a \ldots\ldots$ to m factors, we have already found the following laws to govern the use of positive integral exponents:

$$\text{I. } a^m \times a^n = a^{m+n}.$$

$$\text{II. } \frac{a^m}{a^n} = a^{m-n}, \text{ if } m > n.$$

$$\text{III. } (a^m)^n = a^{mn}.$$

$$\text{IV. } \sqrt[n]{a^{mn}} = a^m.$$

$$\text{V. } (ab)^n = a^n b^n.$$

199. Fractional and Negative Exponents. Just as by using fractions as well as integers, and negative as well as positive quantity, the field of quantity and operation in algebra is greatly extended, some processes made simpler, and others more powerful, so by introducing fractional and negative exponents we get like results.

As fractional and negative exponents have no meaning belonging to them at the outset, it will be most advantageous to suppose that the first and fundamental Index Law, $a^m \times a^n = a^{m+n}$, holds for fractional and negative exponents, and then inquire what meaning must be assigned to these exponents. We limit the fractional and negative exponents here treated to those whose terms are either positive or negative integers, and commensurable; that is, expressible in terms of the unit of quantity used in the given problem.

Thus, exponents like $\sqrt{2}$, as in $a^{\sqrt{2}}$, are not included in the discussion, though the student will find later that the same laws hold for these exponents.

200. I. Meaning of a Fractional Exponent.

Since by Index Law I.,

$$a^{\frac{2}{3}} \times a^{\frac{2}{3}} \times a^{\frac{2}{3}} = a^{\frac{2}{3}+\frac{2}{3}+\frac{2}{3}} = a^2,$$

it follows that $a^{\frac{2}{3}}$ is one of the three equal factors which may be considered as composing a^2; that is, $a^{\frac{2}{3}}$ is the cube root of a^2.

Hence, in the exponent of $a^{\frac{2}{3}}$, the numerator, 2, denotes the power of a to be taken, and the denominator, 3, denotes the root of this power to be extracted.

$$\therefore\ a^{\frac{2}{3}} = \sqrt[3]{a^2}.$$

So, in general,

$$a^{\frac{p}{q}} \times a^{\frac{p}{q}} \times a^{\frac{p}{q}} \times a^{\frac{p}{q}} \ldots\ldots \text{ to } q \text{ factors}$$

$$= a^{\frac{p}{q}+\frac{p}{q}+\frac{p}{q}+\ldots\ldots \text{ to } q \text{ terms}}$$

$$= a^{\frac{p}{q}\times q} = a^p.$$

Hence, in general, *in a fractional exponent the numerator denotes the power of the base that is to be taken, and the denominator denotes the root that is to be extracted.*

Ex. 1. $8^{\frac{2}{3}} = \sqrt[3]{8^2} = \sqrt[3]{64} = 4.$

Ex. 2. $a^{\frac{3}{4}} \times a^{\frac{1}{2}} \times a^{\frac{1}{3}} = a^{\frac{3}{4}+\frac{1}{2}+\frac{1}{3}}$

$= a^{\frac{19}{12}}$, *Product.*

Ex. 3. $x^{\frac{2}{3}}\sqrt[5]{x^4} = x^{\frac{2}{3}} \times x^{\frac{4}{5}}.$

$= x^{\frac{22}{15}}$, *Product.*

Ex. 4. $32^{\frac{6}{5}} = \sqrt[5]{32^6} = 2^6 = 64.$

EXERCISE 73.

Express with radical signs—

1. $a^{\frac{1}{2}}$.
2. $x^{\frac{2}{5}}$.
3. $c^{\frac{5}{3}}$.
4. $2a^{\frac{1}{4}}$.
5. $ax^{\frac{3}{4}}$.
6. $2a^2b^{\frac{2}{7}}$.
7. $a^{\frac{1}{2}}m^{\frac{2}{3}}$.
8. $5x^{\frac{2}{3}}y^{\frac{3}{4}}$.
9. $2c^{\frac{3}{2}}d^{\frac{4}{5}}$.
10. $ax^{\frac{3}{n}}$.
11. $5y^{\frac{n}{4}}$.
12. $x^{\frac{n}{m}}y^{\frac{m}{2n}}$.

Express with fractional exponents—

13. $\sqrt[3]{a^4}$.
14. $\sqrt{c^5}$.
15. $2\sqrt[7]{x^3}$.
16. $a\sqrt{x}$.
17. $b\sqrt[3]{y}$.
18. $2x\sqrt{y^5}$.
19. $\sqrt{x}\sqrt[3]{y^2}$.
20. $2\sqrt[4]{x^3}\sqrt{y}$.
21. $\sqrt[3]{5}\sqrt[7]{a^{10}}$.
22. $\frac{3a\sqrt{x^3}}{4\sqrt[3]{a}\sqrt{c^3}}$.
23. $\frac{2\sqrt[3]{81}\sqrt{3^4}}{3\sqrt{2^3}\sqrt{2^2}}$.

Find the value of—

24. $27^{\frac{2}{3}}$.
25. $25^{\frac{3}{2}}$.
26. $16^{\frac{3}{4}}$.
27. $\sqrt[3]{64^2}$
28. $\sqrt[6]{64^7}$.
29. $(-8)^{\frac{2}{3}}$.
30. $(-27)^{\frac{4}{3}}$.
31. $(-32)^{\frac{6}{5}}$.
32. $(-216)^{\frac{2}{3}}$.
33. $(-243)^{\frac{3}{5}}$.
34. $(\frac{8}{27})^{\frac{4}{3}}$.
35. $(\frac{81}{16})^{\frac{5}{4}}$.

Simplify the following by performing the indicated operations:

36. $a^{\frac{1}{2}} \times a^{\frac{2}{3}}$.
37. $2a \times a^{\frac{1}{3}}$.
38. $a^2x^{\frac{1}{2}} \times a^{\frac{1}{2}}x^{\frac{3}{2}}$.
39. $3x^{\frac{2}{3}} \times a^{\frac{1}{2}}x^{\frac{1}{3}}$.
40. $2^{\frac{1}{2}}x^{\frac{1}{3}} \times 2^{\frac{3}{2}}x^{\frac{5}{3}}$.
41. $a^{\frac{4}{3}}\sqrt{a^3}$.
42. $7\sqrt{a}\sqrt[3]{a^2}$.
43. $2x^2\sqrt[3]{x^2}$.
44. $\sqrt{a^3}\sqrt[3]{a^2}$.
45. $\sqrt[3]{2}\sqrt{2^3}$.
46. $a^{\frac{3}{5}}\sqrt[4]{x^3} \cdot x^{\frac{1}{2}}\sqrt{a}$.
47. $x^{\frac{3}{5}}\sqrt[4]{x^5} \cdot 2x^{\frac{3}{20}}$.
48. $\frac{a^{\frac{1}{2}}\sqrt{x^5}}{\sqrt[4]{c^5}} \times \frac{2x^{\frac{3}{2}}\sqrt{a^5}}{\sqrt[4]{c}}$.
49. $\frac{\sqrt{2}\sqrt[5]{7}}{\sqrt[3]{3}\sqrt[4]{5}} \times \frac{2^{\frac{3}{2}}7^{\frac{4}{5}}}{\sqrt[3]{3^5}\sqrt[8]{5^{14}}}$

201. II. Meaning of the Exponent Zero, or of a^0.

By the Index Law I.,

$$a^0 \times a^m = a^{0+m} = a^m = 1 \times a^m$$

$$\therefore\ a^0 = 1.$$

Hence, a^0 is another symbol for unity.

The student will realize the meaning of a^0 more readily thus,

By direct division, $\dfrac{a^m}{a^m} = 1$

By subtraction of exponents, $\dfrac{a^m}{a^m} = a^0$

$\therefore$ by Ax. 1, $a^0 = 1.$

202. III. Meaning of a Negative Exponent.

If n be an integer or a fraction, by the Index Law,

$$a^n \times a^{-n} = a^{n-n} = a^0 = 1$$

$$\therefore\quad a^{-n} = \frac{1}{a^n},\ \text{or}\ a^n = \frac{1}{a^{-n}}.$$

Ex. 1. $4^{-2} = \dfrac{1}{4^2} = \dfrac{1}{16}$

Ex. 2. $8^{-\frac{2}{3}} = \dfrac{1}{8^{\frac{2}{3}}} = \dfrac{1}{4}.$

203. Transference of Factors in Terms of a Fraction. It follows from the meaning of a negative exponent that *any factor may be transferred from the numerator to the denominator of a fraction, or vice versâ, provided the sign of the exponent of the factor be changed.*

Ex. 1. Transfer to the numerator the factors of the denominator of $\dfrac{ab^{-2}}{xy^{-\frac{1}{2}}}$.

$$\frac{ab^{-2}}{xy^{-\frac{1}{2}}} = ab^{-2}x^{-1}y^{+\frac{1}{2}}, \text{ *Result.*}$$

Ex. 2. Transfer factors in the terms of $\dfrac{2a^{-2}b}{xy^{-\frac{2}{3}}z^{-1}}$, so as to make all exponents positive.

$$\frac{2a^{-2}b}{xy^{-\frac{2}{3}}z^{-1}} = \frac{2by^{\frac{2}{3}}z}{a^2x}, \text{ *Result.*}$$

It will not be a difficult exercise for the student to prove Law II., $\dfrac{a^m}{a^n} = a^{m-n}$, for fractional and negative exponents.

EXERCISE 74.

Transfer to the numerator all factors of the denominator—

1. $\dfrac{a}{c^2}$.
2. $\dfrac{x^3}{z^3}$.
3. $\dfrac{3a}{x^{-4}}$.
4. $\dfrac{ab^3}{cd^{-2}}$.
5. $\dfrac{3}{x^{\frac{1}{2}}}$.
6. $\dfrac{a^3y^{-2}}{xz^{-3}}$.
7. $\dfrac{3a}{2c^3}$.
8. $\dfrac{2a^3}{b^{-2}c^3d^{-\frac{1}{2}}}$.
9. $\dfrac{7x^{-1}}{4m^{-3}n^{-\frac{1}{3}}}$.
10. $\dfrac{1}{2^{-1}x^{\frac{1}{3}}y^{-4}}$.
11. $\dfrac{7}{5^{\frac{1}{3}}x^{-\frac{1}{2}}z^{-\frac{3}{4}}}$.
12. $\dfrac{x^2}{a^{\frac{1}{n}}x^{-\frac{m}{n}}}$.

Express with positive exponents—

13. $7x^{-2}$.
14. $3ab^{-1}$.
15. $a^2b^{-\frac{3}{2}}$.
16. $5a^{-1}b^3$.
17. $3a^{-\frac{1}{2}}b^{-2}$.
18. $ab^{-3}x^2y^{-\frac{1}{2}}$.
19. $\dfrac{5a^2x^{-3}}{z^4}$.
20. $\dfrac{3a^{-2}b}{cd^{-3}}$.

21. $\dfrac{1}{ab^{-2}}$.

22. $\dfrac{x^{-3}y^{-1}}{3a^{-2}}$.

23. $\dfrac{7a^{-2}x^{-\frac{2}{3}}}{3b^{-1}y^{-\frac{4}{5}}}$.

24. $\dfrac{5a^{-2}c^{\frac{1}{3}}}{2^{-1}xy^{-\frac{2}{3}}}$.

25. $\dfrac{3x^{-n}y^{-5}}{5^{-2}cz^{-\frac{1}{n}}}$.

26. $\dfrac{3^{-2}\sqrt{a^{-1}}\sqrt[3]{x^2}}{2\sqrt{a^3}\sqrt[3]{x^{-1}}}$.

Find the numerical value of—

27. $4^{-\frac{1}{2}}$.

28. $27^{-\frac{2}{3}}$.

29. $\sqrt{\dfrac{1}{4^{-5}}}$.

30. $\dfrac{1}{5^{-2}}$.

31. $8^{-\frac{2}{3}}$.

32. $\sqrt{81^{-3}}$.

33. $\dfrac{2}{3^{-3}}$.

34. $3^{-2}\times 4^{\frac{1}{2}}$.

35. $2^{-3}\div 8^{-\frac{1}{3}}$.

36. $\left(\dfrac{1}{36}\right)^{-\frac{3}{2}}$.

37. $\left(\dfrac{8}{27}\right)^{-\frac{4}{3}}$.

38. $\left(-\dfrac{8}{27}\right)^{-\frac{1}{3}}$.

39. $(-125)^{-\frac{4}{3}}$.

40. $\dfrac{(-8)^{-\frac{5}{3}}}{(-27)^{-\frac{2}{3}}}$.

41. $(5\tfrac{4}{9})^{-\frac{3}{2}}$.

42. $\left(\dfrac{1}{64}\right)^{\frac{5}{6}}\cdot\left(\dfrac{1}{4}\right)^{-\frac{5}{2}}$.

43. $\dfrac{2^{-3}\cdot 3^{-2}\cdot 4^{-\frac{3}{2}}}{9^{-1}\cdot 8^{-\frac{5}{3}}}$.

Simplify the following by performing the indicated operations, and reducing the results:

44. $2a^2\times a^{-1}$.

45. $a^3x^{-1}\times a^{-2}x^{-3}$.

46. $5a^{-7}\times 2a^5x^3$.

47. $a^4\div a^{-2}$.

48. $4x^{-2}\div 2x^{-3}$.

49. $xy^{-3}\div x^2y^{-4}$.

50. $a^{\frac{1}{2}}\cdot 3a^{-\frac{1}{3}}$.

51. $c^{-1}d^{\frac{2}{3}}\div c^{-3}d^{\frac{1}{3}}$.

52. $m^{-\frac{1}{2}}n\cdot mn^{-1}$.

53. $6a^{\frac{1}{2}}x^{-\frac{2}{3}}\cdot a^{\frac{1}{2}}x^{\frac{1}{3}}$.

54. $a^{-\frac{1}{3}}\cdot 2a^{\frac{1}{3}}$.

55. $8x^{-\frac{2}{3}}y\div 4x^{\frac{1}{3}}y^3$.

56. $4x^{\frac{2}{3}}\div 3xy^{\frac{1}{2}}$.

57. $7a^3x^{-\frac{3}{2}}\div 5x^{\frac{2}{3}}y$.

58. $a^{\frac{2}{3}}\sqrt{x^{-3}}\cdot x^{\frac{1}{2}}\sqrt[3]{a^{-1}}$.

59. $x^{-\frac{3}{2}}\sqrt[4]{y^{-3}}\div x^{\frac{1}{2}}y^{\frac{1}{4}}$.

60. $\dfrac{x^{\frac{1}{2}}\sqrt[6]{x^{-7}}}{x^{-\frac{1}{3}}\sqrt{x^{-5}}}$.

61. $\dfrac{x^{\frac{1}{2}}\sqrt[3]{x^{-2}}}{\sqrt[6]{x^{-1}}}$.

62. $a^{\frac{1}{3}}b^{-\frac{1}{2}}\sqrt[3]{c^{-2}}\cdot\sqrt[3]{c}\sqrt{b^3}$.

63. $\dfrac{a\sqrt{a^{-2}}}{3\sqrt[3]{a^{-1}}}$.

64. $\dfrac{3x^{-1}\sqrt{x^{-3}}}{2x^{-\frac{5}{2}}}$.

65. $\dfrac{7a^5\sqrt[3]{x^{-5}}}{3a^{-5}\sqrt[3]{x^{-4}}}$.

66. $\dfrac{x^n\sqrt{y^{-m}}}{y^m\sqrt{x^{-2n}}}$.

67. $\dfrac{x^{\frac{n}{2}}\sqrt[n]{y^{-2}}}{y^{\frac{1}{n}}\sqrt{x^{-n}}}$.

68. $\dfrac{\sqrt[2n]{x^3}\cdot\sqrt[n]{x^{-2}}}{x\sqrt[n]{x^{-3}}}$.

204. IV. $(a^m)^n = a^{mn}$ for Fractional and Negative Exponents. It will now be found that using the meanings for fractional and negative exponents which have been determined (Arts. 200, 202), Law III., $(a^m)^n = a^{mn}$ applies to them also.

First, when n is a positive fraction, $\frac{p}{q}$, the terms of the fraction, p and q, being positive integers.

$$[(a^m)^{\frac{p}{q}}]^q = (a^m)^{\frac{p}{q}\times q}$$
$$= a^{mp}.$$

Extracting the qth root of both sides,

$$(a^m)^{\frac{p}{q}} = a^{\frac{mp}{q}}$$

Substitute n for $\frac{p}{q}$, $\quad (a^m)^n = a^{mn}$.

Second, when n is a negative integer or negative fraction; as, $-t$.

$$(a^m)^n = (a^m)^{-t} = \frac{1}{(a^m)^{+t}} = \frac{1}{a^{mt}} = a^{-mt} = a^{mn}.$$

It will not be a difficult exercise for the student to prove also Law V., $(ab)^n = a^n b^n$.

First, when n is a positive fraction.

Second, when n is a negative quantity.

Ex. 1. Find the value of $(4^{-\frac{3}{2}})^{-\frac{5}{3}}$.

$$(4^{-\frac{3}{2}})^{-\frac{5}{3}} = 4^{\frac{5}{2}} = 32, \textit{ Result.}$$

Ex. 2. $\sqrt[3]{(8a^{-\frac{5}{6}})^2} = (8a^{-\frac{5}{6}})^{\frac{2}{3}} = 8^{\frac{2}{3}}a^{-\frac{5}{9}} = \dfrac{4}{a^{\frac{5}{9}}}$, *Result.*

Ex. 3. $\left(\dfrac{16a^{-4}}{81b^3}\right)^{-\frac{3}{4}} = \dfrac{16^{-\frac{3}{4}}a^3}{81^{-\frac{3}{4}}b^{-\frac{9}{4}}} = \dfrac{81^{\frac{3}{4}}a^3b^{\frac{9}{4}}}{16^{\frac{3}{4}}}$

$$= \frac{27a^3b^{\frac{9}{4}}}{8}, \textit{ Result.}$$

EXERCISE 75.

Reduce to the simplest form—

1. $(a^2)^{-3}$.
2. $(x^{-2})^{-1}$.
3. $(a^{-3})^{\frac{1}{2}}$.
4. $(x^{-\frac{3}{2}})^4$.
5. $(c^{\frac{1}{2}})^{-\frac{2}{3}}$.
6. $(a^{-\frac{5}{3}})^{-\frac{3}{5}}$.
7. $(a^2b^{-\frac{1}{2}})^{-2}$.
8. $(x^{-\frac{1}{3}}y^{\frac{1}{2}})^{-6}$.
9. $(8^{-2})^{\frac{2}{3}}$.
10. $(64^{-1})^{-\frac{2}{3}}$.
11. $(9^{\frac{5}{2}})^{-\frac{3}{5}}$.
12. $(3a^{-2})^2$.
13. $(5x^{-\frac{1}{3}})^{-2}$.
14. $(8a^3)^{-\frac{2}{3}}$.
15. $(4x^{-4})^{-\frac{3}{2}}$.
16. $(9a^{-2}x^{\frac{1}{2}}y^{-3})^{-\frac{3}{2}}$.
17. $(-2a^2x^{-\frac{1}{2}})^2$.
18. $(-5x^{-1}y^{\frac{1}{2}})^{-2}$.

19. $(9x^{-3}y^{-2})^{-\frac{5}{2}}$.

20. $(a^2\sqrt{a^{-1}})^{-3}$.

21. $(a^{\frac{2}{3}}\sqrt[4]{a^{-3}})^{-\frac{6}{5}}$.

22. $\sqrt[6]{(a^{-1})^{-3}}$.

23. $(a^2\sqrt{3a^{-1}b^3})^{-2}$.

24. $\left\{\sqrt{(\sqrt{a^{-5}})^{-\frac{5}{2}}}\right\}^{-\frac{6}{5}}$.

25. $\left\{\sqrt[4]{(x^{-\frac{4}{3}}y^2)^6}\right\}^{-\frac{1}{2}}$.

26. $(c^{2a}x^{-3a})^{-\frac{b}{a}}$.

27. $\left(\frac{25\sqrt{x}}{9\sqrt[3]{x^{-2}}}\right)^{-\frac{3}{2}}$.

28. $\left\{\frac{27x\sqrt[3]{y^{-2}}}{8y^{\frac{1}{3}}\sqrt{x^{-2}}}\right\}^{-\frac{2}{3}}$.

29. $\left\{\sqrt[4]{x^{-\frac{2}{3}}y^{\frac{1}{2}}\sqrt{xy}}\right\}^{-\frac{2}{3}}$.

30. $\frac{\sqrt[3]{x^{-1}\sqrt{y^3}}}{\sqrt{y^{-1}\sqrt[3]{x^3}}}$.

31. $\frac{\sqrt[4]{8a^{-2}b^3\sqrt[3]{c^2}}}{\sqrt[3]{8^{-\frac{1}{4}}\sqrt{ab^{\frac{9}{2}}c^{-5}}}}$.

32. $\sqrt{x^{-1}\sqrt[3]{y^2}} \div \sqrt[3]{y\sqrt{x^{-3}}}$.

33. $\sqrt[3]{ab^{-2}c^{-1}} \times \sqrt[6]{ab^4c^2}$.

34. $(x^{\frac{1}{3}}\sqrt[3]{a^{-1}})^{-3} \times \sqrt{a^{-1}x^3}$.

35. $\left(\frac{x^{-1}y^2}{a^3b^{-4}}\right)^{-2} \div \left(\frac{xy^{-1}}{a^{-2}b^3}\right)^3$.

36. $\left(\frac{x\sqrt[4]{a}}{b^2\sqrt{y^{-1}}} \times \frac{\sqrt[4]{a^3y^6}}{x^{-1}b^{-1}}\right)^{-2}$.

37. $\sqrt[4]{\sqrt[3]{(16^{-2})^6}}$.

38. $\sqrt{\left[\sqrt{(\tfrac{16}{25})^{-\frac{1}{2}}}\right]^{-6}}$.

39. $\left[\sqrt[4]{(-\tfrac{27}{64})^{-\frac{2}{3}}}\right]^{-\frac{2}{5}}$.

40. $\left(\frac{x^{n+1}}{x}\right)^{n} \div \left(\frac{x}{x^{1-n}}\right)^{n-1}$.

41. $8^{-\frac{2}{3}} + 9^{\frac{3}{2}} - 2^{-2} + 1^{-\frac{2}{5}} - 7^0$.

42. $\left\{\frac{\sqrt[3]{a^2}}{\sqrt[4]{b^{-1}}} \cdot \frac{\sqrt{c^{-3}}}{a^{\frac{1}{3}}} \cdot \frac{b^{-\frac{1}{4}}\sqrt[3]{a}}{c^{-1}}\right\}^{-2}$.

43. $\left(a^{\frac{3}{4}}b^{-\frac{1}{2}}\sqrt[3]{a^{-2}b^2\sqrt{c^3}}\right)^6$.

44. $\left(\frac{a^{-4}}{b^{-2}c}\right)^{-\frac{3}{4}} \times \left(\frac{a^{-1}b\sqrt{c^{-3}}}{ab^{-1}}\right)^{\frac{1}{2}}$.

45. $\left\{ \sqrt[5]{\dfrac{a^{\frac{1}{2}}x^{-2}}{x^{\frac{1}{2}}a^{-2}}} \times \sqrt{\dfrac{a\sqrt{x}}{x^{-1}\sqrt{a}}} \right\}^{-4}.$

46. $\left(\dfrac{a^2b}{c^2d}\right)^{\frac{1}{2}} \times \left(\dfrac{c^3d}{ab^3}\right)^{\frac{1}{3}} \times \left(\dfrac{a^{\frac{1}{3}}c}{b^{\frac{1}{4}}d^{\frac{5}{12}}}\right)^2.$

47. $\sqrt[5]{\dfrac{x^{\frac{5}{2}}y^{\frac{4}{3}}}{z^{-\frac{5}{4}}} \cdot \dfrac{z^4}{x^{-3}y^{-\frac{5}{3}}} \div \dfrac{y^{-2}z^{\frac{1}{4}}}{x^{-\frac{1}{2}}}}.$

Ex. Expand $(x^{\frac{2}{3}} - 4x^{-\frac{1}{2}})^3$

$$= (x^{\frac{2}{3}})^3 - 3(x^{\frac{2}{3}})^2(4x^{-\frac{1}{2}}) + 3(x^{\frac{2}{3}})(4x^{-\frac{1}{2}})^2 - (4x^{-\frac{1}{2}})^3$$

$$= x^2 - 12x^{\frac{5}{6}} + 48x^{-\frac{1}{3}} - 64x^{-\frac{3}{2}},$$ *Result.*

Expand—

48. $(2x - 3x^{-2})^3.$

49. $(\sqrt{x} - 2\sqrt[3]{x})^4.$

50. $(3\sqrt[3]{x^2} + 2\sqrt{x^3})^5.$

51. $(\frac{1}{2}x^{-\frac{1}{2}} - 2x^{\frac{1}{4}})^4.$

52. $\left(2\sqrt{x^{-1}} + \dfrac{5}{\sqrt[3]{x^2}}\right)^4.$

53. $\left(\sqrt{x} - \dfrac{1}{\sqrt{y}}\right)^4.$

54. $\left(1 + \dfrac{2\sqrt{x^{-3}}}{3\sqrt[3]{y^2}}\right)^6.$

Ex. Solve $x^{-\frac{3}{2}} = 27$. Raise both sides to the power $(-\frac{2}{3})$.

$$(x^{-\frac{3}{2}})^{-\frac{2}{3}} = (27)^{-\frac{2}{3}} = \frac{1}{27^{\frac{2}{3}}} = \frac{1}{9}. \qquad \therefore x = \frac{1}{9}.$$

Find the value of x in each of the following—

55. $x^{\frac{1}{2}} = 2.$

56. $x^{\frac{3}{2}} = -27.$

57. $x^{-\frac{1}{2}} = 3.$

58. $x^{-\frac{2}{3}} = 4.$

59. $x^{-\frac{3}{5}} = -\frac{1}{8}.$

60. $x^{-n} = 1.$

61. $x^{-n} = 2.$

62. $x^{-\frac{1}{n}} = -3.$

63. $x^{-\frac{3}{4}} = -\frac{1}{27}.$

205. Polynomials whose Terms contain Fractional or Negative Exponents.

Ex. 1. Multiply $x + 3x^{\frac{2}{3}} - 2x^{\frac{1}{3}}$ by $3 - 2x^{-\frac{1}{3}} + 4x^{-\frac{2}{3}}$.

$$\begin{array}{l}
x + 3x^{\frac{2}{3}} \quad - 2x^{\frac{1}{3}} \\
\underline{3 - 2x^{-\frac{1}{3}} + 4x^{-\frac{2}{3}}} \\
3x + 9x^{\frac{2}{3}} \quad - 6x^{\frac{1}{3}} \\
\qquad - 2x^{\frac{2}{3}} \quad - 6x^{\frac{1}{3}} + 4 \\
\qquad\qquad\qquad \underline{+ 4x^{\frac{1}{3}} + 12 - 8x^{-\frac{1}{3}}} \\
3x + 7x^{\frac{2}{3}} \quad - 8x^{\frac{1}{3}} + 16 - 8x^{-\frac{1}{3}}, \text{ Product.}
\end{array}$$

Ex. 2. Extract the square root of

$$1 + \frac{4}{\sqrt[3]{x}} - 2x^{-\frac{2}{3}} - \frac{4}{x} + 25x^{-\frac{4}{3}} - 24x^{-\frac{5}{3}} + \frac{16}{x^2},$$

writing the given expression by use of exponents only.

$$\begin{array}{r|l}
 & 1 + 4x^{-\frac{1}{3}} - 2x^{-\frac{2}{3}} - 4x^{-1} + 25x^{-\frac{4}{3}} - 24x^{-\frac{5}{3}} + 16x^{-2} \quad \underline{|\, 1 + 2x^{-\frac{1}{3}} - 3x^{-\frac{2}{3}} + 4x^{-1}} \\
 & \underline{1} \\
2 + 2x^{-\frac{1}{3}} & 4x^{-\frac{1}{3}} - 2x^{-\frac{2}{3}} \\
 & \underline{4x^{-\frac{1}{3}} + 4x^{-\frac{2}{3}}} \\
2 + 4x^{-\frac{1}{3}} - 3x^{-\frac{2}{3}} & - 6x^{-\frac{2}{3}} - 4x^{-1} + 25x^{-\frac{4}{3}} \\
 & \underline{- 6x^{-\frac{2}{3}} - 12x^{-1} + 9x^{-\frac{4}{3}}} \\
2 + 4x^{-\frac{1}{3}} - 6x^{-\frac{2}{3}} + 4x^{-1} & 8x^{-1} + 16x^{-\frac{4}{3}} - 24x^{-\frac{5}{3}} + 16x^{-2} \\
 & \underline{8x^{-1} + 16x^{-\frac{4}{3}} - 24x^{-\frac{5}{3}} + 16x^{-2}}
\end{array}$$

206. Summary of Principles Relating to Exponents.

1. *In a fractional exponent the numerator denotes a power, the denominator a root.* Exs. $a^{\frac{2}{3}} = \sqrt[3]{a^2}$; $32^{\frac{3}{5}} = 8$.

2. $a^0 = 1$. $\qquad 5^0 = 1$.

3. $a^{-n} = \frac{1}{a^n}$, $\qquad 4^{-2} = \frac{1}{16}$.

4. *In the use of exponents, fractional and negative as well as positive, use the rules which govern the use of positive integral exponents;*

That is, in brief,

(1) *To multiply, add the exponents.*

(2) *To divide, subtract the exponent of the divisor from the exponent of the dividend.*

(3) *To raise to a power, multiply the exponents.*

(4) *To extract a root, divide the exponent by the index of the root.*

EXERCISE 76.

Multiply—

1. $a - 2a^{\frac{1}{2}} + 3$ by $2a^{\frac{1}{2}} + 3$.

2. $a^{\frac{2}{3}} - a^{\frac{1}{3}} + 1$ by $a^{\frac{1}{3}} + 1$.

3. $3x^{\frac{2}{3}} - 2x^{\frac{1}{3}}y^{\frac{1}{3}} + 3y^{\frac{2}{3}}$ by $3x^{\frac{1}{3}} + 2y^{\frac{1}{3}}$.

4. $2x^{\frac{2}{3}} - 3x^{\frac{1}{3}} + 4$ by $2 + 3x^{-\frac{1}{3}}$.

5. $a^{-1} - a^{-\frac{1}{2}}b^{\frac{1}{2}} + b$ by $a^{-1} + a^{-\frac{1}{2}}b^{\frac{1}{2}} + b$.

6. $x^2 - xy + 2y^2$ by $2x^{-2} + x^{-1}y^{-1} + y^{-2}$.

7. $2x^{\frac{1}{2}} - 3y^{-1} + x^{-\frac{1}{2}}y^{-2}$ by $2x^{-\frac{1}{3}}y + 3x^{-\frac{5}{6}}$.

8. $2x^{\frac{2}{3}} - 3x^{\frac{1}{3}} - 4 + x^{-\frac{1}{3}}$ by $3x^{\frac{4}{3}} + x - 2x^{\frac{2}{3}}$.

9. $a^{\frac{2}{3}}x^{-\frac{3}{4}} + 2 + a^{-\frac{2}{3}}x^{\frac{3}{4}}$ by $2a^{-\frac{2}{3}}x^{\frac{3}{4}} - 4a^{-\frac{4}{3}}x^{\frac{3}{2}} + 2a^{-2}x^{\frac{9}{4}}$.

10. $2\sqrt{x} + 3x^{\frac{1}{6}}\sqrt[3]{y} + \dfrac{4\sqrt[3]{y^2}}{\sqrt[6]{x}}$ by $\dfrac{3}{\sqrt{x}} - \dfrac{3}{\sqrt[6]{xy^3}} + \dfrac{2\sqrt[6]{x}}{\sqrt[3]{y^2}}$.

Divide—

11. $5x + 2x^{\frac{2}{3}} - 2x^{\frac{1}{3}} + 1$ by $x^{\frac{1}{3}} + 1$.

12. $8x^{-2} + \dfrac{6}{xy} + 3y^{-2} - 18xy^{-3} - 8x^2y^{-4}$ by $2x^{-1} + 3y^{-1}$
$+ 4xy^{-2}$

13. $x^{-\frac{1}{3}} - x^{-\frac{1}{6}} + 5 - 2x^{\frac{1}{6}}$ by $1 + 2\sqrt[6]{x}$.

14. $\sqrt[3]{x} - \sqrt[3]{y}$ by $\sqrt[9]{x} - \sqrt[9]{y}$.

15. $\sqrt{a} + \sqrt[4]{ab} + \sqrt{b}$ by $\sqrt[4]{a} + \sqrt[8]{ab} + \sqrt[4]{b}$.

16. $27a^2 - 30ay^{-1} + 3y^{-2}$ by $3a - 2a^{\frac{1}{2}}y^{-\frac{1}{2}} - y^{-1}$.

17. $x^{-\frac{4}{3}} + x^{-\frac{2}{3}}y^{-2} + y^{-4}$ by $x^{-1} + x^{-\frac{2}{3}}y^{-1} + x^{-\frac{1}{3}}y^{-2}$.

18. $\dfrac{x}{y} - x^{\frac{2}{3}}y^{-\frac{1}{2}} - 4\sqrt[3]{x} - \dfrac{8y}{\sqrt[3]{x}}$ by $\sqrt[3]{x} + 2y^{\frac{1}{2}}$.

19. $\dfrac{9}{a} - \dfrac{3\sqrt{x}}{\sqrt[4]{a^3}} + \dfrac{10x}{\sqrt{a}} + \dfrac{\sqrt{x^3}}{\sqrt[4]{a}} - x^2$ by $\dfrac{3}{\sqrt[4]{a}} + \sqrt{x}$.

20. $4\sqrt[3]{a^2} - 8\sqrt[3]{a} - 5 + \dfrac{10}{\sqrt[3]{a}} + \dfrac{3}{\sqrt[3]{a^2}}$ by $2a^{\frac{5}{12}} - \sqrt[12]{a} - \dfrac{3}{\sqrt[4]{a}}$.

Extract the square root of—

21. $x^{\frac{2}{3}} - 4x^{\frac{5}{6}}y^{\frac{1}{2}} + 4xy$.

22. $9xy^{-2} + 12y^{-1} + 4x^{-1}$.

23. $a^{-1} - 4a^{-\frac{1}{2}}b^{\frac{1}{2}} + 10b - 12a^{\frac{1}{2}}b^{\frac{3}{2}} + 9ab^2$.

24. $x^{-\frac{3}{2}} + 8x^{-1} - 2x^{-\frac{5}{4}} + 16x^{-\frac{1}{2}} - 8x^{-\frac{1}{4}} + 1$.

25. $9x^{-3} - 30x^{-\frac{5}{2}}y + 13x^{-2}y^2 + 20x^{-\frac{3}{2}}y^3 + 4x^{-1}y^4$.

26. $25a^{\frac{4}{3}}b^{-3} - 10a^{\frac{2}{3}}b^{-\frac{3}{2}} - 49 + 10a^{-\frac{2}{3}}b^{\frac{3}{2}} + 25a^{-\frac{4}{3}}b^3$.

27. $\dfrac{9}{x^4} - \dfrac{18\sqrt{y}}{x^3} + \dfrac{15y}{x^2} - \dfrac{6\sqrt{y^3}}{x} + y^2$.

28. $9x^5 - \dfrac{24\sqrt{x^7}}{\sqrt[3]{y}} + \dfrac{4x^2}{\sqrt[3]{y^2}} + 16\sqrt{\dfrac{x}{y^2}} + \dfrac{4}{xy\sqrt[3]{y}}$.

29. $\dfrac{y^2}{4x} - \dfrac{y}{3x^{\frac{3}{4}}} + \dfrac{28}{9\sqrt{x}} - \dfrac{2}{y\sqrt[4]{x}} + \dfrac{9}{y^2}$.

CHAPTER XVII.

RADICALS.

207. Indicated Roots. The root of a quantity may be indicated in two ways:

(1) By the use of a fractional exponent; as $a^{\frac{1}{3}}$.

(2) By the use of a radical sign; as $\sqrt[3]{a}$.

For some purposes, one of these methods is better; for some, the other.

Thus, when we have $a^{\frac{2}{3}} \times a^{\frac{1}{3}} \times a^{-2}$, where the quantities are alike except in their exponents, it is better to use fractional exponents to indicate roots; but if we have $5\sqrt{3} - 7\sqrt{27} + 8\sqrt{12}$, where exponents are alike, but coefficients and bases unlike, it is better to use the radical sign to indicate roots.

In the preceding chapter we considered exponents; we have now to investigate the properties of radicals.

208. A **Radical** is a root of a quantity indicated by the use of the radical sign. Exs. $\sqrt{x}$, $\sqrt[3]{27}$.

209. Surds. An indicated root which may be exactly extracted is said to be **Rational.** Ex. $\sqrt[3]{27}$, since the cube root of 27 is 3.

An indicated root which cannot be exactly extracted is called a **Surd.** Exs. $\sqrt{3}$, $\sqrt[3]{5}$.

210. The **Coefficient** of a radical is the number prefixed to the radical proper, to show how many times the radical is taken.

Ex. The coefficient of $5\sqrt{3}$ is 5; of $6a\sqrt[3]{x}$ is $6a$.

211. Entire Surds. If a surd have unity for its coefficient, it is said to be Entire.

212. The **Degree** of a radical is the number of the indicated root. Ex. $\sqrt[3]{x}$ is a radical of the third degree.

213. Similar Radicals are those which have the same quantity under the radical sign and the same index. (The coefficients and signs of the radicals may be unlike; hence, similar radicals must be alike in two respects, and may be unlike in two other respects.) Ex. $5\sqrt{3}$, $-4\sqrt{3}$ are similar radicals.

214. Fundamental Principle. Since a radical and a quantity affected by a fractional exponent differ only in form, in investigating the properties of radicals we may use all the principles demonstrated concerning fractional exponents.

Thus, since $(ab)^n = a^n b^n$ is true, when n is a fraction, as $\frac{1}{n}$,

$$(ab)^{\frac{1}{n}} = a^{\frac{1}{n}} b^{\frac{1}{n}}$$

$$\therefore \sqrt[n]{ab} = \sqrt[n]{a} \cdot \sqrt[n]{b}.$$

TRANSFORMATIONS OF RADICALS.

215. I. Simplification of a Quantity under Radical Sign. If a factor of the quantity under the radical sign is a perfect power of the same degree with the radical, the root of this factor may be extracted and set outside as a factor of the coefficient.

Ex. 1 Simplify $\sqrt[3]{56}$.

$$\sqrt[3]{56} = \sqrt[3]{8 \times 7} = 2\sqrt[3]{7}, \textit{ Result.} \quad \text{(Art. 214.)}$$

Ex. 2. Simplify $5\sqrt{18a^3b^2c^5}$.

$$5\sqrt{18a^3b^2c^5} = 5\sqrt{9a^2b^2c^4 \times 2ac} = 15abc^2\sqrt{2ac}, \textit{ Result.}$$

Hence, in general,

Separate the quantity under the radical sign into two factors, one of which is the greatest perfect power of the same degree as the radical;

Extract the required root of this factor, and multiply the coefficient of the radical by the result;

The other factor remains under the radical sign.

216. Quantity under Radical Sign a Fraction. To simplify in this case,

Multiply both numerator and denominator of the fraction by such a quantity as will make the denominator a perfect power of the same degree as the radical;

Proceed as in Art. 215.

Ex. 1. Simplify $\sqrt[3]{\frac{40}{9}}$.

$$\sqrt[3]{\tfrac{40}{9}} = \sqrt[3]{\tfrac{40}{9} \times \tfrac{3}{3}} = \sqrt[3]{\tfrac{120}{27}} = \sqrt[3]{\tfrac{8}{27} \times 15} = \tfrac{2}{3}\sqrt[3]{15}, \text{ Result.}$$

Ex. 2. Simplify $\sqrt{\dfrac{5ax^3}{18b}}$.

$$\sqrt{\frac{5ax^3}{18b}} = \sqrt{\frac{5ax^3}{18b} \times \frac{2b}{2b}} = \sqrt{\frac{10abx^3}{36b^2}}$$

$$= \sqrt{\frac{x^2}{36b^2} \times 10ab} = \frac{x}{6b}\sqrt{10ab}, \text{ Result.}$$

217. Meaning of Simplification. By simplication radicals are reduced to their prime form, so that it is made easier to determine, for instance, whether a number of given radicals are similar or not.

Thus, it is difficult to say whether $7\sqrt{18}$, $-5\sqrt{72}$ are similar, but when the given radicals are put in the form $21\sqrt{2}$, $-30\sqrt{2}$, it is easy to see that they are similar.

Again, the radicals $(a-1)\sqrt{\dfrac{a+1}{a-1}}$ and $(a+1)\sqrt{\dfrac{a-1}{a+1}}$, although unlike in present form, may be reduced not only to similar radicals, but to the same expression, $\sqrt{a^2-1}$.

The pupil should show this reduction for himself.

EXERCISE. 77.

Express in the simplest form—

1. $\sqrt{12}$.
2. $\sqrt{18}$.
3. $\sqrt{27}$.
4. $-\sqrt{20}$.
5. $2\sqrt{24}$.
6. $-3\sqrt{28}$.
7. $\frac{1}{2}\sqrt{44}$.
8. $\frac{2}{3}\sqrt{45}$.
9. $\frac{4}{5}\sqrt{50}$.
10. $\sqrt[3]{48}$.
11. $\sqrt[3]{24}$.
12. $\sqrt[3]{54}$.
13. $\frac{1}{2}\sqrt[3]{72}$.
14. $-\frac{5}{8}\sqrt{108}$.
15. $\sqrt[4]{48}$.
16. $\sqrt[4]{128a^5x^3}$.
17. $\sqrt[3]{250a^2b^8}$.
18. $\sqrt{99a}$.
19. $2\sqrt{4a^5x^3}$.
20. $a\sqrt{8a^5x^3}$.
21. $\sqrt{200a^7}$.
22. $\sqrt{147x^5y^8}$.
23. $-2\sqrt{63x^{15}y^{10}}$.
24. $\sqrt[3]{-81a^3x^5}$.
25. $\sqrt{a^2(x-y)^3}$.
26. $\sqrt{49x^3(a+1)^5}$.
27. $10\sqrt{\frac{12a^3c^4n}{25x^4}}$.
28. $\frac{3}{4}\sqrt{\frac{112x^5z^{11}}{9a^2}}$.

Simplify—

29. $\sqrt{\frac{3}{8}}$.
30. $2\sqrt{\frac{5}{2}}$.
31. $3\sqrt{\frac{5}{6}}$.
32. $\sqrt{\frac{24}{5}}$.
33. $\sqrt{\frac{25}{24}}$.
34. $\sqrt{\frac{63a}{8x^2}}$.
35. $4\sqrt{\frac{45c^3}{32a^2x}}$.
36. $\frac{3}{2}\sqrt{\frac{2x}{3a}}$.
37. $\frac{2a}{b}\sqrt{\frac{8b^2}{27a}}$
38. $3\sqrt[3]{\frac{3}{4}}$.
39. $5a\sqrt[3]{\frac{8}{25}}$.
40. $-3\sqrt[3]{\frac{5}{6}}$.
41. $-6\sqrt[3]{\frac{7x^4}{36a^2b^3}}$.
42. $\frac{3}{8}\sqrt[4]{\frac{1}{2}}$.
43. $z^2\sqrt{\frac{48a^2bc^3}{5x^2yz^3}}$.
44. $a\sqrt[5]{\frac{64x^2}{a^3}}$.
45. $\sqrt{\frac{12(x-y)}{5(x+y)}}$.
46. $x\sqrt{\frac{3}{(x+1)^3}}$.
47. $(a+b)^2\sqrt[3]{\frac{-a^4}{(a+b)^5}}$.
48. $4xy\sqrt{\frac{147a^3b^4c}{320xy^3}}$.

EXERCISE 78.

ORAL.

Reduce by inspection—

1. $\sqrt{8}$.
2. $\sqrt{a^3}$.
3. $\sqrt[3]{a^5x^4}$.
4. $\sqrt[3]{16x^5}$.
5. $\sqrt{27x^2y^5}$.
6. $\sqrt{\frac{1}{2}}$.
7. $\sqrt{\frac{1}{3}}$.
8. $\sqrt{\frac{1}{5}}$.
9. $\sqrt{\frac{1}{a}}$.
10. $\sqrt[3]{\frac{1}{4}}$.
11. $\sqrt[3]{\frac{1}{81}}$.
12. $\sqrt[3]{\frac{1}{x}}$.
13. $3\sqrt{\frac{2}{3}}$.
14. $2\sqrt{\frac{3}{8}}$.
15. $\sqrt{\frac{a}{b}}$.
16. $\sqrt[3]{\frac{x}{y}}$.
17. $2\sqrt{4\frac{1}{2}}$.
18. $\frac{3}{5}\sqrt{12\frac{1}{2}}$.
19. $\sqrt{2\frac{2}{3}}$.
20. $\frac{1}{2}\sqrt{3\frac{1}{5}}$.

218. Making Entire Surds. It is sometimes desired to introduce the coefficient of a radical under the radical sign. This may be done by simply reversing the process of Art. 215.

Ex. 1. Express $3\sqrt[3]{5}$ as an entire surd.

$$3\sqrt[3]{5} = \sqrt[3]{3^3 \times 5} = \sqrt[3]{135}.$$

Ex. 2. $$-2\sqrt[5]{3} = -\sqrt[5]{96} = \sqrt[5]{-96}.$$

Ex. 3. $$-2\sqrt[4]{3} = -\sqrt[4]{48}.$$

EXERCISE 79.

Express as entire surds—

1. $2\sqrt{3}$.
2. $3\sqrt{5}$.
3. $6\sqrt[3]{2}$.
4. $-2\sqrt{5}$.
5. $-3\sqrt[3]{2}$.
6. $-2\sqrt[3]{-2}$.
7. $2\sqrt[4]{3}$.
8. $2m\sqrt{3m}$.
9. $m^2\sqrt[3]{4n^2}$.
10. $\frac{1}{8}\sqrt{6}$.
11. $\frac{2}{5}\sqrt{10}$.
12. $\frac{3m}{4n}\sqrt[3]{\frac{2n}{9m^2}}$.
13. $2x\sqrt{\frac{3x-y}{x}}$.
14. $3a\sqrt[3]{\frac{a+2}{6a^2}}$.

15. $(x-1)\sqrt{2x}$.

16. $(a+2y)\sqrt{\dfrac{a-2y}{a+2y}}$.

17. $\dfrac{a-b}{a+b}\sqrt{a^2-b^2}$.

18. $(1-x)\sqrt{\dfrac{1}{x-1}}$.

219. II. **Simplification of Indices.** If the exponent of the quantity under the radical sign and the index of the radical sign have a common factor, this factor may be canceled and the radical thereby simplified.

Ex. 1. $\sqrt[6]{a^2}=a^{\frac{2}{6}}=a^{\frac{1}{3}}=\sqrt[3]{a}$.

Ex. 2. $\sqrt[6]{125}=\sqrt[6]{5^3}=\sqrt{5}$.

EXERCISE 80.

Simplify the indices—

1. $\sqrt[4]{a^2}$.
2. $\sqrt[6]{a^3}$.
3. $\sqrt[4]{a^2y^2}$.
4. $\sqrt[4]{49}$.
5. $\sqrt[6]{27a^3}$.
6. $\sqrt[4]{100a^2x^8}$.
7. $\sqrt[9]{8a^3b^6c^9}$.
8. $\sqrt[12]{81a^4x^8}$.
9. $\sqrt[10]{32a^5x^{10}y^{15}}$.
10. $\sqrt[2n]{9x^2y^4z^{10}}$.
11. $\sqrt[5n]{x^ny^{2n}z^{3n}}$.
12. $6\sqrt[4]{2\frac{7}{9}}$.

220. III. **Reducing Radicals to the Same Index.** Radicals of different degrees may be reduced to equivalent radicals of the same degree.

Ex. 1. Reduce $\sqrt{2}$ and $\sqrt[3]{5}$ to equivalent radicals having the same index.

$$\sqrt{2}=2^{\frac{1}{2}}=2^{\frac{3}{6}}=\sqrt[6]{2^3}=\sqrt[6]{8}$$
$$\sqrt[3]{5}=5^{\frac{1}{3}}=5^{\frac{2}{6}}=\sqrt[6]{5^2}=\sqrt[6]{25}.$$

Ex. 2. Arrange in ascending order of magnitude $\sqrt[4]{5}$, $\sqrt[3]{3}$, $\sqrt{2}$.

We obtain $\sqrt[12]{125}$, $\sqrt[12]{81}$, $\sqrt[12]{64}$;

hence, the ascending order of magnitude is, $\sqrt{2}$, $\sqrt[3]{3}$, $\sqrt[4]{5}$.

EXERCISE 81.

Reduce to equivalent radicals of the same (lowest) degree—

1. $\sqrt{7}$ and $\sqrt[3]{11}$.
2. $\sqrt[3]{5}$ and $\sqrt[4]{3}$.
3. $\sqrt[4]{3}$ and $\sqrt[6]{5}$.
4. $\sqrt{\frac{1}{2}}$ and $\sqrt[3]{\frac{3}{8}}$.
5. $\sqrt[4]{100}$ and $\sqrt[3]{25}$.
6. $\sqrt{6}$ and $\sqrt[6]{200}$.
7. $\sqrt[3]{2}$, $\sqrt[9]{9}$, $\sqrt[6]{5}$.
8. $\sqrt{a}$, $\sqrt[3]{a^2}$, $\sqrt[4]{a^3}$.
9. $\sqrt[3]{3a}$, $\sqrt[4]{2b}$, $\sqrt[6]{5c}$.
10. $\sqrt[4]{x+y}$ and $\sqrt[5]{x-y}$.
11. $\sqrt[n]{x^m}$ and $\sqrt[m]{x^n}$.
12. $\sqrt[x]{c^y}$, $\sqrt[y]{c^z}$, $\sqrt[z]{c^x}$.

Which is greater—

13. $\sqrt{3}$ or $\sqrt[3]{4}$?
14. $\sqrt[3]{15}$ or $\sqrt{6}$?
15. $\sqrt{5}$ or $\sqrt[3]{11}$?
16. $\sqrt[3]{23}$ or $2\sqrt{2}$?
17. $\sqrt[4]{10}$ or $2\sqrt[6]{\frac{1}{2}}$?
18. $\sqrt{2\frac{3}{5}}$ or $\sqrt[3]{4\frac{1}{4}}$?
19. $3\sqrt[3]{6}$ or $2\sqrt{5\frac{3}{8}}$?
20. $\sqrt{\frac{3}{5}}$ or $\sqrt[3]{\frac{7}{15}}$?

Which is the greatest—

21. $\sqrt{3}$, $\sqrt[3]{5}$, or $\frac{1}{2}\sqrt[3]{40\frac{5}{8}}$?
22. $3\sqrt[3]{4\frac{1}{8}}$, $2\sqrt{6}$, $2\sqrt[3]{14\frac{1}{2}}$?

OPERATIONS WITH RADICALS.

I. Addition and Subtraction of Radicals.

221. The **Addition of Similar Radicals** is performed like the addition of similar terms, by *taking the algebraic sum of their coefficients.*

The **Addition of Dissimilar Radicals** *can only be indicated.*

Ex. 1. Add $\sqrt{128} - 2\sqrt{50} + \sqrt{72} - \sqrt{18}$.

$$\sqrt{128} - 2\sqrt{50} + \sqrt{72} - \sqrt{18} = 8\sqrt{2} - 10\sqrt{2} + 6\sqrt{2} - 3\sqrt{2}$$
$$= \sqrt{2}, \textit{ Sum.}$$

Ex. 2.
$$2\sqrt{\tfrac{5}{3}} + \tfrac{1}{6}\sqrt{60} + \sqrt{15} + \sqrt{\tfrac{3}{5}}$$
$$= \tfrac{2}{3}\sqrt{15} + \tfrac{1}{3}\sqrt{15} + \sqrt{15} + \tfrac{1}{5}\sqrt{15}$$
$$= \tfrac{11}{5}\sqrt{15}, \textit{ Sum.}$$

Ex. 3.
$$\sqrt[3]{128} + 2\sqrt[3]{\tfrac{1}{4}} - 3\sqrt[3]{81}$$
$$= 4\sqrt[3]{2} + \sqrt[3]{2} - 9\sqrt[3]{3}$$
$$= 5\sqrt[3]{2} - 9\sqrt[3]{3}, \textit{ Sum.}$$

EXERCISE 82.

Collect—

1. $\sqrt{18} + \sqrt{8}$.
2. $\sqrt{50} - \sqrt{32}$.
3. $2\sqrt{27} + \sqrt{75}$.
4. $3\sqrt{90} - 5\sqrt{40}$.
5. $\sqrt{5} + \sqrt{20} + \sqrt{45}$.
6. $4\sqrt[3]{16} - 2\sqrt[3]{54}$.
7. $3\sqrt[3]{625} - 4\sqrt[3]{135}$.
8. $\sqrt[3]{162} + 3\sqrt[3]{48}$.
9. $2\sqrt[3]{189} - \sqrt[3]{448}$.
10. $\sqrt[3]{24} + \sqrt[3]{81} - \sqrt[3]{375}$.
11. $\sqrt{\tfrac{2}{3}} + \sqrt{\tfrac{3}{2}}$.
12. $2\sqrt{\tfrac{3}{4}} + \sqrt{48}$.
13. $a\sqrt{\dfrac{x}{a}} + x\sqrt{\dfrac{a}{x}}$.
14. $\sqrt[3]{\tfrac{5}{16}} + \tfrac{1}{4}\sqrt[3]{160}$.
15. $\tfrac{2}{3}\sqrt[3]{\tfrac{3}{4}} - \tfrac{3}{4}\sqrt[3]{\tfrac{2}{9}}$.
16. $2\sqrt{25b} + 3\sqrt{4b} - 2\sqrt{36b}$.
17. $3\sqrt[3]{2c} + 3\sqrt[3]{54c} - \sqrt[3]{2000c}$.
18. $\sqrt{12ab^2} + b\sqrt{48a} - 6\sqrt{3ab^2}$.
19. $2\sqrt{a^2bc^2} - 3a\sqrt{16bc^2} + 5c\sqrt{9a^2b}$.
20. $b\sqrt[3]{2a} + \sqrt[3]{250ab^3} - 2b\sqrt[3]{432a}$.
21. $\sqrt{2} + \sqrt{18} - \sqrt{50} + \sqrt{162}$.
22. $\sqrt{75} - 4\sqrt{243} + 2\sqrt{108}$.
23. $6\sqrt{\tfrac{2}{3}} - 5\sqrt{24} + 12\sqrt{\tfrac{3}{2}}$.
24. $5\sqrt{\tfrac{3}{5}} - 12\sqrt{\tfrac{5}{3}} + 6\sqrt{60} - 30\sqrt{\tfrac{1}{15}}$.
25. $3\sqrt{5} - 10\sqrt{\tfrac{1}{5}} + 2\sqrt{45} - 5\sqrt{\tfrac{49}{5}}$.
26. $\sqrt{27} - \sqrt{18} + \sqrt{300} - \sqrt{162} + 6\sqrt{2} - 7\sqrt{3}$.

27. $2\sqrt{63} - 3\sqrt{\tfrac{1}{5}} - \sqrt{\tfrac{9}{7}} + \tfrac{1}{5}\sqrt{45} - \tfrac{4}{7}\sqrt{7}.$

28. $\sqrt{243} + \sqrt{48} - \sqrt{768} + 9\sqrt{\tfrac{1}{3}} + \sqrt{75} - 3\sqrt{33\tfrac{1}{3}}.$

29. $21\sqrt{\tfrac{2}{3}} - 5\sqrt{\tfrac{4}{5}} + 6\sqrt{4\tfrac{1}{6}} - 10\sqrt{3\tfrac{1}{5}} + \tfrac{40}{3}\sqrt{11\tfrac{1}{4}}.$

30. $5a\sqrt{12ab^2} - 3b\sqrt{27a^3} + 2\sqrt{300a^3b^2} - 40ab\sqrt{\tfrac{3}{4}a}.$

II. Multiplication of Radicals.

222. Multiplication of Monomials.

Since by the commutative law,

$$a\sqrt[n]{b} \times c\sqrt[n]{d} = ac\sqrt[n]{b}\sqrt[n]{d}$$
$$= ac\sqrt[n]{bd},$$

we have the general rule,

Reduce the radicals if necessary to the same index;

Multiply the coefficients together for a new coefficient;

Multiply the quantities under the radical sign together for a new quantity under the radical sign;

Simplify the result.

Ex. 1. Multiply $5\sqrt{6}$ by $2\sqrt{3}$.

$$5\sqrt{6} \times 2\sqrt{3} = 10\sqrt{18} = 30\sqrt{2}, \text{ } \textit{Product.}$$

Ex. 2.
$$5\sqrt{2} \times 2\sqrt[3]{3} = 5\sqrt[6]{8} \times 2\sqrt[6]{9}$$
$$= 10\sqrt[6]{72}, \text{ } \textit{Product.}$$

Ex. 3.
$$\sqrt[3]{\tfrac{20}{9}} \times \sqrt[4]{\tfrac{27}{40}} = \sqrt[3]{\frac{2^2 \times 5}{3^2}} \times \sqrt[4]{\frac{3^3}{2^3 \times 5}} = \sqrt[12]{\frac{2^8 \times 5^4}{3^8}} \times \sqrt[12]{\frac{3^9}{2^9 \times 5^3}}$$
$$= \sqrt[12]{\frac{5 \times 3}{2}} = \sqrt[12]{\tfrac{15}{2}}, \text{ } \textit{Product.}$$

223. Multiplication of Polynomials. The Distributive Law applies here as in ordinary algebraic multiplication of polynomials; hence,

Reduce each term of the multiplier and multiplicand to its simplest form;

Multiply each term of the multiplicand by each term of the multiplier;

Simplify each term of the result, and collect.

Ex. 1. Multiply $3\sqrt{2}+5\sqrt{3}$ by $3\sqrt{2}-\sqrt{3}$.

$$\begin{array}{r} 3\sqrt{2}+5\sqrt{3} \\ 3\sqrt{2}-\sqrt{3} \\ \hline 18+15\sqrt{6} \quad\quad \\ -3\sqrt{6}-15 \\ \hline 3+12\sqrt{6}, \text{ \textit{Product.}} \end{array}$$

Ex. 2. Multiply $\sqrt{6}-2\sqrt{12}+5\sqrt{3}$ by $3\sqrt{3}-2\sqrt{2}$.

$$\sqrt{6}-2\sqrt{12}+5\sqrt{3}=\sqrt{6}-4\sqrt{3}+5\sqrt{3}=\sqrt{3}+\sqrt{6}$$

$$\begin{array}{l} \sqrt{3}+\sqrt{6} \\ 3\sqrt{3}-2\sqrt{2} \\ \hline 9+3\sqrt{18}-2\sqrt{6}-2\sqrt{12} \\ \quad =9+9\sqrt{2}-2\sqrt{6}-4\sqrt{3}, \text{ \textit{Product.}} \end{array}$$

EXERCISE 83.

Multiply—

1. $\sqrt{3}$ by $2\sqrt{12}$.
2. $3\sqrt{5}$ by $\sqrt{15}$.
3. $2\sqrt[3]{4}$ by $3\sqrt[3]{6}$.
4. $\sqrt[5]{24}$ by $\sqrt[5]{4}$.
5. $3\sqrt{18}$ by $2\sqrt{12}$.
6. $2\sqrt{15}$ by $3\sqrt{35}$.
7. $\frac{2}{3}\sqrt{28}$ by $\frac{3}{4}\sqrt{35}$.
8. $\frac{1}{3}\sqrt{\frac{5}{6}}$ by $\frac{2}{5}\sqrt{\frac{15}{32}}$.
9. $4\sqrt[3]{\frac{2}{3}}$ by $\frac{1}{2}\sqrt[3]{\frac{4}{3}}$.
10. $\frac{3}{8}\sqrt[3]{\frac{9}{16}}$ by $\frac{8}{9}\sqrt[3]{\frac{9}{4}}$.
11. $\sqrt{a}$ by $\sqrt[3]{a^2b}$.
12. $\sqrt{2}$ by $\sqrt[3]{3}$.
13. $\sqrt[3]{2ax^2}$ by $\sqrt{6x}$.
14. $\sqrt[3]{9}$ by $\sqrt{6}$.
15. $\sqrt[4]{6}$ by $\sqrt[6]{4}$.
16. $\sqrt[3]{18}$ by $\sqrt{12}$.
17. $\sqrt[3]{\frac{4}{9}}$ by $\sqrt[4]{\frac{27}{4}}$.
18. $\sqrt[4]{\frac{3}{5}}$ by $\sqrt[6]{\frac{10}{9}}$.
19. $\sqrt{\frac{2ab^3}{3x^3y}}\times\sqrt[3]{\frac{9x^4y}{4ab^4}}$.

20. $\sqrt[3]{\frac{4}{81}} \times \sqrt[3]{\frac{27}{32}}$.

21. $\sqrt[3]{\frac{9}{28}} \times \sqrt[4]{\frac{2}{3}}$.

22. $\sqrt[3]{\frac{112}{405}} \times \sqrt[6]{\frac{135}{56}}$.

23. $\sqrt[3]{\frac{48}{625}} \times \sqrt[6]{\frac{125}{4}}$.

24. $\sqrt{\frac{35}{24}} \times \sqrt[3]{\frac{48}{175}} \times \sqrt[6]{\frac{10}{21}}$.

25. $\sqrt[4]{\frac{48}{375}} \times \sqrt[3]{\frac{25}{96}} \times \sqrt[6]{360}$.

26. $\sqrt{3} - \sqrt{6} + 2\sqrt{10}$ by $2\sqrt{2}$.

27. $3\sqrt{5} - \sqrt{10} + 2\sqrt{15}$ by $4\sqrt{5}$.

28. $4\sqrt{6} - 3\sqrt{3} + 3\sqrt{2}$ by $2\sqrt{6}$.

29. $\frac{1}{2}\sqrt{\frac{2}{3}} + \frac{3}{4}\sqrt{\frac{1}{3}} - \frac{2}{5}\sqrt{\frac{5}{4}}$ by $20\sqrt{12}$.

30. $10\sqrt{\frac{5}{4}} - 5\sqrt{\frac{9}{2}} + 14\sqrt{\frac{45}{56}}$ by $\frac{2}{5}\sqrt{\frac{8}{15}}$.

31. $3 + \sqrt{2}$ by $2 - 2\sqrt{2}$.

32. $5 - 2\sqrt{3}$ by $4 + 3\sqrt{3}$.

33. $2\sqrt{3} - 3\sqrt{2}$ by $4\sqrt{3} + 5\sqrt{2}$.

34. $3\sqrt{3} + 5\sqrt{5}$ by $5\sqrt{3} - 3\sqrt{5}$.

35. $4\sqrt{2} - 3\sqrt{3}$ by $3\sqrt{2} + 4\sqrt{3}$.

36. $3\sqrt{5} - 2\sqrt{2} + \sqrt{3}$ by $3\sqrt{5} - \sqrt{3}$.

37. $\sqrt{2} + \sqrt{3} - \sqrt{5}$ by $\sqrt{2} - \sqrt{3} + \sqrt{5}$.

38. $2\sqrt{3} - 3\sqrt{6} - 4\sqrt{15}$ by $2\sqrt{3} + 3\sqrt{6} + 4\sqrt{15}$.

39. $3\sqrt{30} + 2\sqrt{5} - 3\sqrt{6}$ by $2\sqrt{5} + 3\sqrt{6}$.

40. $\frac{1}{2}\sqrt{8} + \sqrt{32} - \sqrt{48}$ by $3\sqrt{8} - \frac{1}{4}\sqrt{32} + 2\sqrt{12}$.

41. $12\sqrt{\frac{1}{6}} - 4\sqrt{\frac{1}{2}} + \frac{1}{8}\sqrt{216}$ by $6\sqrt{\frac{2}{3}} - 2\sqrt{\frac{3}{2}} + 3\sqrt{6}$.

42. $\sqrt{2x} + \sqrt{x-1}$ by $\sqrt{3x}$.

43. $\sqrt{3x} + \sqrt{x+1}$ by $\sqrt{x+1}$.

44. $\sqrt{x-1} - 3\sqrt{x+1}$ by $2\sqrt{x+1}$.

45. $a - \sqrt{a-x} + \sqrt{a}$ by $\sqrt{a-x} + \sqrt{a}$.

46. $3\sqrt{2x} - 5\sqrt{x-1}$ by $3\sqrt{2x} + 5\sqrt{x-1}$.

47. $\sqrt{a+x} - \sqrt{a-x}$ by $\sqrt{a+x} + \sqrt{a-x}$.

48. $(2\sqrt{2} + \sqrt{3})(3\sqrt{2} - \sqrt{3})(3\sqrt{3} - \sqrt{2})$.

49. $(2\sqrt{x+2} + 3\sqrt{2})(6x - 5\sqrt{2x+4})(3\sqrt{x+2} - 2\sqrt{2})$.

III. Division of Radicals.

224. Division of Monomials. Reversing the process for multiplication, we have the rule,

If necessary, reduce the radicals to the same index;

Find the quotient of the coefficients for a new coefficient, and the quotient of the quantities under the radical signs for a new quantity under the radical;

Simplify the result.

Ex. 1. Divide $6\sqrt{8}$ by $3\sqrt{6}$.

$$\frac{6\sqrt{8}}{3\sqrt{6}}=2\sqrt{\tfrac{4}{3}}=2\sqrt{\tfrac{12}{9}}=\tfrac{4}{3}\sqrt{3}, \textit{ Quotient.}$$

Ex. 2. Divide $6\sqrt[3]{3}$ by $2\sqrt{2}$.

$$\frac{6\sqrt[3]{3}}{2\sqrt{2}}=\frac{6\sqrt[6]{3^2}}{2\sqrt[6]{2^3}}=3\sqrt[6]{\frac{3^2\times 2^3}{2^3\times 2^3}}=3\sqrt[6]{\frac{72}{2^6}}=\tfrac{3}{2}\sqrt[6]{72}, \textit{ Quotient.}$$

EXERCISE 84.

Divide—

1. $\sqrt{27}$ by $\sqrt{3}$.
2. $4\sqrt{12}$ by $2\sqrt{6}$.
3. $12\sqrt{15}$ by $4\sqrt{5}$.
4. $2\sqrt{60}$ by $3\sqrt{5}$.
5. $8\sqrt{125}$ by $10\sqrt{10}$.
6. $3\sqrt{405}$ by $9\sqrt{45}$.
7. $a^3\sqrt{ab^3}$ by $2a\sqrt{a^3b}$.
8. $4\sqrt{18}$ by $5\sqrt{32}$.
9. $3\sqrt{40}$ by $5\sqrt{28}$.
10. $\sqrt{\tfrac{15}{16}}$ by $\sqrt{\tfrac{5}{8}}$.
11. $4\sqrt{\tfrac{5}{6}}$ by $3\sqrt{\tfrac{20}{27}}$.
12. $5\sqrt{\tfrac{22}{35}}$ by $2\sqrt{\tfrac{7}{11}}$.
13. $\tfrac{3}{5}\sqrt{\tfrac{15}{32}}$ by $\tfrac{1}{10}\sqrt{\tfrac{45}{64}}$.
14. $2\sqrt[3]{3}$ by $3\sqrt{2}$.
15. $\sqrt{54}$ by $\sqrt[4]{36}$.
16. $\sqrt[3]{12}$ by $\sqrt{6}$.
17. $\sqrt{\tfrac{8}{15}}$ by $\sqrt[3]{\tfrac{16}{135}}$.
18. $\sqrt{6\tfrac{1}{2}}$ by $\sqrt[3]{3\tfrac{1}{4}}$.
19. $3\sqrt{\tfrac{45}{56}}$ by $2\sqrt[3]{\tfrac{135}{224}}$.
20. $\sqrt[10]{\tfrac{100}{63}}$ by $2\sqrt[5]{\tfrac{10}{21}}$.

21. $5\sqrt{35} - 7\sqrt{20}$ by $\sqrt{5}$.

22. $3\sqrt{6} + 9\sqrt{3}$ by $3\sqrt{3}$.

23. $12\sqrt{7} - 60\sqrt{5}$ by $4\sqrt{3}$.

24. $6\sqrt{105} + 18\sqrt{40} - 45\sqrt{12}$ by $3\sqrt{15}$.

25. $8\sqrt{45} - 15\sqrt{24} - \sqrt{60}$ by $2\sqrt{30}$.

26. $12\sqrt[3]{45} + 30\sqrt[3]{20} + 42\sqrt[3]{30}$ by $2\sqrt[3]{18}$.

27. $10\sqrt[3]{18} - 4\sqrt[3]{60} + 5\sqrt[3]{100}$ by $3\sqrt[3]{30}$.

225. Rationalizing a Monomial Denominator. If the denominator of a fraction be a surd, in order to make the denominator rational,

Multiply both numerator and denominator by such a number as will make the denominator rational.

Ex. $$\frac{5}{\sqrt[3]{2}} = \frac{5}{\sqrt[3]{2}} \times \frac{\sqrt[3]{4}}{\sqrt[3]{4}} = \frac{5\sqrt[3]{4}}{\sqrt[3]{8}} = \tfrac{5}{2}\sqrt[3]{4}, \textit{ Result.}$$

One object in thus rationalizing the denominator of a fraction is to diminish the labor of finding the approximate value of the fraction. Thus, if we find the approximate numerical value of $\frac{5}{\sqrt[3]{2}}$ directly, we must find the cube root of 2, and divide 5 by the decimal which we obtain. On the other hand, if we find the value of the equivalent expression, $\frac{5}{2}\sqrt[3]{4}$, we extract the cube root of 4, multiply by 5, and divide by 2. In the latter process we therefore avoid the tedious long division, and diminish the labor of the process by nearly one-half.

226. Rationalizing a Binomial or Trinomial Denominator. If the denominator of a fraction be a binomial containing radicals of the second degree, since

$$(\sqrt{a} + \sqrt{b})(\sqrt{a} - \sqrt{b}) = a - b,$$

Multiply both numerator and denominator by the denominator, with one of its signs changed;

For a trinomial denominator repeat the process.

Ex. 1. $\dfrac{2\sqrt{5}+4\sqrt{3}}{3\sqrt{5}-\sqrt{3}}=\dfrac{2\sqrt{5}+4\sqrt{3}}{3\sqrt{5}-\sqrt{3}}\times\dfrac{3\sqrt{5}+\sqrt{3}}{3\sqrt{5}+\sqrt{3}}$

$=\dfrac{42+14\sqrt{15}}{45-3}=\dfrac{42+14\sqrt{15}}{42}=\dfrac{3+\sqrt{15}}{3}$, *Result.*

Ex. 2. $\dfrac{4}{1+\sqrt{3}+\sqrt{2}}=\dfrac{4}{1+\sqrt{3}+\sqrt{2}}\times\dfrac{1+\sqrt{3}-\sqrt{2}}{1+\sqrt{3}-\sqrt{2}}$

$=\dfrac{2(1+\sqrt{3}-\sqrt{2})}{1+\sqrt{3}}\times\dfrac{1-\sqrt{3}}{1-\sqrt{3}}=2+\sqrt{2}-\sqrt{6}$, *Result.*

EXERCISE 85.

Reduce to equivalent fractions having rational denominators—

1. $\dfrac{1}{\sqrt{2}}$.
2. $\dfrac{\sqrt{2}}{2\sqrt{3}}$.
3. $\dfrac{2}{3\sqrt{5}}$.
4. $\dfrac{1-\sqrt{2}}{\sqrt{6}}$.
5. $\dfrac{2+\sqrt{5}}{2\sqrt{7}}$.
6. $\dfrac{3\sqrt{2}-\sqrt{3}}{2\sqrt{6}}$.
7. $\dfrac{\sqrt[3]{5}-1}{3\sqrt[3]{2}}$.
8. $\dfrac{2\sqrt[3]{9}-3\sqrt[3]{4}}{5\sqrt[3]{6}}$.
9. $\dfrac{3-\sqrt{2}}{3+\sqrt{2}}$.
10. $\dfrac{5+\sqrt{3}}{2-\sqrt{3}}$.
11. $\dfrac{3\sqrt{3}-2\sqrt{2}}{2\sqrt{3}+3\sqrt{2}}$.
12. $\dfrac{3\sqrt{6}-2\sqrt{3}}{4\sqrt{6}-3\sqrt{3}}$.
13. $\dfrac{2\sqrt{15}+3\sqrt{10}}{4\sqrt{3}+3\sqrt{2}}$.
14. $\dfrac{3\sqrt{a}-4\sqrt{b}}{2\sqrt{a}-3\sqrt{b}}$.
15. $\dfrac{\sqrt{x+1}+3}{\sqrt{x+1}+2}$.

16. $\dfrac{2\sqrt{2a-1}+3\sqrt{a}}{3\sqrt{2a-1}+2\sqrt{a}}$.

17. $\dfrac{2+\sqrt{6}-\sqrt{2}}{2-\sqrt{6}+\sqrt{2}}$.

18. $\dfrac{\sqrt{5}-\sqrt{6}+1}{\sqrt{6}+\sqrt{5}+1}$.

19. $\dfrac{2\sqrt{6}-3\sqrt{2}-\sqrt{42}}{2\sqrt{6}+3\sqrt{2}+\sqrt{42}}$.

20. $\dfrac{\sqrt{x^2-1}-\sqrt{x^2+1}}{\sqrt{x^2-1}+\sqrt{x^2+1}}$.

21. $\dfrac{\sqrt{a}+\sqrt{b}-\sqrt{a+b}}{\sqrt{a}-\sqrt{b}+\sqrt{a+b}}$.

Find the approximate numerical value of—

22. $\dfrac{3}{\sqrt{2}}$.

23. $\dfrac{2\sqrt{5}}{3\sqrt{2}}$.

24. $\dfrac{12}{\sqrt{7}}$.

25. $\dfrac{1}{\sqrt{300}}$.

26. $\dfrac{3\sqrt{7}}{5\sqrt{5}}$.

27. $\dfrac{1+\sqrt{2}}{2-\sqrt{3}}$.

28. $\dfrac{\sqrt{3}-\sqrt{2}}{\sqrt{2}+\sqrt{3}}$.

29. $\dfrac{5\sqrt{7}-1}{\sqrt{7}+2}$.

30. $\dfrac{3\sqrt{3}-4}{4\sqrt{3}-5}$.

IV. Involution and Evolution of Radicals.

227. The process of raising a radical to a power, or extracting a required root of a radical, is usually performed most readily by the use of fractional exponents.

Ex. 1. Find the square of $3\sqrt[3]{x}$.

$$(3\sqrt[3]{x})^2=(3x^{\frac{1}{3}})^2=9x^{\frac{2}{3}}=9\sqrt[3]{x^2}.$$

Ex. 2. Extract the square root of $4a\sqrt{a^3b^5}$.

$$(4a\sqrt{a^3b^5})^{\frac{1}{2}}=(4a\cdot a^{\frac{3}{2}}b^{\frac{5}{2}})^{\frac{1}{2}}=4^{\frac{1}{2}}a^{\frac{1}{2}}a^{\frac{3}{4}}b^{\frac{5}{4}}$$
$$=2a^{\frac{5}{4}}b^{\frac{5}{4}}=2\sqrt[4]{a^5b^5}=2ab\sqrt[4]{ab},\ \textit{Result.}$$

Ex. 3. Extract the cube root of $\sqrt{a^6b^9}$.

$$(\sqrt{a^6b^9})^{\frac{1}{3}}=(a^3b^{\frac{9}{2}})^{\frac{1}{3}}=ab^{\frac{3}{2}}=a\sqrt{b^3}=ab\sqrt{b}.$$

This process might have been performed by extracting the cube root of a^6b^9 as it stands under the radical sign; thus,

$$\sqrt[3]{\sqrt{a^6b^9}}=\sqrt{a^2b^3}=ab\sqrt{b},\ \textit{Result.}$$

EXERCISE 86.

Perform the operations indicated—

1. $(\sqrt{m})^4$.
2. $(\sqrt[3]{x^2})^6$.
3. $(\sqrt[3]{a^4x^5})^3$.
4. $(\sqrt[6]{36})^3$.
5. $(\sqrt[8]{a^3c^5})^4$.
6. $(\sqrt[6]{x^5})^{15}$.
7. $(\sqrt[3]{2x^4})^9$.
8. $(\sqrt{\sqrt[3]{a^2}})^3$.
9. $(\sqrt[3]{3x^2})^6$.
10. $\sqrt[3]{64a^3\sqrt{8x^3}}$.
11. $(\sqrt[4]{\sqrt[3]{x^4y^8}})^6$.
12. $\sqrt{\sqrt[3]{4\sqrt[4]{64x^{12}}}}$.

V. Square Root of a Binomial Surd.

228. A **Quadratic Surd** is a surd of the second degree.

Exs. $\sqrt{3}$, $\sqrt{ab}$.

A **Binomial Surd** is a binomial expression, at least one term of which contains a surd.

$$\text{Exs. } \sqrt{2}+5\sqrt{3},$$

or

$$a+\sqrt{b}.$$

229. A. *The product of two dissimilar quadratic surds is a quadratic surd.* Thus,

$$\sqrt{2}\times\sqrt{6}=\sqrt{12}=2\sqrt{3},$$

or

$$\sqrt{ab}\times\sqrt{abc}=ab\sqrt{c}.$$

Proof. If the surds are dissimilar, one of them must have under the radical sign a factor which the other has not. This factor must remain under the radical sign in the product.

230. B. *The sum or difference of two dissimilar quadratic surds cannot equal a rational quantity.*

Proof. If $\sqrt{a}\pm\sqrt{b}$ can equal a rational quantity, c,

squaring, $$a\pm 2\sqrt{ab}+b=c^2,$$

$$\pm 2\sqrt{ab}=c^2-a-b.$$

But $\sqrt{ab}$ is a surd by Art. 229; hence we have a surd equal to a rational quantity, which is impossible.

16

231. C. *If* $a+\sqrt{b}=x+\sqrt{y}$, *then* $a=x$, $b=y$.

Proof. If $$a+\sqrt{b}=x+\sqrt{y},$$
transposing, $$\sqrt{b}-\sqrt{y}=x-a.$$

If b does not equal y, we have the difference of two surds equal to a rational quantity, which is impossible; hence,

$$b=y, \quad a=x.$$

In like manner let the student show that if

$$a-\sqrt{b}=x-\sqrt{y}, \quad \text{then } a=x, \quad b=y.$$

232. Extraction of Square Root of a Binomial Surd. If we expand $(\sqrt{2}+\sqrt{3})^2$, we obtain $2+2\sqrt{6}+3$, or $5+2\sqrt{6}$. Hence, the square root of the binomial surd $5+2\sqrt{6}$ is $\sqrt{2}+\sqrt{3}$. Hence, the square root of some binomial surds may be extracted. To investigate a method of doing this, let $a+\sqrt{b}$ be a binomial surd, and let $\sqrt{x}+\sqrt{y}$ be its square root.

$$\therefore \sqrt{a+\sqrt{b}}=\sqrt{x}+\sqrt{y} \quad \ldots \quad (1)$$

Square both sides,

$$a+\sqrt{b}=x+2\sqrt{xy}+y \quad \ldots \quad (2)$$

$$\therefore a=x+y, \quad \sqrt{b}=2\sqrt{xy}. \text{ (See Art. 231.)}$$

Hence, $$a-\sqrt{b}=x+y-2\sqrt{xy} \quad \ldots \quad (3)$$

Extract square root of (3),

$$\sqrt{a-\sqrt{b}}=\sqrt{x}-\sqrt{y} \quad \ldots \quad (4)$$

Multiply (1) by (4),

$$\sqrt{a^2-b}=x-y \quad \ldots \quad (5)$$

But $$a=x+y \quad \ldots \quad (6)$$

Adding (5) and (6) and dividing by 2,

$$x=\frac{a+\sqrt{a^2-b}}{2}.$$

Subtracting (6) from (5) and dividing by 2,

$$y=\frac{a-\sqrt{a^2-b}}{2}$$

$$\therefore \sqrt{x}+\sqrt{y}=\sqrt{a+\sqrt{b}}=\sqrt{\frac{a+\sqrt{a^2-b}}{2}}+\sqrt{\frac{a-\sqrt{a^2-b}}{2}}.$$

Hence, the square root of a binomial surd may always be extracted in form, but we get a result simpler than the original one only when $a^2 - b$ is a perfect square.

Ex. Extract the square root of $5 + 2\sqrt{6}$.

Let $\sqrt{x} + \sqrt{y} = \sqrt{5 + 2\sqrt{6}}$

Then $\sqrt{x} - \sqrt{y} = \sqrt{5 - 2\sqrt{6}}$. {See (4) above.}

Multiplying, $x - y = \sqrt{25 - 24}$

$\therefore x - y = 1$

But $x + y = 5$

$\therefore x = 3$

$y = 2$

$\therefore \sqrt{x} + \sqrt{y} = \sqrt{3} + \sqrt{2}$

$\therefore \sqrt{5 + 2\sqrt{6}} = \sqrt{3} + \sqrt{2}$, *The required root.*

288. Square Root of a Binomial Surd by Inspection.

By actual multiplication we may find,

$$(\sqrt{2} + \sqrt{5})^2 = 2 + 2\sqrt{10} + 5 = 7 + 2\sqrt{10}.$$

In the square, $7 + 2\sqrt{10}$, 7 is the sum of 2 and 5, 10 is the product of 2 and 5. Hence, in extracting the square root of $7 + 2\sqrt{10}$, we are merely required to find two numbers such that their sum is 7, and their product 10; extract the square root of each, take the sum (or difference) of the roots. This may readily be done in all cases where the numbers involved are small. In general,

Transform the surd term so that its coefficient shall be 2;
Find two numbers such that their sum shall equal the rational term, and their product equal the quantity under the radical;
Extract the square root of each of these, and connect the results by the proper sign.

Ex. Find the square root of $18+8\sqrt{5}$.

$$18+8\sqrt{5}=18+2\sqrt{80}.$$

The two numbers whose sum is 18 and product is 80 are 8 and 10.

$$\therefore \sqrt{18+8\sqrt{5}}=\sqrt{8}+\sqrt{10}$$
$$=2\sqrt{2}+\sqrt{10}.$$

EXERCISE 87.

Find the square root of—

1. $17-12\sqrt{2}$.
2. $23+4\sqrt{15}$.
3. $35-12\sqrt{6}$.
4. $9-6\sqrt{2}$.
5. $42+28\sqrt{2}$.
6. $73-12\sqrt{35}$.
7. $26+4\sqrt{30}$.
8. $77-24\sqrt{10}$.
9. $87-36\sqrt{5}$.
10. $14+3\sqrt{3}$.
11. $8-\frac{16}{3}\sqrt{2}$.
12. $5\frac{1}{4}+3\sqrt{3}$.
13. $4\frac{1}{3}-\frac{4}{3}\sqrt{3}$.
14. $2m+2\sqrt{m^2-n^2}$.
15. $10a^2+9+6a\sqrt{a^2+1}$.

Find the fourth root of—

16. $28-16\sqrt{3}$.
17. $49+20\sqrt{6}$.
18. $97-56\sqrt{3}$.
19. $193-132\sqrt{2}$.
20. $\frac{113}{4}-18\sqrt{2}$.
21. $\frac{49}{9}+\frac{20}{9}\sqrt{6}$.

Find by inspection the square root of—

22. $3+2\sqrt{2}$.
23. $9-2\sqrt{14}$.
24. $21+12\sqrt{3}$.
25. $23-6\sqrt{10}$.
26. $18-12\sqrt{2}$.
27. $7+4\sqrt{3}$.
28. Prove that $\sqrt{a}\pm\sqrt{b}$ cannot equal $\sqrt{c}$.
29. Prove that $\sqrt{a}$ cannot equal $b+\sqrt{c}$.

VI. Solution of Equations containing Radicals.

234. Simple Equations containing Radicals.

Ex. 1. Solve $\sqrt{x^2+7}-1=x$.

Transpose terms so that the radical shall be alone on one side of the equation.

$$\sqrt{x^2+7}=x+1.$$

Squaring, $\quad x^2+7=x^2+2x+1$

$$\therefore\ 2x=6$$

$$x=3,\ \textit{Root.}$$

Ex. 2. Solve $\sqrt{x+3}+\sqrt{x}=5$.

Transpose terms so that one radical shall be alone on one side of the equation.

$$\sqrt{x+3}=5-\sqrt{x}.$$

Squaring, $\quad x+3=25-10\sqrt{x}+x$

$$\therefore\ 10\sqrt{x}=22$$

$$5\sqrt{x}=11.$$

Squaring, $\quad 25x=121$

$$x=\tfrac{121}{25},\ \textit{Root.}$$

In general,

Transpose the terms of the given equation so that a single radical shall form one member of the equation;

Raise both members of the equation to the power indicated by the index of this radical;

Repeat the process if necessary.

235. Fractional Equations containing Radicals. If a radical occur in the denominator of a fraction, it is necessary to clear the equation of fractions, being careful to multiply correctly the radical expressions involved.

Ex. 1. $\sqrt{x} - \sqrt{x-8} = \dfrac{2}{\sqrt{x-8}}$.

Multiply by $\sqrt{x-8}$, $\sqrt{x^2-8x} - x + 8 = 2$

$$\therefore \sqrt{x^2-8x} = x - 6$$
$$x^2 - 8x = x^2 - 12x + 36$$
$$4x = 36$$
$$x = 9.$$

Ex. 2. $\dfrac{\sqrt{x}+3}{\sqrt{x}-2} = \dfrac{3\sqrt{x}-5}{3\sqrt{x}-13}$.

Clearing of fractions, $3x - 4\sqrt{x} - 39 = 3x - 11\sqrt{x} + 10$

$$7\sqrt{x} = 49$$
$$\sqrt{x} = 7$$
$$x = 49.$$

EXERCISE 88.

Solve the following equations:

1. $\sqrt{x+1} = 3$.
2. $\sqrt{x} + 1 = 3$.
3. $\sqrt{3x} - 2 = 1$.
4. $5 - \sqrt{2x} = 3$.
5. $1 = \sqrt{3x-5}$.
6. $\sqrt{2x+5} = 2$.
7. $1 = \sqrt[3]{x} - 1$.
8. $\sqrt[3]{2x-1} + 1 = 4$.
9. $3 = 2 - \sqrt[3]{3x+1}$.
10. $x - 1 = \sqrt{x^2+3}$.
11. $\sqrt{x+1} = \sqrt{13}$.
12. $2\sqrt{3x-5} = 3\sqrt{x+1}$.
13. $3\sqrt{x} - 1 = \sqrt{x} + 1$.
14. $\sqrt{x+16} = 8 - \sqrt{x}$.
15. $\sqrt{x} = 3 - \sqrt{x-1}$.
16. $\sqrt{x-15} = 15 - \sqrt{x}$.
17. $\frac{3}{2} + \sqrt{x} = \sqrt{\frac{15}{4} + x}$.
18. $\frac{5}{2} - \sqrt{2x} = \sqrt{2x + \frac{5}{4}}$.
19. $\sqrt{x} + \sqrt{x+8} = 8$.
20. $\sqrt{4x+3} = 2\sqrt{x-1} + 1$.
21. $2\sqrt{x} - \sqrt{4x-22} = \sqrt{2}$.
22. $\sqrt{9x+35} = 7\sqrt{5} - 3\sqrt{x}$.

23. $\sqrt{13 + \sqrt{7 + \sqrt{3 + \sqrt{x}}}} = 4$.

24. $\sqrt{\sqrt{x}+3}-\sqrt{\sqrt{x}-3}=\sqrt{2\sqrt{x}}.$

25. $\sqrt{25x-29}-\sqrt{4x-11}=3\sqrt{x}.$

26. $\sqrt{x}+\sqrt{4a+x}=2\sqrt{b+x}.$

27. $\sqrt{x-1}+\sqrt{x}=\dfrac{2}{\sqrt{x}}.$

28. $\sqrt{3+x}+\sqrt{x}=\dfrac{6}{\sqrt{3+x}}.$

29. $\sqrt{9+2x}=\dfrac{5}{\sqrt{9+2x}}+\sqrt{2x}.$

30. $2\sqrt{x}-\sqrt{4x-3}=\dfrac{1}{\sqrt{4x-3}}.$

31. $3\sqrt{2x+1}-3\sqrt{2x-3}=\dfrac{4}{\sqrt{2x-3}}.$

32. $\dfrac{\sqrt{x}-3}{\sqrt{x}+3}=\dfrac{\sqrt{x}+1}{\sqrt{x}-2}.$

33. $\dfrac{\sqrt{x}+2}{\sqrt{x}-1}=\dfrac{\sqrt{x}+7}{\sqrt{x}+1}.$

34. $\dfrac{6\sqrt{x}-7}{\sqrt{x}-1}-5=\dfrac{7\sqrt{x}-26}{7\sqrt{x}-21}.$

35. $\dfrac{\sqrt{x+a}+\sqrt{x}}{\sqrt{x+a}-\sqrt{x}}=a.$

36. $\dfrac{\sqrt{ax+b}+\sqrt{ax}}{\sqrt{ax+b}-\sqrt{ax}}=1+\dfrac{1}{b}.$

37. $\dfrac{x-a}{\sqrt{x}+\sqrt{a}}=\dfrac{\sqrt{x}-\sqrt{a}}{3}+2\sqrt{a}.$

38. $\dfrac{\sqrt{9x+2}-3\sqrt{x}}{\sqrt{9x+2}+3\sqrt{x}}=4.$

39. $\sqrt[4]{2\frac{1}{2}+\sqrt{1\frac{1}{4}+\sqrt{\frac{1}{2}+\sqrt{x}}}}=\sqrt{2}.$

40. $\sqrt{x}+\sqrt{a-\sqrt{ax+x^2}}=\sqrt{a}.$

41. $\dfrac{a+3}{\sqrt{x}+2}-\dfrac{a-3}{\sqrt{x}-2}=\dfrac{2a}{x-4}.$

42. $\sqrt{\dfrac{a}{x}+1}-\sqrt{\dfrac{a}{x}-1}=1.$

43. $\dfrac{1}{x}+\dfrac{1}{5}=\sqrt{\dfrac{1}{25}+\sqrt{\dfrac{1}{a^2x^2}+\dfrac{1}{x^4}}}.$

44. $\sqrt{x^2+4x+12}+\sqrt{x^2-12x-20}=8.$

45. $\sqrt{4x^2+6x+7}-\sqrt{4x^2+2x+3}=1.$

46. $\sqrt{9x^2+x+5}-\sqrt{9x^2+7x+6}-1=0.$

286. Summary of Principles relating to Radicals. Let the student form a summary of the principles relating to radicals, similar to that given in Art. 206 for exponents. Thus,

TRANSFORMATIONS OF RADICALS.

I. Simplification of Quantity under the Radical Sign.

1. *General Case.* Ex. $\sqrt{8a^3b}=2a\sqrt{2ab}$

2. *Fraction under Radical.* Ex. $\sqrt{\frac{2}{3}}=\frac{1}{3}\sqrt{6}.$

Etc., Etc.

EXERCISE 89.

REVIEW.

1. Write in the simplest form—

$\dfrac{1}{\sqrt{2}},\ \dfrac{1}{\sqrt{3}},\ \dfrac{2}{\sqrt{8}},\ \dfrac{1}{\sqrt{a}},\ \dfrac{3}{\sqrt{5}},\ \sqrt{\frac{1}{6}},\ \sqrt{\frac{3}{8}},\ \sqrt{\frac{5}{6}},\ \dfrac{6}{\sqrt{8}},\ \dfrac{\sqrt{2}+1}{\sqrt{2}},$

$\dfrac{1}{[illegible]-1},\ \dfrac{2}{2-\sqrt{3}},\ \dfrac{1}{\sqrt{5}-2},\ \dfrac{7}{\sqrt{3}+\sqrt{2}},\ \dfrac{8}{\sqrt{7}-1},\ \dfrac{12}{3\sqrt{2}-2\sqrt{3}}.$

Collect—

2. $2\sqrt{12} - \sqrt{75} + \sqrt{48} - \sqrt{27} + \sqrt{3}$.

3. $\frac{2}{3}\sqrt{45} - 30\sqrt{\frac{1}{5}} + \frac{3}{4}\sqrt{20} - 6\sqrt{\frac{5}{9}} + \sqrt{500}$.

4. $4\sqrt{147} - 3\sqrt{75} - 6\sqrt{\frac{1}{3}} + 18\sqrt{\frac{1}{27}} - 24\sqrt{\frac{3}{4}}$.

5. $\frac{2a}{3b}\sqrt{\frac{45b^2x}{16a^4}} - \frac{10b}{a^2}\sqrt{\frac{a^2x}{5b^2}} + \frac{14x}{a^2b}\sqrt{\frac{5a^2b^2}{49x}} - \frac{5x}{2a}\sqrt{\frac{1}{20x}}$.

Multiply—

6. $2 + \sqrt{3} - \sqrt{5}$ by $2 + \sqrt{3} + \sqrt{5}$.

7. $\sqrt{a} + \sqrt{a+x} - \sqrt{x}$ by $\sqrt{a} - \sqrt{a+x} - \sqrt{x}$.

8. $\frac{1}{2}\sqrt{\frac{6}{5}}$ by $3\sqrt[3]{\frac{25}{6}}$; and $2\sqrt[4]{\frac{27}{13}}$ by $\frac{1}{3}\sqrt[6]{\frac{225}{56}}$.

Divide—

9. $6\sqrt{12} + 3\sqrt{8} - 6\sqrt{30} + 4\sqrt{15}$ by $2\sqrt{6}$.

10. $\frac{3}{5}\sqrt{70} - \frac{2}{4}\sqrt{28} + 3\sqrt{105}$ by $\frac{3}{4}\sqrt{42}$.

11. $x^2 - x + 1$ by $x + \sqrt{x-1}$.

Rationalize the denominator of—

12. $\frac{2+\sqrt{3}}{3-\sqrt{3}}$.

13. $\frac{2\sqrt{3} - 3\sqrt{2}}{3 - 2\sqrt{6}}$.

14. $\frac{6\sqrt{3} - 4\sqrt{6}}{5\sqrt{3} + 3\sqrt{6}}$.

15. $\frac{2\sqrt{15}+8}{5+\sqrt{15}} + \frac{8\sqrt{3} - 6\sqrt{5}}{5\sqrt{3} - 3\sqrt{5}}$.

Find the numerical value of—

16. $\frac{5}{\sqrt{3}}$.

17. $\frac{\sqrt{7} - \sqrt{2}}{2\sqrt{7} + \sqrt{2}}$.

18. $\frac{2\sqrt{5} - 3\sqrt{2}}{3\sqrt{5} - 4\sqrt{2}}$.

19. $\frac{3\sqrt{6} - 2\sqrt{3}}{4\sqrt{2} + 5}$.

Which is the greater—

20. $3\sqrt{3}$ or $2\sqrt{7}$?

21. $\sqrt{8}$ or $\sqrt[3]{23}$?

22. $2\sqrt{6\frac{1}{2}}$ or $3\sqrt[3]{5}$?

23. $\sqrt[3]{6}$ or $\sqrt[4]{11}$?

Find the square root of—

24. $33 + 20\sqrt{2}$.

25. $35 - 12\sqrt{6}$.

26. $80 - 32\sqrt{6}$.

27. $107 + 12\sqrt{77}$.

Simplify—

28. $\dfrac{x-1}{x+1}\left\{\dfrac{x-1}{\sqrt{x}-1} + \dfrac{1-x}{x+\sqrt{x}}\right\}$.

29. $\dfrac{\sqrt{1-x} + \dfrac{1}{\sqrt{1+x}}}{1 + \dfrac{1}{\sqrt{1-x^2}}}$.

30. $\left\{\sqrt{\dfrac{x}{3} - 2} - \sqrt{\dfrac{x}{3} + 2}\right\}^2$.

31. $x\sqrt{\dfrac{x-y}{x+y}} + y\sqrt{\dfrac{x+y}{x-y}} - (3y^2 - x^2)\sqrt{\dfrac{1}{x^2-y^2}}$.

Solve—

32. $2 + \sqrt{x+3} = \sqrt{x-2} + 3$.

33. $\dfrac{1}{\sqrt{x}} - \sqrt{x+2} = \sqrt{x}$.

34. $\sqrt{x} + \sqrt{x-4} = \dfrac{4}{\sqrt{x-4}}$.

35. $\dfrac{\sqrt{x}+4}{2\sqrt{x}-1} = \dfrac{\sqrt{x}+2}{2\sqrt{x}-3}$.

36. $2\sqrt{3x+4} = \dfrac{28-9x}{\sqrt{3x+4}} + 5\sqrt{3x-8}$.

37. $\sqrt{\dfrac{2x}{3}+1} + \sqrt{\dfrac{2x}{3}-1} = \sqrt{\dfrac{8x}{3}-2}$.

CHAPTER XVIII.

IMAGINARY QUANTITIES.

237. An **Imaginary Quantity** is an indicated even root of a negative quantity.

$$\text{Exs. } \sqrt{-4}, \quad \sqrt[4]{-3}, \quad \sqrt{-a}.$$

The term "imaginary" is used because so long as we confine ourselves to plus quantity, and to its direct opposite, minus quantity, there is no number which multiplied by itself will give a negative number, as -4, for instance. All the quantity considered hitherto that is plus or minus quantity, whether it be rational or irrational, is called **real** quantity.

If we extend the realm of quantity considered, outside of plus quantity and its direct opposite, minus quantity, imaginary numbers are as real as any others, as will be shown in the next article.

A number part real and part imaginary is called a **Complex Number.**

$$\text{Exs. } 3+2\sqrt{-1}, \quad a+b\sqrt{-1}.$$

If an imaginary number exist by itself, it is called a **Pure Imaginary.** Thus, $3\sqrt{-1}$ is a pure imaginary.

238. Meaning of $\sqrt{-1}$.

Let us consider the simplest imaginary, $\sqrt{-1}$, and, by a geometrical illustration, try to discover how it has a meaning if we extend the realm of quantity outside of plus quantity and its direct opposite, minus quantity. If $OA = +1$, and OA' be of the same length, but lying in the opposite direction from O, $OA' = -1$.

Hence, we regard the operation of converting a plus quantity into negative quantity as equivalent to a rotation through an angle of 180°. If we divide this rotation into two equal rotations, each of these will be a rotation through 90°. But in seeking a square root of -1 we seek a factor, $\sqrt{-1}$, which multiplied by itself will give -1. The result of a rotation of $+1$ through 90°, rotated again through 90°, gives -1.

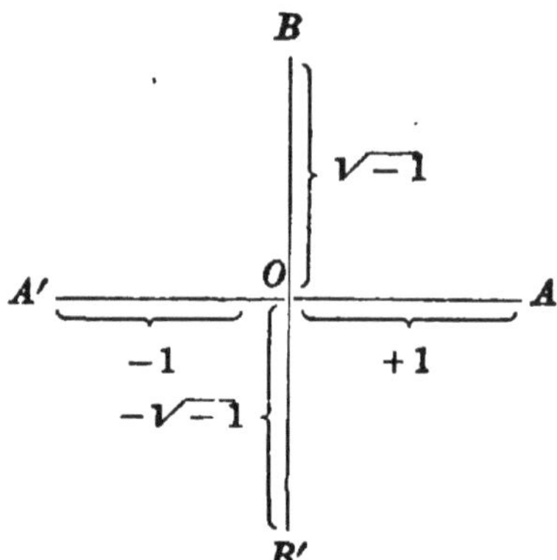

Hence, $\sqrt{-1}$ must be equivalent (geometrically) to the result of rotating the plus unit of quantity through 90°. Hence, $\sqrt{-1}$ on our figure will be represented by *OB*.

Hence, it is easy to see, also, that $\sqrt{-1} \times \sqrt{-1} = -1$. We thus perceive that the introduction of imaginary quantity enlarges the field of quantity considered in algebra from mere quantity in a line to quantity in a plane. This gives a vast extension to power of algebraic processes and introduces many economies in them, as will be found by the student who pursues the study of mathematics extensively.

In taking up the subject for the first time we consider only a few of the first properties of imaginaries, so called.

239. Fundamental Principle. We regard all the ordinary laws of algebra as applying to imaginaries, but, owing to the nature of a square root, one modification of these laws as ordinarily stated must be made.

This fundamental new principle is that $\sqrt{-1} \times \sqrt{-1} = -1$.

Besides the above geometrical illustration (Art. 238), it is important to state formally the algebraic reason for this principle.

The square of $\sqrt{-1}$ must be such a number that when its square root is extracted we shall have the original quantity, $\sqrt{-1}$.

If we use the law of signs in the most general form,

$$(\sqrt{-1})^2 = \sqrt{-1} \times \sqrt{-1} = \sqrt{1} = \pm 1.$$

Now, if we extract the square root of $+1$, we shall not have $\sqrt{-1}$. But if we extract the square root of -1, we shall have $\sqrt{-1}$.

Hence, we must limit the product $\sqrt{-1}\times\sqrt{-1}$ to -1.

Likewise $\sqrt{-a}\times\sqrt{-b}=\sqrt{a}\sqrt{-1}\times\sqrt{b}\sqrt{-1}$

$$=\sqrt{a}\sqrt{b}(\sqrt{-1})^2=-\sqrt{ab},$$

or, in general,

The product of two minus signs under a radical sign of the second degree is a minus sign outside of the radical.

240. Reductions to the Typical Form $A+B\sqrt{-1}$. It may be shown that any imaginary expression of any degree of complexity can be reduced to the form $a+b\sqrt{-1}$. We will limit ourselves at this time to showing that any combination of the sum, difference, product, or quotient of imaginary expressions of the second degree reduces to this form.

First. Sum or difference of two complex numbers.

$$a+b\sqrt{-1}\pm(c+d\sqrt{-1})=(a\pm c)+(b\pm d)\sqrt{-1}$$
$$=A+B\sqrt{-1}$$

Second. Product of two complex numbers.

$$(a+b\sqrt{-1})(c+d\sqrt{-1})=ac-bd+(bc+ad)\sqrt{-1}$$
$$=A'+B'\sqrt{-1}$$

Third. Quotient of two complex numbers.

$$\frac{c+d\sqrt{-1}}{a+b\sqrt{-1}}=\frac{c+d\sqrt{-1}}{a+b\sqrt{-1}}\times\frac{a-b\sqrt{-1}}{a-b\sqrt{-1}}$$
$$=\frac{ac+bd+(ad-bc)\sqrt{-1}}{a^2+b^2}$$
$$=\frac{ac+bd}{a^2+b^2}+\frac{ad-bc}{a^2+b^2}\sqrt{-1}$$
$$=A''+B''\sqrt{-1}$$

Hence, any combination of the sum, difference, product, and quotient of complex numbers may be reduced by successive steps to the typical form.

This serves to illustrate the fundamental value of the imaginary unit, $\sqrt{-1}$.

241. Powers of $\sqrt{-1}$.

$(\sqrt{-1})^1 = \sqrt{-1}$, $\quad\therefore i = \sqrt{-1}$.

$(\sqrt{-1})^2 = -1$, $\quad\therefore i^2 = -1$.

$(\sqrt{-1})^3 = (\sqrt{-1})^2\sqrt{-1} = -\sqrt{-1}$, $\quad\therefore i^3 = -i$. (See OA' of figure

$(\sqrt{-1})^4 = (\sqrt{-1})^3\sqrt{-1} = +1$, $\quad\therefore i^4 = 1$. in Art. 238.)

The symbol i is used for $\sqrt{-1}$.

Thus the first four powers of $\sqrt{-1}$ are $\sqrt{-1}$, -1, $-\sqrt{-1}$, $+1$; and for the higher powers, as the fifth, sixth, etc., these four results recur regularly. The same fact is plain from the figure in Art. 238.

242. Equational Properties of Imaginaries. I. *If an imaginary expression, $x + y\sqrt{-1}$, equals zero, then $x = 0$, $y = 0$.*

Proof.

$$x + y\sqrt{-1} = 0$$
$$y\sqrt{-1} = -x$$
$$-y^2 = x^2.$$
$$\therefore x^2 + y^2 = 0,$$

which, since x and y are both real, can be true only when $x = 0$, $y = 0$.

II. *If two imaginary expressions be equal, the real part of one equals the real part of the other, and the imaginary part of the one equals the imaginary part of the other.*

Proof. Let $\quad x + y\sqrt{-1} = a + b\sqrt{-1}$.

$$\therefore x - a + (y - b)\sqrt{-1} = 0.$$

$\therefore$ by I. of this Art., $x - a = 0$, $\quad\therefore x = a$,

$y - b = 0$, $\quad\therefore y = b$.

243. Conjugate Imaginaries. If two complex numbers

differ only in the sign of their imaginary part, they are called *conjugate imaginaries.*

Exs. $3-\sqrt{-1}$ and $3+\sqrt{-1}$;
$a+b\sqrt{-c}$ and $a-b\sqrt{-c}$.

244. Operations with Imaginaries. It follows from Art. 239 that, in performing operations with imaginaries, we

Use all the ordinary laws of algebra, with the exception of a limitation in use of signs, which may be mechanically stated as follows: *The product of two minus signs under the radical sign of the second degree gives a minus sign outside the radical sign.* But in dividing *first indicate the division and afterwards rationalize the denominator.*

Ex. 1. Add $\sqrt{-9}, -3+2\sqrt{-1}, 7-2\sqrt{-16}$.

$$\begin{aligned} \sqrt{-9} &= 3\sqrt{-1} \\ -3+2\sqrt{-1} &= -3+2\sqrt{-1} \\ 7-2\sqrt{-16} &= 7-8\sqrt{-1} \\ \hline & \quad 4-3\sqrt{-1}, \textit{ Sum.} \end{aligned}$$

Ex. 2. Multiply $2\sqrt{-3}+3\sqrt{-6}$ by $3\sqrt{-3}-5\sqrt{-2}$.

$$\begin{array}{l} 2\sqrt{-3}+3\sqrt{-6} \\ 3\sqrt{-3}-5\sqrt{-2} \\ \hline 6(-3)-9\sqrt{18} \\ \qquad +10\sqrt{6}+15\sqrt{12} \\ \hline -18-27\sqrt{2}+10\sqrt{6}+30\sqrt{3}, \textit{ Product.} \end{array}$$

Ex. 3. Divide $-2\sqrt{6}$ by $\sqrt{-2}$.

$$\frac{-2\sqrt{6}}{\sqrt{-2}} = \frac{-2\sqrt{6}}{\sqrt{-2}} \times \frac{\sqrt{-2}}{\sqrt{-2}}$$
$$= \frac{-2\sqrt{-12}}{-2} = 2\sqrt{-3}, \textit{ Quotient.}$$

Ex. 4. Extract the square root of $1+4\sqrt{-3}$.

$$1+4\sqrt{-3} = 1+2\sqrt{-12}.$$

The two numbers which multiplied together give -12, and added together give 1, are 4 and -3.

$$\therefore \sqrt{1+4\sqrt{-3}} = \sqrt{4}+\sqrt{-3} = 2+\sqrt{-3}, \textit{ Result.}$$

EXERCISE 90.

Reduce to the form $a\sqrt{-1}$.

1. $\sqrt{-9}$.
2. $\sqrt{-25}$.
3. $\sqrt{-100}$.
4. $-\sqrt{-81}$.
5. $2\sqrt{-4}$.
6. $-3\sqrt{-49}$.
7. $-\frac{2}{3}\sqrt{-36}$.
8. $\frac{5}{9}\sqrt{-324}$.

Collect—

9. $7\sqrt{-4}+3\sqrt{-49}-10\sqrt{-9}$.
10. $2\sqrt{-1}-3\sqrt{-121}+5\sqrt{-64}$.
11. $\sqrt{-400}+2\sqrt{-900}-5\sqrt{-144}$.
12. $\sqrt{-12}-\sqrt{-27}+2\sqrt{-3}+\sqrt{-75}$.
13. $5\sqrt{-\frac{1}{2}}-3\sqrt{-\frac{9}{2}}+4\sqrt{-50}-\sqrt{-200}$.
14. $2\sqrt{-a^2}-3a\sqrt{-4}+\frac{2}{a}\sqrt{-16a^4}-\frac{2}{3}\sqrt{-36a^2}$.
15. $a+b\sqrt{-1}-b-a\sqrt{-1}-\sqrt{-b^2}+2\sqrt{-a^2}-a$.
16. $(a-2b)\sqrt{-1}-(2a+b)\sqrt{-1}$.

Multiply—

17. $\sqrt{-1}$ by $\sqrt{-2}$.
18. $2\sqrt{-3}$ by $3\sqrt{-3}$.
19. $\sqrt{-5}$ by $-2\sqrt{-5}$.
20. $-\sqrt{-7}$ by $-\sqrt{-7}$.
21. $-\sqrt{-12}$ by $-\sqrt{-18}$.
22. $\sqrt{-6}$ by $-3\sqrt{-3}$.
23. $2\sqrt{-14}$ by $-2\sqrt{-21}$.
24. $-5\sqrt{-2}$ by $-2\sqrt{-5}$.
25. $-\sqrt{x-y}$ by $\sqrt{y-x}$.
26. $-a\sqrt{1-a}$ by $-\sqrt{(a-1)^3}$.
27. $\sqrt{-1}+\sqrt{-2}$ by $\sqrt{-1}-2\sqrt{-2}$.
28. $3\sqrt{-3}-2\sqrt{-2}$ by $2\sqrt{-3}+3\sqrt{-2}$.
29. $2\sqrt{2}-2\sqrt{-2}$ by $3\sqrt{2}+3\sqrt{-2}$.
30. $3\sqrt{-5}-2\sqrt{-7}$ by $4\sqrt{-7}+5\sqrt{-5}$.

31. $1-3\sqrt{-3}$ by $1+5\sqrt{-3}$.

32. $\sqrt{-1}-\sqrt{-2}+\sqrt{-3}$ by $\sqrt{-1}+\sqrt{-2}+\sqrt{-3}$.

33. $x-2+\sqrt{-3}$ by $x-2-\sqrt{-3}$.

34. $a\sqrt{-a}+b\sqrt{-b}$ by $a\sqrt{-a}-b\sqrt{-b}$.

35. $x-1-\sqrt{-1}$ by $x-1+\sqrt{-1}$.

36. $x-\dfrac{1-\sqrt{-3}}{2}$ by $x-\dfrac{1+\sqrt{-3}}{2}$.

Divide—

37. $-\sqrt{18}$ by $\sqrt{-6}$.

38. $-\sqrt{-12}$ by $-\sqrt{-2}$.

39. $-6\sqrt{-15}$ by $2\sqrt{-3}$.

40. $8\sqrt{-a^3}$ by $-2\sqrt{a}$.

41. $2\sqrt{-18}-4\sqrt{-15}+10\sqrt{30}$ by $-2\sqrt{-3}$.

42. $a\sqrt{-a}-2a\sqrt{-6a}-a^2\sqrt{3a^3}$ by $-a\sqrt{-a}$.

Express with rational denominators—

43. $\dfrac{1}{3-\sqrt{-2}}$.

44. $\dfrac{2-\sqrt{-3}}{2+\sqrt{-3}}$.

45. $\dfrac{3\sqrt{2}-\sqrt{-2}}{2\sqrt{2}+\sqrt{-2}}$.

46. $\dfrac{a+b\sqrt{-1}}{a-b\sqrt{-1}}$.

47. $\dfrac{3\sqrt{-5}-\sqrt{-2}}{2\sqrt{-5}-3\sqrt{-2}}$.

48. $\dfrac{\sqrt{x-1}+\sqrt{1-x}}{2\sqrt{1-x}-\sqrt{x-1}}$.

49. $\dfrac{(1-\sqrt{-1})^2}{2-3\sqrt{-1}}$.

50. $\dfrac{3\sqrt{2}+2\sqrt{-7}-\sqrt{-10}}{3\sqrt{2}-2\sqrt{-7}+\sqrt{-10}}$.

Find the square root of—

51. $3-6\sqrt{-6}$.

52. $1-2\sqrt{-6}$.

53. $12\sqrt{-6}-6$.

54. $12\sqrt{10}-38$.

55. $-29-24\sqrt{-5}$.

56. $7+40\sqrt{-2}$.

17

57. Find the value of, $(\sqrt{-1})^5$; $(-\sqrt{-1})^4$; $(\sqrt{-1})^6$; $(-\sqrt{-1})^7$; $(\sqrt{-1})^{-3}$; $(-\sqrt{-1})^{-9}$; $(\sqrt{-1})^{\frac{2}{3}}$.

In expanding binomials containing imaginaries, the labor is perhaps lessened by the substitution of a letter, as i, for the imaginary, $\sqrt{-1}$, and after simplifying this derived expression, in i, return to the imaginary, $\sqrt{-1}$.

For example, let it be required to simplify the expression, $3(\sqrt{-1}+2)^2 - (2\sqrt{-1}-1)^2$. Substitute i for $\sqrt{-1}$.

$$\begin{aligned} 3(i+2)^2 - (2i-1)^2 &= 3i^2 + 12i + 12 - 4i^2 + 4i - 1 \\ &= -i^2 + 16i + 11 \\ &= -(\sqrt{-1})^2 + 16\sqrt{-1} + 11 \\ &= 12 + 16\sqrt{-1}, \textit{ Result.} \end{aligned}$$

Simplify—

58. $(\sqrt{-1}-1)^3 - (\sqrt{-1}-1)^2 + 2(\sqrt{-1}-1)$.

59. $(\sqrt{-1}-2)(3\sqrt{-1}+1) - (\sqrt{-1}-3)^2 - (\sqrt{-2})^2$.

60. $(\sqrt{-1}-1)^4 + 3(\sqrt{-1}-1)^3 + 4(\sqrt{-1}-1)^2$.

61. Prove that the sum and the product of any pair of conjugate imaginaries are real. (See Art. 243.)

62. If $x = \dfrac{1-\sqrt{-1}}{2}$, find the value of, $3x^2 - 6x + 7$. Of $x^3 - 5x^2 + 2x - 1$.

63. If $x = \dfrac{2-3\sqrt{-1}}{5}$, find the value of, $10x^2 - 8x + 3$.

CHAPTER XIX.

QUADRATIC EQUATIONS OF ONE UNKNOWN QUANTITY.

245. General Problem. The relation of the square of some unknown number (as well as of its first power) to known numbers may be given in the form of a more or less complex equation. It then is often required to reduce this complex relation to some simple relation from which the value of the unknown number may be at once recognized.

246. A Quadratic Equation of one unknown quantity is an equation containing the second power of the unknown quantity, but no higher power.

A **Pure Quadratic Equation** is one in which the second power of the unknown quantity occurs, but not the first power.

Ex. $5x^2 - 12 = 0$.

An **Affected Quadratic Equation** is one in which both the first and second powers of the unknown quantity occur.

Ex. $3x^2 - 7x + 12 = 0$.

PURE QUADRATIC EQUATIONS.

247. Solution of Pure Quadratics. Since only the second power, x^2, of the unknown quantity occurs in a pure quadratic equation,

Reduce the given equation to the form $x^2 = c$;
Extract the square root of both members.

Ex. 1. Solve $\frac{x^2-12}{3}=\frac{x^2-4}{4}$.

Clearing of fractions, $4x^2-48=3x^2-12$

Hence, $x^2=36$

Extracting the square root of each member,

$$x=+6, \quad \text{or } -6.$$

That is, since the square of $+6$ is 36, and also the square of -6 is 36, x has two values, either of which satisfies the original equation. These two values of x are best written together. Thus,

$$x=\pm 6, \textit{ Roots.}$$

Ex. 2. Solve $\frac{a}{x^2-b}=\frac{b}{x^2-a}$.

$$ax^2-a^2=bx^2-b^2$$

$$ax^2-bx^2=a^2-b^2$$

$$x^2=a+b$$

$$x=\pm\sqrt{a+b}, \textit{ Roots.}$$

EXERCISE 91.

Solve—

1. $5x^2=80$.
2. $3x^2-5=x^2+3$.
3. $\frac{7}{8}x^2-1=\frac{1}{8}-3x^2$.
4. $1-\frac{2}{5}x^2=x^2-4\frac{3}{5}$.
5. $\frac{x^2}{8}-\frac{1}{2}+x^2=0$.
6. $\frac{x^2-5}{7}=\frac{\frac{2}{7}-x^2}{5}$.
7. $\frac{3-x^2}{11}+\frac{x^2+5}{6}=3$.
8. $\frac{3}{4x^2}-\frac{1}{3x^2}=\frac{5}{12}$.
9. $\frac{1}{2x-1}-\frac{1}{2x+1}=3\frac{5}{9}$.
10. $ax^2+a^3=5a^3-3ax^2$.
11. $ax^2+c=b$.
12. $\frac{x+2a}{x-2a}+\frac{x-2a}{x+2a}=\frac{26}{5}$.
13. $\frac{2c}{x-c}+\frac{5x+2c}{3x}+1=0$.

14. $(ax + b)^2 + (ax - b)^2 = 10b^2$.

15. $(x + a)(x - b) + (x - a)(x + b) = 2(a^2 + b^2 + ab)$.

16. $3(2x - 5)(x + 1) - 2(3x + 2)(2x - 3) = x - 9$.

AFFECTED QUADRATIC EQUATIONS.

248. Completing the Square. An affected quadratic equation may in every instance be reduced to the form

$$x^2 + px = q.$$

An equation in this form may then be solved by a process called *completing the square.* This process consists in adding such a number to both members of the equation as will make the left-hand member a perfect square. The use of familiar elementary processes then gives the values of x.

Thus, to solve $\qquad x^2 + 6x = 16,$

take half the coefficient of x (that is, 3), square it, and add the result (that is, 9) to both members of the original equation. We obtain

$$x^2 + 6x + 9 = 25.$$

Extract the square root of both members,

$$x + 3 = \pm 5,$$

Hence, $\qquad x = -3 \pm 5,$

That is, $\qquad x = -3 + 5 = 2,$

Also, $\qquad x = -3 - 5 = -8,$ } *Roots.*

Hence we have the general rule:

By clearing the given equation of fractions and parentheses, transposing terms, and dividing by the coefficient of x^2, reduce the given equation to the form $x^2 + px = q$;

Add the square of half the coefficient of x to each member of the equation;

Extract the square root of each member;

Solve the resulting simple equations.

Ex. 1. Solve $6x^2 - 14x = 12$.

Dividing by 6, $x^2 - \frac{7}{3}x = 2$

Completing the square, $x^2 - \frac{7}{3}x + (\frac{7}{6})^2 = 2 + \frac{49}{36} = \frac{121}{36}$

Extracting square root, $x - \frac{7}{6} = \pm \frac{11}{6}$

$x = \frac{7}{6} \pm \frac{11}{6}$

$x = 3$, or $-\frac{2}{3}$, *Roots.*

Ex. 2. Solve $3x^2 = 2(1 + 2x)$.

Clearing, $3x^2 = 2 + 4x$

Transposing, $3x^2 - 4x = 2$

Hence, $x^2 - \frac{4}{3}x = \frac{2}{3}$

$x^2 - \frac{4}{3}x + (\frac{2}{3})^2 = \frac{10}{9}$

$x = \frac{2}{3} \pm \sqrt{\frac{10}{9}} = \frac{2 \pm \sqrt{10}}{3}$, *Roots.*

EXERCISE 92.

Solve—

1. $x^2 + 14x = 32$.
2. $x^2 + 10x = 24$.
3. $x^2 - 8x - 20 = 0$.
4. $x^2 - 5x = 6$.
5. $x^2 + 11x + 24 = 0$.
6. $3x^2 + 4x = 7$.
7. $5x^2 - 6x = 8$.
8. $2x^2 - 5x = 7$.
9. $3x^2 + 7x = 26$.
10. $4x^2 + 8x - 5 = 0$.
11. $6x^2 - 5x - 6 = 0$.
12. $2x + 3\frac{3}{4}x^2 = 4$.
13. $x^2 + 5 = \frac{14x}{3}$.
14. $35 = 2x^2 - 3x$.
15. $3x + 77 = 2x^2$.
16. $6x^2 - x - 35 = 0$.
17. $3x^2 + \frac{1}{2}x = 1\frac{2}{3}$.
18. $3x^2 = \frac{1}{3}x + 2\frac{2}{3}$.
19. $\frac{1}{2} = \frac{7x}{6} + x^2$.
20. $\frac{x-1}{2} + \frac{2}{x-1} = 2\frac{1}{2}$.
21. $\frac{9}{5x} - 1 + \frac{2}{x+2} = 0$.
22. $\frac{3x+5}{x+4} = 3 - \frac{2x-5}{x-2}$.
23. $\frac{2x-1}{x+3} - \frac{x+1}{2x-3} = -\frac{1}{2}$.

24. $x-\frac{1}{x}=4-\frac{x}{2}+\frac{1}{2x}.$

25. $\frac{x}{5-x}-\frac{5-x}{x}=\frac{15}{4}.$

26. $\frac{1}{x+2}-\frac{x}{x-2}=\frac{4}{3}.$

27. $\frac{2x+1}{x-1}+x=\frac{x-1}{3}.$

28. $\frac{2}{3x-1}+\frac{3x}{2x-5}=0.$

29. $\frac{3x+1}{5x+4}=\frac{x+3}{4x+5}.$

30. $\frac{3}{2x+3}+\frac{4}{3x-4}=\frac{17}{15}.$

31. $\frac{1}{2}(x+1)-\frac{x}{3}(2x-1)=-12.$

32. $\frac{x-3}{2x}+\frac{3x-1}{3}=1-\frac{1-x}{4x}+\frac{3+2x}{6}.$

33. $\frac{2x-1}{x+1}-\frac{3x-4}{x-1}=1-\frac{4x-14}{1-x^2}.$

34. $\frac{x-3}{3x-2}+\frac{5x-7}{4-9x^2}=\frac{2x+5}{2+3x}-1.$

35. $\frac{2x-1}{x+1}-\frac{1-3x}{x+2}-\frac{x-7}{x-1}=4.$

36. $x^2+2x=1.$

37. $3x^2-5x=-1.$

38. $9x^2-18x+4=0.$

39. $5x^2+3x=1.$

40. $11x^2-12x=-3.$

41. $2x^2+5x=-4.$

42. $3x^2-7x=-5.$

43. $9x^2-6x+5=0.$

44. $3x(x+1)-(x-2)(x+3)=2+(1-x)^2.$

45. $(x+1)(x-5)-2x(x-1)=1-(1-2x)^2.$

46. $\frac{x^2+x-1}{x^2+x+1}+\frac{x^2-x-1}{x^2-x+1}=2.$

N. B. The verification of equations having irrational roots is a profitable exercise.

249. Literal Quadratic Equations are solved by the same

methods employed in solving quadratic equations with numerical coefficients.

Ex. 1. Solve $\frac{x^2}{2}-\frac{ax}{6}=\frac{a}{6}(x+a)$.

Clearing, $3x^2-ax=ax+a^2$

Hence, $3x^2-2ax=a^2$

$$x^2-\frac{2a}{3}x=\frac{a^2}{3}$$

$$x^2-(\ )+\left(\frac{a}{3}\right)^2=\frac{4a^2}{9}$$

$$x=a,\quad -\frac{a}{3},\ \textit{Roots.}$$

Ex. 2. Solve $(a-b)^2x^2-(a^2-b^2)x=-ab$.

$$x^2-\frac{a+b}{a-b}x=-\frac{ab}{(a-b)^2}$$

$$x^2-(\ )+\tfrac{1}{4}\left(\frac{a+b}{a-b}\right)^2=\tfrac{1}{4}\left(\frac{a+b}{a-b}\right)^2-\frac{4ab}{4(a-b)^2}$$

$$x-\frac{a+b}{2(a-b)}=\pm\frac{a-b}{2(a-b)}$$

$$x=\frac{a}{a-b},\quad \frac{b}{a-b},\ \textit{Roots.}$$

EXERCISE 93.

Solve—

1. $x^2+4ax=12a^2$.
2. $x^2+4bx=21b^2$.
3. $x^2+3cx-10c^2=0$.
4. $x^2+5abx=6a^2b^2$.
5. $6x^2-bx=12b^2$.
6. $3x^2+4cdx=15c^2d^2$.
7. $2a^2x^2+ax=3$.
8. $7c^2x^2-10acx+3a^2=0$.
9. $x^2+\frac{5x}{a}=\frac{6}{a^2}$.
10. $2a^2x^2+abx=15b^2$.
11. $x^2-(a+1)x=-a$.
12. $x^2+\frac{bx}{a}=\frac{3b^2}{4a^2}$.
13. $x^2+(2-3a)x=6a$.
14. $3a^2x^2+a(3b-5)x=5b$.
15. $abx^2+(a^2+b^2)x+ab=0$.
16. $ax^2-(a^2-1)x=a$.
17. $\frac{x+1}{x^2}=\frac{a+1}{a^2}$.

18. $x - \dfrac{1}{x-a} = a.$ 19. $4(x^2-1) = b(4x-b).$

20. $(a+b)x^2 - (a-b)x - \dfrac{ab}{a+b} = 0.$

21. $abx^2 = \dfrac{1}{ab}\left[x(a+b) - \dfrac{1}{ab}\right].$

22. $\dfrac{x^2+1}{x} = \dfrac{a+b}{c} + \dfrac{c}{a+b}.$

23. $a(x^2-b^2) + b(x^2-b^2+c) + cx = 0.$

24. $(a+c)x^2 - (2a+c)x + a = 0.$

FACTORIAL METHOD OF SOLVING EQUATIONS.

250. Factorial Method for Solving Quadratic Equations. If any factor of a product equals zero, the entire product equals zero. Hence, if a quadratic equation be reduced to the form $ax^2 + bx + c = 0$, and the left-hand member be factored, and each factor be made equal to zero, the values of x thus obtained will satisfy the equation, and therefore be its roots.

Ex. Solve $x^2 + 5x - 24 = 0.$

Factoring, $(x+8)(x-3) = 0.$

Hence, letting $x + 8 = 0$, and $x - 3 = 0$,

$x = -8$, $x = +3$, *Roots.*

251. Factorial Solution of Equations of Higher Degrees. Since the principle of Art. 250 applies to the product of any number of factors, equations of degree higher than the second may often be readily solved by this method.

Ex. 1. Solve $x(x-1)(x+3)(x-5) = 0.$

The roots are $x = 0, 1, -3, 5.$

Ex. 2. Solve $x^3+1=0$.

Factoring, $(x+1)(x^2-x+1)=0$

$x+1=0$, gives $x=-1$, *Root.*

Also, $x^2-x+1=0$

Whence, $x^2-x=-1$

$x=\frac{1}{2}\pm\frac{1}{2}\sqrt{-3}$, *Roots.*

Ex. 3. Solve $x^3+x^2-4(x^2-1)=0$.

Factoring, $x^2(x+1)-4(x^2-1)=0$

$(x+1)(x^2-4x+4)=0$

$(x+1)(x-2)(x-2)=0$

$x=-1,\ 2,\ 2$, *Roots.*

EXERCISE 94.

Solve by factoring—

1. $x^2+8x+7=0$.
2. $x^2-5x=84$.
3. $6x^2-x=15$.
4. $6x^2+7x=90$.
5. $12x^2-5x=3$.
6. $3x^2-10x+3=0$.
7. $24x^2=2x+15$.
8. $3a^2x^2+10ax=8$.
9. $x^4=16$.
10. $x^3=8$.
11. $x^4-5x^2+4=0$.
12. $x^5-x^4-x+1=0$.
13. $(2x-1)(6x^2-x-2)=0$
14. $3(x^2-1)-2(x+1)=0$
15. $5(x^2-4)=3(x-2)$.
16. $7(x^4-16)-53x(x^2-4)=0$.
17. $3x(x^2-1)+2(x-1)=0$.
18. $x^3-27=13x-39$.
19. $2x^3+2x^2=x+1$.
20. $2x^3+6x^2=3x^2+8x-3$.
21. Find the six roots of $x^6-1=0$.

EQUATIONS IN THE QUADRATIC FORM.

252. Simple Unknown Quantity. An equation containing but two powers of the unknown quantity, the index of one power being twice the index of the other power, is an equation of the quadratic form. It may be solved by the methods already given for affected quadratic equations.

Ex. 1. Solve $x^4 - 5x^2 = -4$.

Adding $(\frac{5}{2})^2$ to both members will make the left-hand member a perfect square, giving

$$x^4 - 5x^2 + (\tfrac{5}{2})^2 = \tfrac{9}{4}$$

Hence, $$x^2 - \tfrac{5}{2} = \pm \tfrac{3}{2}$$

$$x^2 = 4, \quad \text{or } 1$$

$$x = \pm 2, \quad \pm 1, \textit{ Roots.}$$

This equation might also have been solved by the factorial method.

Ex. 2. Solve $2\sqrt[3]{x^{-2}} - 3\sqrt[3]{x^{-1}} = 2$.

Using fractional exponents,

$$2x^{-\frac{2}{3}} - 3x^{-\frac{1}{3}} = 2$$

Whence $$x^{-\frac{2}{3}} - \tfrac{3}{2}x^{-\frac{1}{3}} = 1$$

$$x^{-\frac{2}{3}} - (\ ) + \tfrac{9}{16} = \tfrac{25}{16}$$

$$x^{-\frac{1}{3}} - \tfrac{3}{4} = \pm \tfrac{5}{4}$$

$$x^{-\frac{1}{3}} = 2, \quad -\tfrac{1}{2}$$

Whence $$x^{\frac{1}{3}} = \tfrac{1}{2}, \quad -2$$

$$x = \tfrac{1}{8}, \quad -8, \textit{ Roots.}$$

253. Compound Unknown Quantity. A polynomial may be used in the place of a single quantity as an unknown quantity.

Ex. 1. Solve $(2x - 3)^2 - 6(2x - 3) = 7$.

Let $2x - 3 = y$, and substitute.

We obtain $y^2 - 6y = 7$

Whence $y = 7, \quad -1.$

Hence, $2x - 3 = 7, \quad$ also, $\quad 2x - 3 = -1$

$\therefore x = 5$, *Root.* $\quad x = 1$, *Root.*

Ex. 2. Solve $\sqrt{x+12}+\sqrt[4]{x+12}=6.$

This equation may be written, $(x+12)^{\frac{1}{2}}+(x+12)^{\frac{1}{4}}=6.$

Let $\quad (x+12)^{\frac{1}{4}}=y;\quad$ then $(x+12)^{\frac{1}{2}}=y^2$

Hence, substituting, $\quad y^2+y=6.$

Whence $\quad y=2,\ \text{or}\ -3.$

$\therefore\ \sqrt[4]{x+12}=2,\qquad$ Also, $\sqrt[4]{x+12}=-3,$

$x+12=16,\qquad x+12=81,$

$x=4,$ *Root.* $\qquad x=69,$ *Root.*

Ex. 3. Solve $x^2-7x+\sqrt{x^2-7x+18}=24.$

Add 18 to both sides,

$$x^2-7x+18+\sqrt{x^2-7x+18}=42$$

Let $\quad \sqrt{x^2-7x+18}=y;\quad$ then $y^2+y=42$

$$y=6,\ \text{or}\ -7.$$

Hence, $\sqrt{x^2-7x+18}=6,\qquad$ Also, $\sqrt{x^2-7x+18}=-7,$

$x^2-7x+18=36,\qquad x^2-7x+18=49,$

$x=9,\ -2,$ *Roots.* $\qquad x=\frac{1}{2}(7\pm\sqrt{173}),$ *Roots.*

EXERCISE 95.

Solve—

1. $x^4-17x^2+16=0.$
2. $4x^4-13x^2+9=0.$
3. $27x^6=35x^3-8.$
4. $3x^{\frac{1}{2}}-5x^{\frac{1}{4}}=2.$
5. $27x^3+19x^{\frac{3}{2}}=8.$
6. $3x^{\frac{2}{3}}=4x^{\frac{1}{3}}+4.$
7. $2\sqrt[3]{x}=\sqrt[6]{x}+1.$
8. $3x^{-\frac{2}{3}}+5x^{-\frac{1}{3}}=2.$
9. $6x^{-1}-x^{-\frac{1}{2}}=12.$
10. $9x^{-\frac{4}{5}}+4=13x^{-\frac{2}{5}}.$
11. $3\sqrt[4]{x}-5\sqrt{x}=-2.$
12. $5\sqrt[3]{x^2}=8\sqrt[3]{x}+4.$
13. $7\sqrt[5]{x^{-6}}-4\sqrt[5]{x^{-3}}=3.$
14. $3\sqrt{2x}-2\sqrt[4]{2x}=1.$
15. $(x-1)^2+4(x-1)=21.$
16. $2(x^2-3)^2-7(x^2-3)=30.$
17. $6(x^2+1)^2+13(x^2+1)=28.$
18. $2\sqrt{2x-3}+5\sqrt[4]{2x-3}=7.$

19. $3\left(\frac{x}{6}+\frac{2}{x}\right)^2-2\left(\frac{x}{6}+\frac{2}{x}\right)=2\frac{2}{3}$.

20. $5(4x+1)-27\sqrt{4x+1}=-10$.

21. $3(3x^2-2x+1)-4\sqrt{3x^2-2x+1}=15$.

22. $2(2x^2+3x-4)^2-3(2x^2+3x-4)=-1$.

23. $x^2+7x-3\sqrt{x^2+7x+1}=17$.

24. $6(x^2+x)-7\sqrt{3x(x+1)-2}=8$.

25. $3x^2-7+3\sqrt{3x^2-16x+21}=16x$.

26. $3x^{-\frac{1}{2}}-7x^{\frac{1}{2}}=4$.

27. $3x^{\frac{2}{3}}=8x^{-\frac{2}{3}}-10$.

28. $16x^{\frac{3}{5}}-22=3x^{-\frac{3}{5}}$.

29. $2x^2-\sqrt{x^2-2x-3}=4x+9$.

30. $5(2x^2-1)^{\frac{1}{3}}-4=\sqrt[3]{(2x^2-1)^{-1}}$.

31. $3(x^2+1)^{-\frac{1}{2}}+5=2(x^2+1)^{\frac{1}{2}}$.

RADICAL EQUATIONS.

254. Radical Equations resulting in Affected Quadratic Equations. If an equation be cleared of radicals by the methods given in Art. 235, the result is often a quadratic equation.

Ex. Solve $\sqrt{3x+10}+\sqrt{x+2}=\sqrt{10x+16}$.

Squaring, $3x+10+2\sqrt{(3x+10)(x+2)}+x+2=10x+16$

Hence, $\sqrt{(3x+10)(x+2)}=3x+2$

Squaring again, $3x^2+16x+20=9x^2+12x+4$

$$6x^2-4x=16$$

$$x=2,\quad -\tfrac{4}{3}.$$

Substituting these values in the original equation, the only value that verifies is $x=2$, which is the root. The other value, $x=-\frac{4}{3}$, is not a root of the original equation, but is introduced by squaring in the process of clearing the equation of radical signs. It satisfies the equation,

$$\sqrt{3x+10}-\sqrt{x+2}=\sqrt{10x+16}.$$

EXERCISE 96.

Solve—

1. $x-1=\sqrt{3x-5}$.
2. $2x+1=\sqrt{7x+2}$.
3. $x-\sqrt{3x}=6$.
4. $3x-2\sqrt{6x}=6$.
5. $\sqrt{3x+1}-2\sqrt{2x}=-3$.
6. $2+\sqrt{2x+7}=\sqrt{5x+4}$.
7. $\sqrt{3x+7}=\sqrt{x+1}+2\sqrt{x-2}$.
8. $\sqrt{2x+1}=2\sqrt{x}-\sqrt{x-3}$.
9. $\sqrt{x-a^2}+\sqrt{x+2a^2}=\sqrt{x+7a^2}$.
10. $\sqrt{x+2a}+2\sqrt{x-2a}=3a-1$.
11. $3\sqrt{x^3+17}-2\sqrt{5x^3+41}+\sqrt{x^3+1}=0$.
12. $\sqrt{2x+3}-\frac{3}{2}\sqrt{7-x}=\frac{1}{2}\sqrt{11x-33}$.
13. $\sqrt{x+4}+\sqrt{3x+1}-\sqrt{9x+4}=0$.
14. $2\sqrt{5x}-\sqrt{2x-1}=\dfrac{4x+1}{\sqrt{2x-1}}$.
15. $\dfrac{3\sqrt{2x}-5}{3+\sqrt{2x}}=\dfrac{9-2\sqrt{2x}}{\sqrt{2x}-3}$.
16. $\sqrt{x+\frac{2}{3}}+\sqrt{4x-\frac{1}{3}}=\sqrt{5x+\frac{7}{3}}$.
17. $\sqrt{x+5}+\sqrt{3x+4}-\sqrt{12x+1}=0$.
18. $\sqrt{12x^2-x-6}+\sqrt{12x^2+x-6}=\sqrt{24x^2-12}$.
19. $\dfrac{x+\sqrt{x^2-a^2}}{x-\sqrt{x^2-a^2}}-\dfrac{x-\sqrt{x^2-a^2}}{x+\sqrt{x^2-a^2}}=8\sqrt{x^2-a^2}$.
20. $\sqrt{4x+3}+\sqrt{2x+3}=\sqrt{5x+1}+\sqrt{x+5}$.

OTHER METHODS OF SOLVING QUADRATIC EQUATIONS.

255. I. **Completing the Square when the Coefficient of x^2 is a Square Number or can be Readily made One.** If in a simplified quadratic equation the coefficient of x^2 is a square number, we may readily complete the square by *dividing the second term by twice the square root of the first term, and adding the square of the quotient thus obtained to both members of the equation.*

That this process gives a perfect square is readily seen from the fact that

$$(ax + b)^2 = a^2x^2 + 2abx + b^2.$$

Hence, given $a^2x^2 + 2abx$, the term b^2 with which to complete the square may be obtained by dividing $2abx$ by $2ax$, and squaring the quotient.

Ex. 1. Solve $9x^2 + 4x = 5$.

To complete the square, take the square root of $9x^2$ (that is, $3x$), and divide $4x$ by twice $3x$; this gives as a quotient $\frac{2}{3}$. Add the square of this, $\frac{4}{9}$, to both members.

$$\therefore\ 9x^2 + 4x + \tfrac{4}{9} = \tfrac{49}{9}.$$

Whence $$3x + \tfrac{2}{3} = \pm \tfrac{7}{3}$$

$$3x = \tfrac{5}{3}, \text{ or } -3$$

$$x = \tfrac{5}{9}, \text{ or } -1, \textit{ Roots.}$$

Ex. 2. Solve $8x^2 + 3x = 26$.

To make the coefficient of x^2 a square number, multiply both members of the equation by 2.

$$\therefore\ 16x^2 + 6x = 52$$

Completing the square, $$16x^2 + 6x + \tfrac{9}{16} = 52 + \tfrac{9}{16} = \tfrac{841}{16}$$

$$4x + \tfrac{3}{4} = \pm \tfrac{29}{4}$$

$$x = \tfrac{13}{8},\ -2, \textit{ Roots.}$$

256. II. Hindoo Method to Avoid Fractions in Completing the Square. After simplifying the equation, *multiply through by four times the coefficient of x^2, and add to both sides the square of the coefficient of x in the simplified equation.*

The reason for this process is evident, since if $ax^2 + bx = c$ be multiplied by $4a$, we obtain

$$4a^2x^2 + 4abx = 4ac.$$

The addition of b^2 gives on the left-hand side $4a^2x^2 + 4abx + b^2$, which is a perfect square.

Ex. Solve $3x^2 - 2x = 8$ by the Hindoo method.

Multiply by 4×3, or 12,

$$36x^2 - 24x = 96.$$

Add the square of the coefficient of x in the original equation; that is, $(-2)^2$, or 4.

$$36x^2 - 24x + 4 = 100$$
$$6x - 2 = \pm 10$$
$$6x = 12, \quad -8$$
$$x = 2, \quad -\tfrac{4}{3}, \textit{ Roots.}$$

EXERCISE 97.

Solve—

1. $x^2 + 5x = 6.$
2. $3x^2 - x = 2.$
3. $6x^2 + 5x = 4.$
4. $7x^2 + 11x = 6.$
5. $8x^2 - 2x = 3.$
6. $4x^2 + 4x = 35.$
7. $9x^2 - 3x = 30.$
8. $16x^4 - 40x^2 + 9 = 0.$
9. $6x^2 - ax = 2a^2.$
10. $4a^2x^2 + 5ax = 21.$
11. $(x^2 + 3)^2 - 7(x^2 + 3) = 60.$
12. $4x^{-4} - 101x^{-2} + 25 = 0.$
13. $6x^{\frac{1}{2}} - 5x^{\frac{1}{4}} = 6.$
14. $4x^{\frac{2}{3}} + 4x^{\frac{1}{3}} = 3.$
15. $6\sqrt[3]{x^{-2}} - 11\sqrt[3]{x^{-1}} = 10.$
16. $3(x-2)^2 + 5(x-2) = 12.$
17. $x + \dfrac{1}{2x} = a + \dfrac{1}{2a}.$
18. $(a-1)x^2 + (a+1)x = -2.$
19. $(a^2 - b^2)x^2 + (a^2 + b^2)x = ab.$
20. $8abx^2 - (6b^2 + 4a^2)x = -3ab.$

257. III. Use of Formula. Any quadratic equation can be reduced to the form

$$ax^2 + bx + c = 0.$$

Solving this equation by use of Art. 248,

$$x = \frac{-b \pm \sqrt{b^2 - 4ac}}{2a}.$$

By substituting in this result, as a formula, the values of a, b, c in any given equation, the values of x may be at once obtained.

Ex. Solve $5x^2 + 3x - 2 = 0$ by use of the formula.

Here $a = 5, \quad b = 3, \quad c = -2.$

Substituting for a, b, c in the above formula,

$$x = \frac{-3 \pm \sqrt{9 + 40}}{10}$$

$$= \frac{-3 \pm 7}{10} = \tfrac{2}{5}, \quad -1, \textit{ Roots.}$$

EXERCISE 98.

Solve by the formula—

1. $2x^2 + 5x = 7.$
2. $4x^2 - 3x = 7.$
3. $6x^2 + 7x = 10.$
4. $4x^2 - 11x = 3.$
5. $12x^2 + 8x - 15 = 0.$
6. $6x^2 + 13x + 6 = 0.$
7. $33x^2 - 17x = 36.$
8. $12x^2 - x = 6.$
9. $2x^2 + ax = 6a^2.$
10. $12b^2 = 3a^2x^2 - 5abx.$
11. $ax^2 = (1 + a^2)x - a.$
12. $2x^{\frac{2}{3}} - 3x^{\frac{1}{3}} = 9.$
13. $6x - 7\sqrt{x} = 20.$
14. $8x^{-3} + 19x^{-\frac{3}{2}} = 27.$
15. $4x^{-4} - 73x^{-2} = -144.$
16. $3(x^2 - 1)^2 = 7(x^2 - 1) + 6.$
17. $\frac{2}{3}(2x - 3) - \frac{3}{x}(x - 2)^2 = \frac{1}{6}(18 - 5x).$
18. $\dfrac{x}{cd(c + d)}\left[\dfrac{x}{c + d} - 1\right] = \dfrac{1}{(c - d)^2}.$
19. $\dfrac{2x + 3}{2(2x - 1)} - \dfrac{7 - x}{2x + 2} = \dfrac{7 - 3x}{4 - 3x}.$
20. $(a + 3)^2x^2 - (a^2 - 9)x = 3a.$

Find the approximate numerical values of x to four decimal places.

21. $x^2 - 4x + 1 = 0.$
22. $9x^2 - 12x = 1.$
23. $2x^2 + 3 = 10x.$
24. $5x^2 + 2x = 2.$

EXERCISE 99.

REVIEW.

Solve—

1. $6x^2 + x = 1.$
2. $3x^2 + \frac{25}{6}x = \frac{25}{3}.$
3. $\sqrt{x} - \sqrt[4]{x} = 2.$
4. $x^3 - 16x = 0.$
5. $3\sqrt{x} - 12x^{-\frac{1}{2}} = 5.$
6. $\dfrac{x-1}{x} + \dfrac{x}{x-1} = 2\frac{1}{6}.$
7. $\dfrac{3}{x-7} + \dfrac{2x+1}{3x} = \frac{17}{16}.$
8. $\dfrac{x+1}{\sqrt{x}} = \dfrac{a+1}{\sqrt{a}}.$
9. $x + \dfrac{a}{b} = \dfrac{1}{x} + \dfrac{b}{a}.$
10. $\sqrt{4x-3} = 1 + \sqrt{x+1}.$
11. $3x^{-\frac{2}{3}} - 7x^{-\frac{1}{3}} = 6.$
12. $x^4 - 27x = 0.$
13. $5x^{-1} + 6x^{-\frac{1}{2}} = 11$
14. $\dfrac{x-3}{x-4} - \dfrac{x+3}{x+4} = \frac{6}{7}.$
15. $2x^2 + 2x + 1 = 0.$
16. $5(x+2)^4 = 3(x+2)^2 + 2.$
17. $3\sqrt{x+1} - 5\sqrt[4]{x+1} = 2.$
18. $3 - \left(\sqrt{x} - \dfrac{1}{\sqrt{x}}\right) = \sqrt{x} - \dfrac{1}{\sqrt{x}}.$
19. $20 - 2\left(x + \dfrac{3}{x}\right)^2 = 3\left(x + \dfrac{3}{x}\right).$
20. $\dfrac{abx^2}{a^2 - b^2} + 1 = \dfrac{(a^2 + b^2)x}{a^2 - b^2}.$
21. $\dfrac{8}{\sqrt{x^3}} + \dfrac{7}{\sqrt[4]{x^3}} = 1.$
22. $\dfrac{1}{x + a + b} = \dfrac{1}{x} + \dfrac{1}{a} + \dfrac{1}{b}.$
23. $\dfrac{3\sqrt{x} - 4\sqrt{2}}{4\sqrt{x} - 2\sqrt{2}} = \dfrac{2\sqrt{3x} - \sqrt{6}}{3\sqrt{3x} - 5\sqrt{6}}.$
24. $4x^2 - 7\sqrt{2x^2 + 3x - 2} = 19 - 6x.$
25. $\dfrac{1}{x}\left\{\dfrac{4x-3}{3} - \dfrac{3x-4}{4}\right\} - \frac{2}{3}\left\{\dfrac{x-2}{x+1} - \dfrac{2x-1}{x-1}\right\} + \frac{25}{36} = 0.$
26. $(x^2 - 5x)^2 - 8(x^2 - 5x) = 84.$
27. $(x^2 + 6x)^2 - 2(x^2 + 6x) = 35.$
28. $\frac{3}{2}n^2 = 10 - \frac{1}{4}n.$
29. $9n^4 = 23n^2 + 12.$
30. $3y^2 = 36 + 4\sqrt{y^2 - 7}.$
31. $104 + 11\sqrt{y^3} = 3y^3.$
32. $y^2 + a^2 + y = 2ay + a + 6.$
33. $x^2 - 1 = (1 - x)\sqrt{2} - 4x.$

EXERCISE 100.

1. Find two consecutive numbers the sum of whose squares is 61.

2. There are two consecutive numbers, such that if the larger be added to the square of the less the sum will be 57. Find the numbers.

3. There are two numbers whose difference is 3, and if twice the square of the larger be added to 3 times the smaller, the sum is 56. Find them.

4. Seven times a certain number is one less than the square of the number next larger than the original number. Find the number.

5. A gentleman is 3 years older than his brother, but twice the product of their ages is 17 more than 21 times the sum of their ages. How old is he?

6. If a train had traveled 6 miles an hour faster, it would have required 1 hour less to run 180 miles. How fast did it travel?

7. Two numbers when added produce 5.7, and when multiplied produce 8. What are they?

8. What are the two parts of 18 whose product exceeds 8 times their difference by 1?

9. A gentleman distributed among some boys $9; if he had begun by giving each boy 5 cents more, 6 of them would have received nothing. How many boys were there?

10. A cistern is filled by two pipes in 18 minutes; by the greater alone it can be filled in 15 minutes less than by the smaller. Find the time required to fill it by each.

11. A certain number of eggs cost a dollar, but if there had been 10 more eggs at the same price, they would have cost 6 cents a dozen less. What was the price of a dozen eggs?

12. One number is $\frac{2}{3}$ of another, and their product, plus their sum, is 69. Find the numbers.

13. Find two numbers whose product is 90 and quotient $2\frac{1}{2}$.

14. Find two numbers whose difference is 4 and the sum of whose squares is 170.

15. Find two numbers whose product is 42, such that if the larger be divided by the less, the quotient is 4 and the remainder 2.

[Let x and $\frac{42}{x}$ represent the numbers.]

16. A number of boys bought a boat, each paying as many dollars as there were boys in the party; had there been 5 boys more, and each paid $\frac{3}{5}$ as much as he did pay, they would have lacked $10 of the price of the boat. How many boys were there?

17. A company of gentlemen agreed to buy a boat for $7200, but 3 of their number died, and each survivor was obliged to contribute $400 more than he otherwise would have done. How many men were there?

18. Divide the number 12 into two parts, such that the sum of the fractions obtained by dividing 12 by the parts shall be $\frac{64}{15}$.

19. The length of a certain rectangle is twice its width, and it has the same area as another, $1\frac{1}{3}$ times as wide, and shorter by $4\frac{1}{2}$ feet. Find its length.

20. A rectangular lot is 8 rods long and 6 rods wide, and is surrounded by a drive of uniform width which occupies $\frac{2}{3}$ as much area as the lot. Required the width of the drive.

21. A rectangular lot 20 by 15 rods is surrounded by a fence, within which is a drive occupying as much area as the rest of the lot. Find its width.

22. A number of two figures has the units' digit double the tens' digit, but the product of this number and the one obtained by inverting the order of the figures is 1008. Find the number.

23. A cistern can be filled by 2 pipes in 1 hour and $33\frac{3}{4}$

minutes, but the larger alone can fill it in 1 hour and 40 minutes less than the smaller one. Find the time required by the less.

24. The left-hand digit of a certain number of two figures is $\frac{3}{2}$ of the right digit. If the product of this number and the number obtained by inverting the order of the digits be increased by twice the original number, the sum is 800. Find the number.

25. A man can row down a stream 16 miles and back in 10 hours. If the stream runs 3 miles an hour, find his rate of rowing in calm water.

26. Two trains run at uniform rates over the same 120 miles of rail; one of them goes 5 miles an hour faster than the other, and takes 20 minutes less time to run this distance. Find the rate of the faster train.

27. A and B accomplish a certain task in a certain time, but if each were to do half the work, A would work $2\frac{1}{2}$ days more, and B, $1\frac{1}{2}$ days less than if they work together till the work is completed. Find the time required for each to do it.

28. If a carriage wheel 11 feet in circumference took $\frac{1}{12}$ of a second less to revolve, the rate of the carriage would be 1 mile more per hour. At what rate is the carriage traveling?

Solve—

29. $\frac{x}{a-x} + \frac{a+b}{x} = \frac{a}{a-x}.$

30. $2cx^2 + 2a^2(x+c) = ax(x+5c).$

31. $a(b-c)x^2 + b(c-a)x + c(a-b) = 0.$

32. $(4a^2 - 9b^2)(x^2+1) = 2x(4a^2 + 9b^2).$

33. $\frac{a-b+x}{a+b+x} + \frac{a+b}{x+b} = 2.$

34. $\frac{a+4b}{x+2b} - \frac{a-4b}{x-2b} = \frac{4b}{a}.$

35. $\frac{x^2}{a+b} - \left(1 + \frac{1}{ab}\right)x + \frac{1}{a} + \frac{1}{b} = 0.$

CHAPTER XX.

SIMULTANEOUS QUADRATIC EQUATIONS.

258. The General Problem. If the relations of two unknown numbers to known numbers be given in the shape of two quadratic equations, the problem is to combine these relations so as to obtain simpler ones which will show directly the values of the unknown numbers. This can be done in certain special cases only, if we limit the work to methods already given for solving quadratic equations.

259. A Homogeneous Equation is one in which all the terms, containing an unknown quantity, are of the same degree. Thus,

$$3x^2y - 5xy^2 + y^3 = 18$$

is homogeneous, and of the third degree.

CASE I.

260. When One Equation is of the First Degree, the Other of the Second.

Two simultaneous equations of the kind just specified *may always be solved by the method of substitution.*

Ex. Solve
$$\begin{cases} 2x - 3y = 2 \quad \ldots\ldots\ldots\ldots \quad (1) \\ x^2 - 2xy = -7 \quad \ldots\ldots\ldots \quad (2) \end{cases}$$

From (1),
$$y = \frac{2x-2}{3} \quad \ldots\ldots\ldots \quad (3)$$

Substitute for y in (2),

$$x^2 - 2x\left(\frac{2x-2}{3}\right) = -7,$$

Hence,
$$3x^2 - 4x^2 + 4x = -21$$
$$x^2 - 4x = 21$$
$$x = -3,\ 7$$

Substitute for x in (3),
$$y = -\tfrac{8}{3},\ 4$$
Roots.

It is to be observed that corresponding values of x and y must be used together. Thus, when $x = -3$, y must $= -\frac{8}{3}$, and not 4. Likewise the values, 7 and 4, go together.

EXERCISE 101.

Find the values of x and y—

1. $3x^2 - 2y^2 = -5.$
   $x + y - 3 = 0.$

2. $x - 2y = 3.$
   $x^2 + 4y^2 = 17.$

3. $2x^2 + xy = 2.$
   $3x + y = 3.$

4. $x^2 - 3y^2 = 1.$
   $x + 2y = 4.$

5. $x - 3y = 1.$
   $7xy - x^2 = 12.$

6. $2x + y + 3 = 0.$
   $3x^2 - 7y^2 = 5.$

7. $2x + 5y = 1.$
   $2x^2 + 3xy = 9.$

8. $\frac{1}{3}x - \frac{1}{2}y = \frac{1}{3}.$
   $(x - y)^2 = y^2 - 7.$

9. $x^2 - 3xy + 2y^2 = 0.$
   $2x + 3y = 7.$

10. $4y^2 + 4y = 4x - 13.$
    $10y - 2x = 1.$

11. $9y^2 - 6y - 5 = 3x.$
    $9y + x + 5 = 0.$

12. $\frac{3}{y} - \frac{2}{x} = \frac{9}{xy}.$
    $\frac{2x}{y} + \frac{10}{xy} - \frac{3y}{x} = 5.$

13. $\frac{3y}{2x} - \frac{4x}{3y} = 1.$
    $3y + 4x = 6.$

14. $3x - 5y - 1 = 0.$
    $2x^2 + 3xy - 5y^2 - 6x + 7y = 4.$

15. $4x^2 - 4xy = y^2 + x + 3y - 1.$
    $4x - 2 - 5y = 0.$

CASE II.

261. When both Equations are Homogeneous and of the Second Degree.

Two simultaneous quadratic equations of this kind can *always be solved by the substitution $y = vx$.vx.*

Ex. Solve
$$x^2 - xy + y^2 = 21.$$
$$y^2 - 2xy = -15.$$

Substitute $y = vx$, $x^2 - vx^2 + v^2x^2 = 21 \ldots\ldots\ldots (1)$

$$v^2x^2 - 2vx^2 = -15 \ldots\ldots\ldots (2)$$

From (1), $$x^2 = \frac{21}{1 - v + v^2} \ldots\ldots\ldots (3)$$

From (2), $$x^2 = \frac{-15}{v^2 - 2v} \ldots\ldots\ldots (4)$$

Equate the values of x^2 in (3) and (4),

$$\frac{21}{1 - v + v^2} = \frac{-15}{v^2 - 2v}$$

Hence, $$21v^2 - 42v = -15 + 15v - 15v^2$$
$$36v^2 - 57v = -15$$
$$12v^2 - 19v = -5 \qquad \therefore v = \tfrac{5}{4}, \tfrac{1}{3}$$

Substitute for v its values in (3),

$$x^2 = \frac{21}{1 - \frac{5}{4} + \frac{25}{16}}, \quad \text{or} \quad \frac{21}{1 - \frac{1}{3} + \frac{1}{9}}$$

Hence, $$x = \pm 4, \quad \text{or} \quad \pm 3\sqrt{3}$$

Since $y = vx$, multiply each value of x by the corresponding value of v.

$$\therefore y = (\pm 4)\tfrac{5}{4} = \pm 5,$$
$$y = (\pm 3\sqrt{3})\tfrac{1}{3} = \pm\sqrt{3}.$$

EXERCISE 102.

Find the values of x and y—

1. $x^2 + 3xy = 28.$
   $xy + 4y^2 = 8.$
2. $2x^2 + xy = 15.$
   $x^2 - y^2 = 8.$
3. $x^2 + 3xy = 7.$
   $y^2 + xy = 6.$
4. $2x^2 - 3y^2 = 6.$
   $3xy - 4y^2 = 2.$
5. $2x^2 - y^2 = 46.$
   $xy + y^2 = 14.$
6. $3x^2 + y^2 = 12.$
   $5xy - 4x^2 = 11.$
7. $2y^2 - 4xy + 3x^2 = 17.$
   $y^2 - x^2 = 16.$
8. $x^2 + xy + 2y^2 = 74.$
   $2x^2 + 2xy + y^2 = 73.$

9. $2x^2 + 3xy + y^2 = 14.$
$3x^2 + 2xy - 4y^2 = 9.$

10. $4xy - x^2 = 5.$
$13x^2 - 31xy + 16y^2 = 2\frac{1}{2}.$

11. $x^2 + xy + 2y^2 = 44.$
$2x^2 - xy + y^2 = 16.$

12. $2x^2 - 7xy - 2y^2 = 5.$
$3xy - x^2 + 6y^2 = 44.$

SPECIAL METHODS OF SOLVING SIMULTANEOUS QUADRATICS.

262. The methods of Cases I. and II. are the only general methods which can be used in solving all simultaneous quadratic equations of a given class. Besides these, however, there are certain special methods which enable us to solve important particular examples.

Examples which come directly under Cases I. and II. are often solved more advantageously by one of these special methods.

The special methods apply with particular advantage to what are called symmetrical equations.

263. A **Symmetrical Equation** is one in which, if y be substituted for x, and x for y, the resulting equation is identical with the original equation.

Thus, each of the following is a symmetrical equation:

$$x^3 + 3x^2y^2 + y^3 = 18,$$

$$x + y = 12, \qquad xy = 6.$$

264. I. **Addition and Subtraction Method** (often in connection with multiplication and division). In this method the object is to find, *first, the values of $x + y$, and $x - y$, and then the values of x and y themselves.*

Ex. 1. Solve
$$\begin{cases} x + y = 7 & \dots\dots\dots (1) \\ xy = 12 & \dots\dots\dots (2) \end{cases}$$

Here we have the value of $x + y$ given, and the object is to find the value of $x - y$.

Square (1), $\qquad x^2 + 2xy + y^2 = 49 \quad \dots\dots\dots (3)$

Multiply (2) by 4, $\qquad 4xy = 48 \quad \dots\dots\dots (4)$

[illegible] . . . . (5)

[illegible] . . . . . . (6)

[illegible] Roots.

[illegible] . . . . (1)

[illegible] . . . . (2)

[illegible] . . . . (3)

[illegible] . . . . (4)

[illegible] . . . . (5)

[illegible] Roots.

Ex. [illegible] Solve [illegible] . . . . (1)

[illegible] $= 61$ . . . . (2)

[illegible] . . . . (3)

[illegible] . . . . (4)

[illegible] $= 1$ . . . . (5)

Hence, $\frac{1}{x} - \frac{1}{y} = \pm 1$

But, from (1), $\frac{1}{x} + \frac{1}{y} = 11$

Hence, adding, $\frac{2}{x} = 12, 10$

$$\therefore x = \tfrac{1}{6}, \tfrac{1}{5} \quad y = \tfrac{1}{5}, \tfrac{1}{6} \quad \textit{Roots.}$$

EXERCISE 103.

Find the values of x and y—

1. $x+y=13.$
   $xy=36.$
2. $x^2+y^2=25.$
   $x+y=1.$
3. $x+y=-10.$
   $xy=21.$
4. $x^2+xy+y^2=21.$
   $x+y=-1.$
5. $x^2-xy+y^2=37.$
   $x^2+xy+y^2=79.$
6. $x^2+y^2=2\frac{1}{2}.$
   $3xy=2\frac{1}{4}.$
7. $x+y+1=0.$
   $xy+3\frac{1}{9}=0.$
8. $x^3+y^3=9.$
   $x+y=3.$
9. $x^3+y^3=37.$
   $x+y=1.$
10. $x^3+y^3-218=0.$
    $x^2-xy+y^2=109.$
11. $x^2+3xy+y^2=-2\frac{3}{4}.$
    $x^2-xy+y^2=12\frac{1}{4}.$
12. $xy-6a^2=0.$
    $x^2+y^2=xy+7a^2.$
13. $x^3+y^3=2a^3+6a.$
    $x^2-xy+y^2=a^2+3.$
14. $\frac{x}{y}+\frac{y}{x}=2\frac{1}{6}.$
    $x+y=5.$
15. $x^3+y^3=224.$
    $x^2y+xy^2=96.$
16. $\frac{1}{x^2}+\frac{1}{y^2}=13.$
    $\frac{1}{xy}-6=0.$
17. $\frac{1}{x^3}+\frac{1}{y^3}=3\frac{1}{2}.$
    $\frac{1}{x}+\frac{1}{y}=2.$
18. $x^3+y^3=-\frac{7}{8}xy.$
    $x+y=\frac{2}{3}.$
19. $x^4+x^2y^2+y^4=4\frac{4}{27}.$
    $x^2+xy+y^2=1\frac{1}{3}.$

Solve also by the same method—

20. $x^2+y^2=\frac{17}{2}.$
    $x-y=4.$
21. $x^3-y^3=98.$
    $x-y=2.$
22. $x^2+y^2=5(a^2+b^2).$
    $y-x=a+3b.$
23. $3x^2+5xy+3y^2=13.$
    $5x^2+3xy+5y^2=27.$

Subtract (4) from (3), $x^2 - 2xy + y^2 = 1 \ldots\ldots\ldots\ldots (5)$

Extract square root of (5), $x - y = \pm 1 \ldots\ldots\ldots (6)$

Add (1) and (6), divide by 2. $x = 4$ or 3

Subtract (6) from (1), divide by 2, $y = 3$ or 4 } *Roots.*

Ex. 2. Solve

$$\begin{cases} x^3 + y^3 = 65 \ldots\ldots\ldots\ldots (1) \\ x + y = 5 \ldots\ldots\ldots\ldots (2) \end{cases}$$

Divide (1) by (2), $x^2 - xy + y^2 = 13 \ldots\ldots\ldots\ldots (3)$

Square (2), $x^2 + 2xy + y^2 = 25 \ldots\ldots\ldots\ldots (4)$

Subtract (3) from (4), $3xy = 12$

Hence, $xy = 4 \ldots\ldots\ldots\ldots (5)$

Subtract (5) from (3), $x^2 - 2xy + y^2 = 9$

$\therefore x - y = \pm 3$

But $x + y = 5$

Hence, $x = 4, 1$; $y = 1, 4$ } *Roots.*

Ex. 3. Solve

$$\begin{cases} \dfrac{1}{x} + \dfrac{1}{y} = 11 \ldots\ldots\ldots\ldots (1) \\ \dfrac{1}{x^2} + \dfrac{1}{y^2} = 61 \ldots\ldots\ldots\ldots (2) \end{cases}$$

Squaring (1), $\dfrac{1}{x^2} + \dfrac{2}{xy} + \dfrac{1}{y^2} = 121 \ldots\ldots\ldots\ldots (3)$

Subtracting (2) from (3), $\dfrac{2}{xy} = 60 \ldots\ldots\ldots\ldots (4)$

Subtracting (4) from (2), $\dfrac{1}{x^2} - \dfrac{2}{xy} + \dfrac{1}{y^2} = 1 \ldots\ldots\ldots\ldots (5)$

Hence, $\dfrac{1}{x} - \dfrac{1}{y} = \pm 1$

But, from (1), $\dfrac{1}{x} + \dfrac{1}{y} = 11$

Hence, adding, $\dfrac{2}{x} = 12, 10$

$\therefore x = \frac{1}{6}, \frac{1}{5}$; $y = \frac{1}{5}, \frac{1}{6}$ } *Roots.*

EXERCISE 103.

Find the values of x and y—

1. $x + y = 13.$
   $xy = 36.$
2. $x^2 + y^2 = 25.$
   $x + y = 1.$
3. $x + y = -10.$
   $xy = 21.$
4. $x^2 + xy + y^2 = 21.$
   $x + y = -1.$
5. $x^2 - xy + y^2 = 37.$
   $x^2 + xy + y^2 = 79.$
6. $x^2 + y^2 = 2\frac{1}{2}.$
   $3xy = 2\frac{1}{4}.$
7. $x + y + 1 = 0.$
   $xy + 3\frac{1}{9} = 0.$
8. $x^3 + y^3 = 9.$
   $x + y = 3.$
9. $x^3 + y^3 = 37.$
   $x + y = 1.$
10. $x^3 + y^3 - 218 = 0.$
    $x^2 - xy + y^2 = 109.$
11. $x^2 + 3xy + y^2 = -2\frac{3}{4}.$
    $x^2 - xy + y^2 = 12\frac{1}{4}.$
12. $xy - 6a^2 = 0.$
    $x^2 + y^2 = xy + 7a^2.$
13. $x^3 + y^3 = 2a^3 + 6a.$
    $x^2 - xy + y^2 = a^2 + 3.$
14. $\frac{x}{y} + \frac{y}{x} = 2\frac{1}{6}.$
    $x + y = 5.$
15. $x^3 + y^3 = 224.$
    $x^2y + xy^2 = 96.$
16. $\frac{1}{x^2} + \frac{1}{y^2} = 13.$
    $\frac{1}{xy} - 6 = 0.$
17. $\frac{1}{x^3} + \frac{1}{y^3} = 3\frac{1}{2}.$
    $\frac{1}{x} + \frac{1}{y} = 2.$
18. $x^3 + y^3 = -\frac{7}{8}xy.$
    $x + y = \frac{2}{3}.$
19. $x^4 + x^2y^2 + y^4 = 4\frac{4}{27}.$
    $x^2 + xy + y^2 = 1\frac{1}{3}.$

Solve also by the same method—

20. $x^2 + y^2 = \frac{17}{2}.$
    $x - y = 4.$
21. $x^3 - y^3 = 98.$
    $x - y = 2.$
22. $x^2 + y^2 = 5(a^2 + b^2).$
    $y - x = a + 3b.$
23. $3x^2 + 5xy + 3y^2 = 13.$
    $5x^2 + 3xy + 5y^2 = 27.$

265. II. Solution by the Substitutions, $x = u + v$ and $y = u - v$.

Ex. Solve
$$\begin{cases} x^5 + y^5 = 242 \quad \ldots\ldots\ldots (1) \\ x + y = 2 \quad \ldots\ldots\ldots (2) \end{cases}$$

Substitute in (1) and (2), $\quad x = u + v, \quad y = u - v$

From (1), $\quad 2u^5 + 20u^3v^2 + 10uv^4 = 242 \ldots\ldots\ldots (3)$

From (2), $\quad 2u = 2 \ldots\ldots\ldots (4)$

Divide (3) and (4) by 2, and substitute in (3) for u, i. e., $u = 1$,

$$1 + 10v^2 + 5v^4 = 121$$

Hence, $\quad v^4 + 2v^2 = 24$

$$v = \pm 2, \ \pm\sqrt{-6}$$

But $\quad u = 1$

Hence,
$$\left.\begin{aligned} x &= u + v = 3, \ -1, \ 1 \pm \sqrt{-6} \\ y &= u - v = -1, \ 3, \ 1 \mp \sqrt{-6} \end{aligned}\right\} \textit{Roots.}$$

266. III. Use of Compound Unknown Quantities. It is often expedient to *consider some expression, as the sum, difference, or product of the unknown quantities, as a single unknown quantity, and find its value, and hence the value of the unknown quantities themselves.*

Ex. Solve
$$\begin{cases} x^2 + y^2 = 18 - x - y \quad \ldots\ldots (1) \\ xy = 6 \quad \ldots\ldots\ldots (2) \end{cases}$$

Add $2xy = 12$ to (1).

Then $\quad x^2 + 2xy + y^2 = 30 - x - y \ldots\ldots (3)$

Let $\quad x + y = v,$

Then from (3), $\quad v^2 = 30 - v$

$$v^2 + v = 30$$

$$v = -6, \ 5$$

Hence, $\quad x + y = -6,$

$$xy = 6.$$

$$\therefore x = -3 \pm \sqrt{3},$$

$$y = -3 \mp \sqrt{3}.$$

also $\quad x + y = 5,$

$$xy = 6.$$

$$\therefore x = 3, \ 2,$$

$$y = 2, \ 3.$$

EXERCISE 103 (A).

Find the values of x and y.

1. $x^5+y^5=244$; $x+y=4$. $x=3, 1, 2\pm 3\sqrt{-1}$; $y=1, 3, 2\mp 3\sqrt{-1}$.
2. $x^2+y^2+x+y=24$; $xy=-12$. $x=3, -4, \pm 2\sqrt{3}$; $y=-4, 3, \mp 2\sqrt{3}$.
3. $x+y+\sqrt{x+y}=6$; $xy=3$. $x=1, 3, \frac{1}{2}(9\pm\sqrt{69})$; $y=3, 1, \frac{1}{2}(9\mp\sqrt{69})$.
4. $x^2y^2+xy=6$; $x+2y=-5$. $x=1, -6, -4, -1$; $y=-3, \frac{1}{2}, -\frac{1}{2}, -2$.
5. $x+y=25$; $\sqrt{x}+\sqrt{y}=x-y$. $x=9, 16$; $y=16, 9$.
6. $y+\sqrt{x^2-9}=6$; $\sqrt{x+3}-\sqrt{x-3}=\sqrt{y}$. $x=3, 5$; $y=6, 2$.

7. $x^4+y^4=97$.
$x+y=5$.

8. $x^2+y^2=x-y+50$.
$xy=24$.

9. $x^2y^2+7xy=-6$.
$5x^2+xy=4$.

10.* $(x-y)^2-3(x-y)=40$.
$x^2y^2-3xy=54$.

11. $x^2+y^2+x+5y=6$.
$xy-2y=-2$.

12.† $x^{\frac{1}{2}}y^{\frac{1}{2}}\left(x^{\frac{1}{2}}+y^{\frac{1}{2}}\right)=70$.
$x^{\frac{1}{2}}y^{\frac{1}{2}}+x^{\frac{1}{2}}+y^{\frac{1}{2}}=17$.

EXERCISE 104.

GENERAL EXERCISE.

Find the values of x and y—

1. $2x-5y=0$.
$x^2-3y^2=13$.

2. $x+y=2$.
$\frac{2}{x}+\frac{3}{y}=6$.

3. $2x^2-xy=28$.
$x^2+2y^2=18$.

4. $3x^2-xy-5y^2=5$.
$3x-5y=1$.

5. $x^3+y^3=91$.
$x+y=1$.

6. $x^2+3y^2=28$.
$x^2+xy+2y^2=16$.

7. $xy+2x=5$.
$2xy-y=3$.

8. $3x^2+xy+y^2=15$.
$31xy-5y^2-3x^2=45$.

9. $\frac{4x}{3y}+\frac{2y}{5x}=\frac{34}{15}$.
$2x-5y=-4$.

---

* There are eight roots for x and eight for y.

† Consider, first, that the unknown quantities are
$u=\sqrt{xy}$ and $v=\sqrt{x}+\sqrt{y}$.

10. $x^2 + 2x - y = 5.$
$2x^2 - 3x + 2y = 8.$

11. $(x+2)(2y-1) = 35.$
$xy - x - y = 7.$

12. $x^2y^2 - 5xy + 6 = 0.$
$5x + 3y = 14.$

13. $(x+y)^2 - (x+y) = 20.$
$2x^2 - 3x + 4y = 14.$

14. $x^2 + y^2 + x + 3y = 18.$
$xy - y = 12.$

15. $x^3 - y^3 = -3xy.$
$x - y = 2.$

16. $\frac{x^2}{y} + \frac{y^2}{x} = \frac{9}{2}.$
$\frac{1}{x} + \frac{1}{y} = 4\frac{1}{2}.$

17. $x^{-1} + y^{-1} = 2.$
$x^{-2} + y^{-2} = 2\frac{1}{2}.$

18. $x^2 + y^2 + x + y = 14.$
$xy + x + y = -5.$

19. $3x^2 - 35 = 5xy - 7y^2.$
$2x^2 - 35 = y^2 - xy.$

20. $x^4 + y^4 = 17.$
$x + y = 3.$

21. $x^5 + y^5 = 211.$
$x + y = 1.$

22. $x^4 + y^4 = 82.$
$x + y = 2.$

23. $x^4 + y^4 = 257.$
$x - y = 5.$

24. $x^4 + y^4 = 17.$
$xy = 2.$

25. $x^2 + y^2 = xy + 7.$
$x - y = xy - 5.$

26. $x^2 + y^2 = xy + 13.$
$x + y = xy - 5.$

27. $x^2 = 4(a^2 + b^2 - y^2).$
$xy = 2ab.$

28. $\frac{1}{x} + \frac{1}{y} = 5.$
$\frac{1}{x+1} + \frac{1}{y+1} = \frac{17}{12}.$

29. $\frac{x-y}{x+y} - \frac{x+y}{x-y} = \frac{5}{6}.$
$2x + 5y = 5.$

30. $3x - 4y = a.$
$2x^2 - 3y^2 + ay = 8a^2.$

31. $x - 4 = y(x-2).$
$y - 8 = x(y-2).$

32. $2x^{-1} - 5y^{-1} = 4.$
$x^{-2} + 2x^{-1}y^{-1} - y^{-2} = 1.$

33. $xy + x + y = 5.$
$x^2y + xy^2 = -84.$

34. $ax + by = 0.$
$(ax-2)(by+3) = -2.$

35. $\frac{x}{2a} + \frac{y}{3b} = 3\frac{1}{2}.$
$\frac{4a}{x} + \frac{5b}{y} = 4.$

36. $a(x-a)=b(y-b)$.
$xy=ax+by$.

37. $x^4+a^4y^4=\frac{17}{4}a^2x^2y^2$.
$3x+ay=5$.

38. $x+y=65$.
$\sqrt[3]{x}+\sqrt[3]{y}=5$.

39. $x-y=\sqrt{x}+\sqrt{y}$.
$x^{\frac{3}{2}}-y^{\frac{3}{2}}=37$.

40. $x^3+x=9y$.
$x^2+1=6y$.

41. $x^2+y^2=3xy-4$.
$x^4+y^4=272$.

42. $x+\sqrt{xy}+y=14$.
$x^2+xy+y^2=84$.

43. $x^2+4y^2+80=15x+30y$.
$xy-6=0$.

EXERCISE 105.

1. The sum of the squares of two numbers is 58, and their product is 21. Find the numbers.

2. Find two numbers whose sum increased by three times their product is 83, and of which 3 times the less exceeds the larger by 1.

3. The area of a rectangle is 84 square feet, and the distance around it (perimeter) is 38 feet. Find the length and breadth (dimensions) of the rectangle.

4. If the dimensions of a rectangular field were each increased by 3 rods, its area would be 140 sq. rds.; but if its width were increased by 8 rods and length diminished by 2, its area would be 135 sq. rds. Find its actual dimensions.

5. A rectangular lot containing 270 square rods is surrounded by a road 1 rod wide; the area of the road is 70 square rods. Find the dimensions of the field.

6. A certain number of two figures when multiplied by the left digit becomes 56; but if by the right digit, 224. Required the number.

7. A hall of 90 square yards can be paved with 720 rectangular tiles of a certain size, but if each tile were 3 inches shorter and 3 inches wider, it would require 648 tiles. What is the size of each tile?

8. A merchant bought a number of yards of cloth for $140; he kept 8 yards and sold the remainder at an advance of 1\frac{1}{2}$ a yard, and gained $20. How many yards did he buy?

9. Two farmers, A and B, have together 30 calves, which they sell for $336, A receiving as many dollars for each of his as B had calves; if they had each sold his calves for as many dollars apiece as the other received for each of his, they would have received only $324. How many calves had A, and at what price did he sell them?

10. The sum of the numerator and denominator of a certain fraction is 8, and if 2$\frac{1}{2}$ be added to each term of the fraction, its value will be increased by $\frac{2}{15}$. What is the fraction?

11. Two trains traveling toward each other left, at the same time, two stations 240 miles apart; each reached the station from which the other started, the one 3$\frac{3}{5}$ hours, and the other 1$\frac{3}{5}$ hours, after they met. Required their rates of running.

12. A crew rowing at $\frac{3}{5}$ their usual rate took 32 hours to row down stream 48 miles and back to starting-place; had they rowed at their usual rate it would have taken 18 hours for same circuit. Find their rate and that of the stream.

13. Two square plots contain together 610 square feet, but a third plot, which is a foot shorter than a side of the larger square, and a foot wider than the less, contains 280 square feet. What are the sides of the two squares?

14. The fore wheel of a carriage makes 28 revolutions more than the hind wheel in going 560 yards, but if the circumference of each wheel were increased by 2 feet, the difference would be only 20 revolutions. What is the circumference of each wheel?

15. A number of foot-balls cost $100, but if they had cost $1 apiece less, I should have had as many more for the money as the number of dollars paid for each ball. Find the cost of each.

16. Find two fractions whose sum is equal to their product and the difference of whose squares is $\frac{5}{6}$ of their product.

CHAPTER XXI.

GENERAL PROPERTIES OF QUADRATIC EQUATIONS.

267. Two General Forms of the Quadratic Equation. Any quadratic equation may be reduced to the general form

$$ax^2 + bx + c = 0 \quad . \quad . \quad . \quad . \quad . \quad . \quad . \quad . \quad \text{I.}$$

Factoring, this becomes

$$a\left(x^2 + \frac{b}{a}x + \frac{c}{a}\right) = 0.$$

Dividing by a and denoting $\frac{b}{a}$ by p, and $\frac{c}{a}$ by q, we obtain

$$x^2 + px + q = 0 \quad . \quad . \quad . \quad . \quad . \quad . \quad . \quad . \quad \text{II.}$$

When a, b, c, or p and q are given, we can often infer at once, without the labor of solving the equation, important facts concerning the roots of an equation. Or if, on the other hand, the roots only of an equation be given, or some property of them, we can at once infer what the equation will be.

PROPERTIES OF $x^2 + px + q = 0$.

268. Relation of the Roots of $x^2 + px + q = 0$ to the Coefficients p and q.

Solving $x^2 + px + q = 0$, and denoting its roots by r_1, r_2, we obtain

$$r_1 = -\frac{p}{2} + \frac{\sqrt{p^2 - 4q}}{2}$$

$$r_2 = -\frac{p}{2} - \frac{\sqrt{p^2 - 4q}}{2}$$

Adding, $\quad r_1 + r_2 = -p.$

Multiplying, $\quad r_1 r_2 = q.$

Hence,

(1) *The sum of the roots of* $x^2 + px + q = 0$ *equals* $-p$, *or the coefficient of* x *with the sign changed;*

(2) *The product of the roots equals the known term* q.

$$\text{Ex. In } x^2 - 5x + 6 = 0$$

the roots are found to be 3, 2.

The sum of these with the sign changed is -5, the coefficient of x; the product is 6, the known term.

This relation is used in the factorial method of solving quadratic equations (See Art. 250.)

269. Formation of a Quadratic Equation, the Roots Only being Given.

If the two roots of a quadratic equation be given, the equation may at once be written out by the use of the relation between the roots and coefficients determined in Art. 268.

Ex. Form the quadratic equation whose roots are 5, and -2.

The sum of 5 and -2 is $+3$; hence, the coefficient of x in the required equation is -3.

The product of 5 and -2 is -10, the third term; hence, $x^2 - 3x - 10 = 0$ is the required equation.

This equation might have been formed also by subtracting each root from x, multiplying together the binomials thus formed, and letting the product $= 0$.

Thus,
$$(x-5)(x+2) = 0,$$
or
$$x^2 - 3x - 10 = 0.$$

270. Factoring a Quadratic Expression. Any quadratic expression may be factored by letting the given expression equal zero, and using the property stated in Art. 268.

Ex. 1. Factor $x^2 - 4x + 2$.

Solving the equation, $x^2 - 4x + 2 = 0$

$$x = 2 \pm \sqrt{2}$$

$\therefore\ x^2 - 4x + 2 = (x - 2 - \sqrt{2})(x - 2 + \sqrt{2})$, *Factors.*

Ex. 2. Factor $3x^2 - 4x + 5$.

Take $3(x^2 - \frac{4}{3}x + \frac{5}{3}) = 0$

Solve $x^2 - \frac{4}{3}x + \frac{5}{3} = 0$,

Whence $x = \dfrac{2 \pm \sqrt{-11}}{3}$.

Hence, the factors of $3x^2 - 4x + 5$ are

$$3\left(x - \frac{2 + \sqrt{-11}}{3}\right)\left(x - \frac{2 - \sqrt{-11}}{3}\right).$$

EXERCISE 106.

Find by inspection the sum and product of the roots in each of the following equations:

1. $x^2 + 3x + 5 = 0$.
2. $x^2 - x + 7 = 0$.
3. $x^2 - 5x = 10$.
4. $2x^2 - 6x - 3 = 0$.
5. $6x^2 - x = 1$.
6. $a^2x^2 - ax + 2 = 0$.
7. $5x - 4x^2 = 1$.
8. $3 - 7x = 11x^2$.
9. $4x^2 - ax + x = a^2$.
10. $1 - 2cx - 2ax^2 = 3c$.

Form the equations whose roots are—

11. 2, 3.
12. 2, -1.
13. 3, -2.
14. -1, -5.
15. $\frac{1}{2}$, 6.
16. -1, $-\frac{2}{3}$.
17. -2, $-\frac{3}{4}$.
18. $\frac{3}{2}$, $-\frac{2}{3}$.
19. $1\frac{1}{2}$, $-2\frac{1}{3}$.
20. $1 + a$, $1 - a$.
21. ab, $-a$.
22. $\dfrac{a}{b}$, $-\dfrac{b}{a}$.
23. $1 + \sqrt{2}$, $1 - \sqrt{2}$.
24. $-3 \pm \sqrt{3}$.
25. $\dfrac{2 \pm \sqrt{2}}{2}$.
26. $\dfrac{1 \pm \sqrt{-1}}{2}$.
27. $\dfrac{-2 \pm \sqrt{-2}}{2}$.
28. $\frac{1}{2}a \pm c\sqrt{-b}$.

Factor—

29. $3x^2 - 10x - 8$.	34. $x^2 + 14 - 6x$.
30. $24x^2 + 2x - 15$.	35. $25x^2 + 2 - 30x$.
31. $x^2 + 2x - 1$.	36. $4x^2 - 8x + 7$.
32. $x^2 - 4x + 1$.	37. $5x^2 + 6x + 7$.
33. $x^2 - x - 1$.	38. $3x - 3x^2 - 1$.

PROPERTIES OF $ax^2 + bx + c = 0$.

271. Character of the Roots Inferred from the Coefficients. It is important to be able to infer at once from the nature of the given coefficients, a, b, c, of an equation in the form $ax^2 + bx + c = 0$, whether the roots of the equation be equal or unequal, real or imaginary, positive or negative.

Solving $ax^2 + bx + c = 0$, and denoting the roots by r_1, r_2, we obtain

$$r_1 = \frac{-b + \sqrt{b^2 - 4ac}}{2a}, \quad r_2 = \frac{-b - \sqrt{b^2 - 4ac}}{2a}.$$

From these expressions we infer that

I. *If* $b^2 - 4ac > 0$, *the roots are real and unequal.*

For if $b^2 - 4ac$ is a positive quantity (greater than zero), the radical $\sqrt{b^2 - 4ac}$ is real and not imaginary, and since the fraction of which it is a numerator is added to $-\frac{b}{2a}$ to form one root, and subtracted from $-\frac{b}{2a}$ to form the other root, the two roots are unequal.

The roots are also rational or irrational, according as $b^2 - 4ac$ is or is not a perfect square.

The roots are also rational if $b^2 - 4ac = 0$.

II. *If* $b^2 - 4ac = 0$, *the roots are real and equal,* since each root reduces to $-\frac{b}{2a}$.

III. *If* $b^2 - 4ac < 0$, *the two roots are imaginary.*

Since the character of the roots is thus determined by the value of $b^2 - 4ac$, this expression is termed the *discriminant* of $ax^2 + bx + c = 0$.

Ex. 1. Determine the character of the roots of the equation, $2x^2 + 7x - 15 = 0$.

We have $\qquad a = 2, \quad b = 7, \quad c = -15.$

$$\therefore b^2 - 4ac = 49 + 120 = 169.$$

Hence, the roots are *real*, *rational*, and *unequal*.

Ex. 2. Of $9x^2 - 12x + 4 = 0$.

Here $\qquad a = 9, \quad b = -12, \quad c = 4$

$$\therefore b^2 - 4ac = 144 - 144 = 0.$$

Hence, the roots are *real* and *equal*.

Ex. 3. Of $3x^2 - 4x + 2 = 0$.

Here $\qquad a = 3, \quad b = -4, \quad c = 2$

$$\therefore b^2 - 4ac = 16 - 24 = -8.$$

Hence, the roots are *imaginary*.

272. Determining Coefficients so that the Roots shall satisfy a Given Condition. It is often possible so to determine the coefficients of an equation that the roots shall satisfy a given condition.

Ex. Find the value of m for which the equation $(m-1)x^2 + mx + 2m - 3 = 0$ shall have equal roots.

By Art. 271, II., in order that the roots be equal, $b^2 - 4ac = 0$.

In the given equation, $\quad a = m - 1, \quad b = m, \quad c = 2m - 3.$

$$\therefore m^2 - 4(m-1)(2m-3) = 0$$

$$m^2 - 8m^2 + 20m - 12 = 0$$

$$7m^2 - 20m = -12$$

$$m = 2, \tfrac{6}{7}.$$

Proof. Substituting these values for m in the original equation,

$$x^2 + 2x + 1 = 0, \qquad x^2 - 6x + 9 = 0,$$

of each of which equations the roots are equal.

EXERCISE 107.

Determine, without solution, the character of the roots in each equation.

1. $x^2-5x+6=0.$
2. $3x^2-7x-2=0.$
3. $4x^2=4x-1.$
4. $3x^2+2x+1=0.$
5. $2x^2-5x+3=0.$
6. $9x^2+12x+4=0.$
7. $2x^2+5x+4=0.$
8. $2x^2+3x=5.$
9. $3x^2-1=x.$
10. $6x^2+\frac{25}{6}=10x.$
11. $x=\frac{2}{5}(x^2+1).$
12. $35x+18+12x^2=0.$
13. $\frac{1}{3}x^2=2x-3.$
14. $7x^2+1=5x.$

Determine the value of m for which the roots of each equation will be equal.

15. $2x^2-2x+m=0.$
16. $2x^2+m+x=0.$
17. $x^2+m=3x.$
18. $mx^2-5x+2=0.$
19. $5x^2+8x-m=0.$
20. $2x^2-mx+12\frac{1}{2}=0.$
21. $18x^2+6x=m.$
22. $4x^2+\frac{1}{9}=mx.$
23. $(m+1)x^2+mx=1.$
24. $(m+1)x^2+3m=12x.$
25. $(m+1)x^2+(m-1)x+m+1=0.$
26. $2mx^2+3mx-7=3x-2m-x^2.$

27. If r_1 and r_2 represent the roots of $3x^2-8x+5=0$ find without determining the actual roots, the values of:

$$r_1+r_2;\ r_1r_2;\ r_1^2+r_2^2;\ r_1-r_2;\ r_1^2-r_2^2;$$
$$r_1^3+r_2^3;\ \frac{1}{r_1}+\frac{1}{r_2};\ \frac{1}{r_1}-\frac{1}{r_2};\ \frac{1}{r_1^2}+\frac{1}{r_2^2}.$$

28. Find the values of the same expressions for the equation $2x^2-9x+7=0$. Also for the equation $6x^2-x-12=0$.

29. Find the values of the same expressions for the equation $ax^2+bx+c=0$. Also for the equation $x^2+px+q=0$.

30. If m and n represent the roots of the equation $10x^2+9x-7=0$, form that equation whose roots shall be mn and $m+n$. Form that equation whose roots shall be $m-n$ and $\frac{1}{m}+\frac{1}{n}$.

CHAPTER XXII.

RATIO AND PROPORTION.

RATIO.

273. THE **Ratio** of two algebraic quantities is their exact relation of magnitude. Thus, also, it is the indicated quotient of the one divided by the other, expressed either in the form of a fraction or by the symbol : placed between the two quantities. Thus, the ratio of a to b is expressed as $\frac{a}{b}$, or as $a : b$.

274. The **Terms** of a ratio are the two quantities compared. The first term is called the **Antecedent**. The second term is called the **Consequent**.

275. Kinds of Ratio. An **inverse** ratio is one obtained by interchanging antecedent and consequent.

Thus, the direct ratio of a to b is $a : b$; the inverse ratio of the same quantities is $b : a$.

A **compound** ratio is one formed by taking the product of the corresponding terms of two given ratios. Thus, $ac : bd$ is the ratio compounded of $a : b$ and $c : d$.

A **duplicate** ratio is formed by compounding a ratio with itself. Thus, the duplicate ratio of $a : b$ is $a^2 : b^2$.

In like manner the **triplicate** ratio of $a : b$ is $a^3 : b^3$.

276. Fundamental Property of Ratios. *If both antecedent and consequent of a ratio be multiplied or divided by the same quantity, the value of the ratio is not changed.*

For, since $$\frac{a}{b} = \frac{ma}{mb},$$

$a : b$ has the same value as $ma : mb$.

PROPORTION.

277. A **Proportion** is an expression of the equality of two or more equal ratios.

$$\text{Ex. } \frac{a}{b} = \frac{c}{d}, \text{ or } a : b = c : d.$$

278. Terms of a Proportion. The four quantities used in a proportion are called its **terms**, or **proportionals**.

The first and third terms are called the *antecedents*.

The second and fourth terms are called the *consequents*.

The first and last terms are called the *extremes*.

The second and third terms are called the *means*.

In $a : b = c : d$, d is called a *fourth proportional* to a, b, and c.

279. A **Continued Proportion** is one in which each consequent and the next antecedent are the same. Thus,

$$a : b = b : c = c : d = d : e.$$

In the simple continued proportion $a : b = b : c$, b is called a **mean proportional** between a and c; c is called a **third proportional** to a and b.

280. Fundamental Property of Proportion. For algebraic purposes the fundamental property of a proportion consisting of four quantities is, that

The product of the means is equal to the product of the extremes.

For, if $$a : b = c : d,$$

then $$\frac{a}{b} = \frac{c}{d}.$$

Multiplying by bd, $$ad = bc.$$

In like manner, if $$a : b = b : c,$$

$$b^2 = ac. \quad \therefore b = \sqrt{ac}.$$

This property enables us to convert a proportion into an equation, and to solve a given proportion by solving the equation thus obtained. (See Art. 291, Ex. 1.)

Before converting a given proportion into an equation it is important, however, first to simplify the given proportion as far as possible. For this purpose we have the following transformations, which are possible in dealing with proportions:

If four quantities are in proportion, they are in proportion by

281. I. **Alternation**; *that is, the first term is to the third as the second is to the fourth.*

For if $a : b = c : d,$

$$\frac{a}{b} = \frac{c}{d}.$$

Multiplying by $\frac{b}{c}$, $\frac{a}{c} = \frac{b}{d},$

$$\therefore\ a : c = b : d.$$

282. II. **Inversion**; *that is, the second term is to the first as the fourth is to the third.*

Given, $a : b = c : d,$

Then $\frac{a}{b} = \frac{c}{d},$

Hence, $1 \div \frac{a}{b} = 1 \div \frac{c}{d},$

$$\therefore\ \frac{b}{a} = \frac{d}{c},$$

Or, $b : a = d : c.$

283. III. **Composition**; *that is, the sum of the first and second terms is to the second as the sum of the third and fourth is to the fourth.*

Given, $a : b = c : d,$

Then $\frac{a}{b} = \frac{c}{d}.$

Add 1 to each side, $\frac{a}{b} + 1 = \frac{c}{d} + 1.$

That is, $\frac{a+b}{b} = \frac{c+d}{d},$

Or, $a + b : b = c + d : d.$

284. IV. Division; *that is, the difference of the first and second terms is to the second term as the difference of the third and fourth is to the fourth.*

Given, $a:b=c:d$,

Then $\frac{a}{b}=\frac{c}{d}$,

And $\frac{a}{b}-1=\frac{c}{d}-1.$

That is, $\frac{a-b}{b}=\frac{c-d}{d}$,

Or, $a-b:b=c-d:d.$

285. V. Composition and Division; *that is, the sum of the first two terms is to their difference as the sum of the last two terms is to their difference.*

Given, $a:b=c:d.$

By composition (Art. 283), $\frac{a+b}{b}=\frac{c+d}{d}$ . . . . . (1)

By division (Art. 284), $\frac{a-b}{b}=\frac{c-d}{d}$ . . . . . (2)

Divide (1) by (2), $\frac{a+b}{a-b}=\frac{c+d}{c-d}.$

That is, $a+b:a-b=c+d:c-d.$

286. VI. Composition of Several Equal Ratios; *that is, in a series of equal ratios, the sum of all the antecedents is to the sum of all the consequents as any one antecedent is to its consequent.*

Given, $\frac{a}{b}=\frac{c}{d}=\frac{e}{f}=\frac{g}{h}.$

Let each of the equal ratios equal r.

Then $\frac{a}{b}=r,\quad \frac{c}{d}=r,\quad \frac{e}{f}=r,\quad \frac{g}{h}=r.$

$\therefore\ a=br,\quad c=dr,\quad e=fr,\quad g=hr.$

Adding the last series of equalities,

$$a+c+e+g=(b+d+f+h)r.$$

$$\therefore \frac{a+c+e+g}{b+d+f+h}=r=\frac{a}{b}.$$

$$\therefore a+c+e+g : b+d+f+h = a : b.$$

287. VII. Product of Corresponding Terms. *In two or more proportions the products of the corresponding terms are in proportion.*

Given, $$a:b=c:d,$$
$$e:f=g:h,$$
$$j:k=l:m.$$

Then $$\frac{a}{b}=\frac{c}{d},\quad \frac{e}{f}=\frac{g}{h},\quad \frac{j}{k}=\frac{l}{m}.$$

Taking the product of corresponding members of these equations,

$$\frac{aej}{bfk}=\frac{cgl}{dhm}.$$

$$\therefore aej : bfk = cgl : dhm.$$

288. VIII. Powers and Roots. *In any proportion like powers or like roots of the terms are in proportion.*

Given, $$a:b=c:d.$$

Then $$\frac{a}{b}=\frac{c}{d}.$$

Hence, $$\frac{a^n}{b^n}=\frac{c^n}{d^n}.$$ Also, $$\frac{a^{\frac{1}{n}}}{b^{\frac{1}{n}}}=\frac{c^{\frac{1}{n}}}{d^{\frac{1}{n}}}.$$

That is, $$a^n : b^n = c^n : d^n.$$

And $$a^{\frac{1}{n}} : b^{\frac{1}{n}} = c^{\frac{1}{n}} : d^{\frac{1}{n}}.$$

289. IX. Cancellation of Factors of Terms. From Arts. 276 and 281 it is evident that *if four quantities be in proportion,* *and if the first two terms or the last two, or the first and third, or second and fourth, be multiplied or divided by the same quantity, the resulting quantities are in proportion.*

Thus, if $a : b = c : d,$

Then $ma : mb = nc : nd,$

And $ma : pb = mc : pd.$

290. Equal Products made into a Proportion; *that is, if the product of two quantities is equal to the product of two other quantities, either two may be made the means, and the other two the extremes of a proportion.*

For, if $ad = bc,$

Dividing by $bd,$ $\frac{a}{b} = \frac{c}{d}.$

$\therefore a : b = c : d.$

This property is evidently the converse of the principle stated in Art. 280.

291. Application of these Principles. The use of proportion in solving algebraic problems and determining the properties of algebraic quantities may be reduced essentially to the following:

I. *By taking the product of the means equal to the product of the extremes, a proportion may be converted into an equation, and the proportion solved by solving the equation.*

Ex. 1. Find the value of x which satisfies the proportion,

$$4x - 1 : x + 1 = 3x + 1 : 2x - 1.$$

Taking the product of the means equal to the product of the extremes,

$$(4x - 1)(2x - 1) = (x + 1)(3x + 1)$$

$$\therefore 5x^2 - 10x = 0$$

$$x = 0,\ 2.$$

292. II. *Before converting a proportion into an equation it is important to simplify the proportion, as far as possible, by use of the properties of a proportion, as Alternation, Composition, Division, etc.*

Ex. 1. Solve $x^2-2x+3:x^2+2x-3=2x^2-x-3:2x^2+x+3$.

By Composition and Division, $\dfrac{2x^2}{4x-6}=\dfrac{4x^2}{2x+6}$

Divide by $2x^2$, $\dfrac{1}{2x-3}=\dfrac{2}{x+3}$

$$\therefore x+3=4x-6$$
$$x=3$$

The factor $2x^2$ divided out also gives the roots $x=0, 0$.

Ex. 2. Solve $\dfrac{\sqrt{x+1}+\sqrt{x-1}}{\sqrt{x+1}-\sqrt{x-1}}=\dfrac{4x-1}{2}$.

By Composition and Division, $\dfrac{\sqrt{x+1}}{\sqrt{x-1}}=\dfrac{4x+1}{4x-3}$

Squaring, $\dfrac{x+1}{x-1}=\dfrac{16x^2+8x+1}{16x^2-24x+9}$

By Composition and Division, $\dfrac{x}{1}=\dfrac{16x^2-8x+5}{16x-4}$

Hence, $16x^2-4x=16x^2-8x+5$

$$x=\tfrac{5}{4}.$$

293. III. *Given some proportion (or equality of several equal ratios), as* $a:b=c:d$, *a required proportion is often readily proved by taking* $\dfrac{a}{b}=\dfrac{c}{d}=x$ *(hence,* $a=bx$, $c=dx$*), and substituting for* a *and* c *in the required proportion.*

Ex. Given, $a:b=c:d$,

Prove $2a^3+3ab^2:2a^3-3ab^2=2c^3+3cd^2:2c^3-3cd^2$.

Let $\dfrac{a}{b}=\dfrac{c}{d}=x$, $\quad\therefore a=bx,\ c=dx.$

Substitute in each ratio the values $a=bx$, $c=dx$.

I. $\dfrac{2a^3+3ab^2}{2a^3-3ab^2}=\dfrac{2b^3x^3+3b^3x}{2b^3x^3-3b^3x}=\dfrac{b^3x(2x^2+3)}{b^3x(2x^2-3)}=\dfrac{2x^2+3}{2x^2-3}$;

II. $\dfrac{2c^3+3cd^2}{2c^3-3cd^2}=\dfrac{2d^3x^3+3d^3x}{2d^3x^3-3d^3x}=\dfrac{d^3x(2x^2+3)}{d^3x(2x^2-3)}=\dfrac{2x^2+3}{2x^2-3}$.

$$\therefore \frac{2a^3+3ab^2}{2a^3-3ab^2}=\frac{2c^3+3cd^2}{2c^3-3cd^2},$$

since they are each equal to the same expression.

EXERCISE 108.

Find the ratio of x to y—

1. $7x - 3y = 4x + y$.
2. $4x - 5y : 5x - 4y = \frac{2}{3}$.
3. $\dfrac{3x - 2y}{4x - 3y} = \dfrac{a}{b}$.
4. $x^2 + 6y^2 = 5xy$.

Find a mean proportional between—

5. $3ab^2$ and $12a^3$.
6. $3\frac{1}{6}$ and $2\frac{2}{3}$.
7. $(a - x)^3$ and $(a + x)^2$.
8. $\dfrac{3x^2 - 5x - 12}{3x^2 + 5x}$ and $\dfrac{3x^2 + 4x}{3x^2 - 4x - 15}$.
9. $\dfrac{2\sqrt{6} + 5\sqrt{3}}{3\sqrt{2} - 4}$ and $\dfrac{3\sqrt{6} - 4\sqrt{3}}{8\sqrt{2} + 20}$.

Find a fourth proportional to—

10. $2a$, $3b$, $4ac$.
11. x^2, xy, $3x^2$.
12. $\frac{2}{3}$, $\frac{5}{7}$, $\frac{2}{15}$.
13. $a - 1$, a, 1.

Find a third proportional to—

14. x and 5.
15. $1\frac{9}{16}$ and $7\frac{1}{2}$.
16. $(a + 1)^2$ and $a^2 - 1$.
17. $a - \dfrac{1}{a}$ and $\dfrac{1}{a} - 1$.

Solve the equations—

18. $2x + 3 : 3x - 1 = 3x + 1 : 2x + 1$.
19. $x + 5 : 3 - x = 10 + 3x : x - 10$.
20. $3x + 5 : 5x + 11 = 7 - x : -3x$.
21. $x^2 - 4 : x^2 - x + 3 = x + 2 : 2x + 3$.
22. $x^2 + 2x - 1 : x^2 + 2x + 5 = 2x + 1 : 2x - 5$.
23. $x^3 - 3x^2 + 5 : x^3 + 3x^2 - 5 = x^2 + 2 : x^2 - 2$.
24. $2x^3 - 8x^2 - 3x + 1 : 2x^3 - 10x^2 + 3x - 1 = x^2 + 11 : x^2 - 11$.
25. $\sqrt{3x + 1} : 2\sqrt{2x - 1} = \sqrt{x - 1} : \sqrt{x + 4}$.
26. $\dfrac{3 + \sqrt{2x + 3}}{5 - \sqrt{2x + 3}} = \dfrac{4 + \sqrt{x + 1}}{4 - \sqrt{x + 1}}$.

27. $\dfrac{3a+\sqrt{4x-3a^2}}{5a-\sqrt{4x-3a^2}}=\dfrac{a+\sqrt{x+a^2}}{3a-\sqrt{x+a^2}}$.

28. $8y-6x:x+y-1=5-3x:4-y=7:4$.

29. $\begin{cases} x+y:y-x=a:1. \\ xy-3:x-1=a+2:1. \end{cases}$

If $a:b=c:d$, prove—

30. $a^2:c^2=ab:dc$.

31. $a^2:b^2=a^2+c^2:b^2+d^2$.

32. $ac:bd=(a+c)^2:(b+d)^2$.

33. $(a-c)^2:(b-d)^2=a^2+c^2:b^2+d^2$.

34. $a:b=\sqrt{a^2+3c^2}:\sqrt{b^2+3d^2}$.

35. $2a^2+3ab:3ab-4b^2=2c^2+3cd:3cd-4d^2$.

36. $a^2-ab+b^2:\dfrac{a^3-b^3}{a}=c^2-cd+d^2:\dfrac{c^3-d^3}{c}$.

If a, b, c, d are in continued proportion, prove—

37. $a:c-d=b^2:bd-cd$.

38. $a:c=a^2+b^2+c^2:b^2+c^2+d^2$.

39. $a:d=a^3+2b^3+3c^3:b^3+2c^3+3d^3$.

Prove that $a:b=c:d$, it being given that—

40. $(a+b)(c-d)+(b+c)(d-a)=cd-ab$.

41. $(a+b-3c-3d)(2a-2b-c+d)=(2a+2b-c-d)$ $(a-b-3c+3d)$.

42. Find two numbers in the ratio of 2 to 5, such that when each is increased by 5 they shall be as 3 to 5.

43. Find two numbers, such that if 7 be added to each they will be in the ratio of 2 to 3; and if 2 be subtracted from each, they will be in the ratio of 1 to 3.

44. Separate 32 into two parts, such that the greater diminished by 11 shall be to the less, increased by 5, as 4 to 9.

45. Separate 12 into two parts, such that their product shall be to the sum of their squares as 2 to 5.

CHAPTER XXIII.

INDETERMINATE EQUATIONS. VARIATION.

294. Indeterminate Equations. If a single equation containing two unknown quantities be given, this equation is called an *indeterminate equation*, for the unknown quantities may have an indefinite number of different values which satisfy the equation.

Thus, given $\qquad 3x + 2y = 5.$

When $\qquad x = 0, \quad y = \frac{5}{2},$

$x = 1, \quad y = 1,$

$x = 2, \quad y = -\frac{1}{2},$

$x = 3, \quad y = -2,$

etc.

In an indeterminate equation some limitation in the character of the values of x and y may be imposed. Very frequently the values of x and y are limited to *positive integers*. Of the values obtained for x and y in the above equation, the only set that satisfies this condition is $x = 1$, $y = 1$.

In like manner, if in a group of given simultaneous equations the number of unknown quantities be greater than the number of the equations, the equations are said to be indeterminate.

The treatment of the subject here made will be limited to indeterminate equations of the first degree.

295. The Solution of Indeterminate Equations is best explained in connection with illustrative examples.

Ex. 1. Solve in positive integers $5x - 7y = 11$.

Divide through by 5, *the smaller of the two coefficients.*

$$x - y - \frac{2y}{5} = 2 + \tfrac{1}{5}$$

$$\therefore x - y - 2 = \frac{2y + 1}{5}$$

Since x and y are integers, $x - y - 2$ must be an integer. Hence, $\frac{2y + 1}{5}$ must be an integer.

$$\therefore \frac{3(2y + 1)}{5}, \text{ or } \frac{6y + 3}{5}, \text{ must be an integer.}$$

(The particular multiplier 3 is used in this case so that on dividing the resulting numerator by the denominator 5, the coefficient of y in the remainder is unity.)

$\frac{6y + 3}{5}$; hence, $y + \frac{y + 3}{5}$; hence, $\frac{y + 3}{5}$ must be an integer.

Let $$\frac{y + 3}{5} = p.$$

$$\therefore y = 5p - 3 \quad \ldots\ldots\ldots \quad (1)$$

Substitute in the original equation for y,

$$5x - 35p + 21 = 11$$

$$\therefore x = 7p - 2 \quad \ldots\ldots\ldots \quad (2)$$

In equations (1) and (2) p must have some integral value.

If $p = 1$, then $x = 5$, $y = 2$.

If $p = 2$, $x = 12$, $y = 7$.

Etc. etc.

It is seen that there are an indefinite number of positive integral values of x and y.

Ex. 2. A number consists of two digits; if the number be divided by the number formed by reversing the digits, the quotient is 2, and the remainder 2. Find the number.

Let $x =$ the tens' digit

$y =$ the units' digit.

Then $$\frac{10x + y - 2}{x + 10y} = 2$$

$$\therefore 8x - 19y = 2.$$

Dividing by 8, $x - 2y - \frac{3y}{8} = \frac{2}{8}$

$$\therefore x - 2y = \frac{3y + 2}{8}$$

$$\therefore \frac{3y + 2}{8} \text{ must be an integer.}$$

Hence, $\frac{3(3y + 2)}{8}$, or $\frac{9y + 6}{8}$, and $\frac{y + 6}{8}$ must be integers.

Let $\frac{y + 6}{8} = p$

$$\begin{cases} y = 8p - 6 \\ x = 19p - 14 \end{cases}$$

The values of x as digits in a number are limited to positive integers, lowest 0, highest 9.

$$\therefore x = 5,\ y = 2 \text{ is the only result allowable.}$$

$$\therefore \text{the number is } 52.$$

Ex. 3. In how many ways can the sum of $5.10 be paid with half-dollars, quarters, and dimes, the whole number of coins used being 20?

Let $x =$ number of half-dollars.
$y =$ number of quarter-dollars.
$z =$ number of dimes.

Then $\frac{x}{2} + \frac{y}{4} + \frac{z}{10} = \frac{51}{10}.$

Or, $10x + 5y + 2z = 102 \quad \ldots\ldots\ldots\ldots \quad (1)$

Also, $x + y + z = 20 \quad \ldots\ldots\ldots\ldots \quad (2)$

Multiply (2) by 2, and subtract from (1),

$$8x + 3y = 62$$

Solving, $\begin{cases} x = 3p + 1. \\ y = 18 - 8p. \\ z = 5p + 1. \end{cases}$

Let $p = 0$, then $x = 1$, $y = 18$, $z = 1$.
$p = 1$, $x = 4$, $y = 10$, $z = 6$.
$p = 2$, $x = 7$, $y = 2$, $z = 11$.

Any other values of p give negative results for one or more of the quantities x, y, z.

Hence, there are three ways of making the required payment.

EXERCISE 109.

Solve in positive integers—

1. $7x + 4y = 63$.
2. $3x + 11y = 31$.
3. $5x + 7y = 82$.
4. $7x + 12y = 111$.
5. $15x + 8y = 101$.
6. $10x + 17y = 199$.
7. $5x - 7y = 11$.
8. $13x - 30y = 61$.
9. $16x - 11y = 26$.
10. $13x - 35y = -64$.

11. Divide the number 107 into two such parts that one is divisible by 3, and the other by 8.

12. Divide 321 into two such parts that one is divisible by 9, and the other by 13.

13. Find two fractions whose denominators are 5 and 12 respectively, and whose sum is $4\frac{3}{20}$.

14. A farmer sold a number of sheep and calves for $194; for each sheep he received $6, and for each calf $11. How many of each did he sell?

15. In how many ways can the sum of $5.80 be paid with dimes and quarters?

16. Find all possible ways of paying three dollars with five-, ten-, and twenty-five-cent pieces, so that half the coins used are five-cent pieces.

17. There is a number which, when divided by 17 gives a remainder of 6, and when divided by 23 gives a remainder of 21. Find it. How many such numbers are there?

VARIATION.

296. Variables and Constants. A **Variable** is a quantity which has an indefinite number of different values.

A **Constant** is a quantity which has a single fixed value.

297. Relation of Variables. Variations. One variable (called the *function*) may depend on another variable for its value in a definite manner. Thus, if a man be hired to work for a certain sum per day, the number of dollars he will receive as wages will vary as the number of days he works.

Thus, if $x =$ number of dollars in his wages,

$t =$ number of days he works,

$x \propto t$. (The symbol $\propto$ reads "varies as.")

This expression is called a variation.

This variation may also be expressed thus,

$$x = mt,$$

where m denotes the number of dollars in one day's wages.

Or,
$$\frac{x}{t} = m.$$

Thus, if the ratio of two variables is always constant, their relation may be expressed in any one of three ways:

(1) As a ratio.

(2) As an equation.

(3) As a variation.

KINDS OF ELEMENTARY VARIATIONS.

298. I. **Simple Direct Variations.** The case considered in Art. 297,

$$x \propto y, \quad \text{or } x = my,$$

is called a *direct variation*.

II. **Inverse Variations.** If x varies inversely as y (that is, as x increases y decreases, and *vice versâ*), then x and $\frac{1}{y}$ have a constant ratio,

$$x \propto \frac{1}{y}, \quad \text{or } x = \frac{m}{y}.$$

This is called an *inverse variation*.

Thus, the number of days required in which to do a given

piece of work varies inversely as the number of workmen employed. Also, in triangles of a given area the altitude varies inversely as the length of the base.

III. **Joint Variation.** If x varies as the product of two or more other variables, as of y and z, for instance, then x and yz have a constant ratio, and

$$x \propto yz, \quad \text{or } x = myz.$$

This is called a *joint variation.*

IV. **Direct and Inverse Variation.** x may also vary directly as one variable, as y, and inversely as another, as z; then

$$x \propto \frac{y}{z}, \quad \text{or } x = \frac{my}{z}.$$

299. Compound Variations. The sum or difference of two or more variations may be taken, the result being termed a *compound variation.*

Thus, if y equals the sum of u and v, and u varies directly as x^2, and v, inversely as x,

$$u = mx^2, \quad v = \frac{n}{x},$$

$$y = u + v.$$

$$\therefore y = mx^2 + \frac{n}{x}.$$

300. Fundamental Property of Variations. *A variation may be converted into an equation by the use of a coefficient which is afterward to be determined, and the properties of variations derived and problems solved by the use of the properties of equations.*

301. Elementary Properties of Variations.

I. If $x \propto y$, and $y \propto z$, then $x \propto z$.

For $x = my, \quad y = nz.$

$$\therefore x = mnz.$$

$$\therefore x \propto z.$$

II. If $x \propto z$, $y \propto z$, then $x \pm y \propto z$ and $\sqrt{xy} \propto z$.

For $x = mz$, $y = nz$.

$$\therefore x \pm y = (m \pm n)z,$$

And $\sqrt{xy} = \sqrt{mz \cdot nz} = \sqrt{mn \cdot z^2} = z\sqrt{mn}.$

Hence, $x \pm y \propto z$.

$$\sqrt{xy} \propto z.$$

III. If $x \propto z$, and $y \propto u$, then $xy \propto uz$.

For $x = mz$. $y = nu$.

$$\therefore xy = mnuz.$$

$$\therefore xy \propto uz.$$

IV. If $x \propto y$, then $x^n \propto y^n$.

For $x = my$.

$$\therefore x^n = m^n y^n.$$

$$\therefore x^n \propto y^n.$$

302. Examples.

Ex. 1. If x varies inversely as y^2, and $x = 4$, when $y = 1$, find x when $y = 2$.

Since $x \propto \frac{1}{y^2}$,

we have $x = \frac{m}{y^2}$ . . . . (1)

Substitute $x = 4$, $y = 1$, in (1), $4 = m$.

Substitute for m its value in (1),

$x = \frac{4}{y^2}$ . . . . (2)

Let $y = 2$ in (2), then $x = 1$, *Result.*

Ex. 2. If y equals the sum of two quantities, one of which varies directly as x^2, the other inversely as x; and $y=5$ when $x=1$, $y=1$ when $x=-1$, find y when $x=2$.

Since $y = u + v$, and $u \propto x^2$, $v \propto \frac{1}{x}$.

[Art. 299.] $$y = mx^2 + \frac{n}{x} \quad (1)$$

Substituting the given pairs of values for x and y in (1),

$$5 = m + n.$$

$$1 = m - n.$$

$$\therefore m = 3, \quad n = 2.$$

Substitute in (1) for m and n,

$$y = 3x^2 + \frac{2}{x} \quad (2)$$

Let $x = 2$ in (2), $y = 13$, *Result.*

Ex. 3. The area of a circle varies as the square of its diameter. Find the diameter of a circle whose area shall be equivalent to the sum of the areas of two circles whose diameters are 6 and 8 inches respectively.

Let A denote the area of a circle, and D the diameter.

Then $A \propto D^2$,

And $A = mD^2$,

denote the areas of the two given circles by A' and A''.

Then $A' = m \times 6^2 = 36m.$

$$A'' = m \times 8^2 = 64m.$$

Adding, $A' + A''$, or $A = 100m$,

Hence, since $A = mD^2$, and also $100m$,

$$mD^2 = 100m$$

$$D^2 = 100$$

$$D = 10$$

Thus, the required diameter is 10.

The student should review examples 1, 2, and 3 thoroughly, until he understands every step taken in their solution, before he undertakes a single example of the following exercise.

EXERCISE 110.

1. If x varies as y, and x is 10 when y is 2, find x when y is 3.

2. If $x \propto y$, and $x = 8$ when $y = 6$, find y when $x = 3$.

3. If $x + 1 \propto y - 5$, and $x = 2$ when $y = 6$, find x when $y = 7$.

4. If $x^2 \propto y^3$, and $x = 4$ when $y = 2$, find y when $x = 32$.

5. If $x^2 \propto y^3 + 8$, and $x = \frac{3}{2}\sqrt{3}$ when $y = 1$, find y when $x = 3$.

6. If x varies inversely as y, and equals 2 when y is 4, find y when $x = 5$.

7. If x varies inversely as y^2, and is 6 when y is $\frac{1}{2}$, find y when $x = 1\frac{1}{3}$.

8. If x varies jointly as y and z, and is 6 when y is 3 and z is 2, find x when y is 5 and z, 7.

9. If x varies jointly as y and z, and equals 2 when $y = \frac{1}{2}$ and $z = \frac{2}{3}$, find x when $y = 3$, $z = \frac{2}{9}$.

10. x varies directly as y and inversely as z, and $= 10$ when $y = 15$ and $z = 6$. Find y when $x = 16$ and $z = 2$.

11. If $2x - 3y \propto 5x + 9y$, and when $y = -2$, $x = 4$, find the equation connecting x and y.

12. If $x^2 - 2x + 1 \propto y^2 - 2y - 1$, and $x = \frac{3}{2}$ when $y = -\frac{1}{2}$, find the equation between x and y.

13. One quantity varies directly as x and another varies inversely as x. If their sum is equal to 10 when $x = 2$, and to -2 when $x = -1$, find each quantity when $x = \frac{3}{2}$.

14. Two quantities vary directly as x^2 and inversely as x respectively. If their sum is $3\frac{1}{2}$ when $x = 2$, and $-3\frac{1}{2}$ when $x = 1$, find the quantities in terms of x.

15. Given that w is equal to the sum of two quantities which vary as x and x^3, respectively. If $w = -2$ when $x = 2$, and -5 when $x = -1$, what is w when $x = \frac{1}{2}$?

16. Given that w is equal to the sum of two quantities which vary as x and x^2 respectively. If $w = -2$ when $x = -1$, and $w = -11\frac{1}{2}$ when $x = -2$, what is the value of w when $x = -\frac{2}{3}$?

17. Given that w is equal to the sum of three quantities, one of which is constant and the others vary directly as x^3 and inversely as x^2 respectively. If $w = 8$ when $x = \frac{1}{2}$, $w = 8$ when $x = -1$, and $w = 16\frac{2}{3}$ when $x = -\frac{3}{2}$, find the equation between w and x.

18. The distance fallen by a body from a position of rest varies as the square of the time during which it falls. If a body falls $144\frac{3}{4}$ feet in 3 seconds, how far will it fall in 8 seconds?

19. The area of a circle varies as the square of its diameter. Find the diameter of a circle equivalent to two circles whose diameters are 5 and 12 inches respectively.

20. The volume of a sphere varies as the cube of its diameter. If three spheres whose diameters are respectively 6, 8, and 10 inches be formed into a single sphere, find its diameter.

21. The volume of a cone of revolution whose altitude is 7, and the radius of whose base is 3, is 66. Find the volume of a cone of revolution of altitude 6 and radius 5.

NOTE. The volume of a cone of revolution (or cylinder of revolution) varies jointly as the altitude and the square of the radius of the base.

22. Find the altitude of a cone of revolution the radius of whose base is 7, and which is equivalent to two cones with altitudes 5 and 11 and radii 2 and 4 respectively.

23. If the illumination from a source of light varies inversely as the square of the distance, how much farther from a candle must a book which is now 18 inches away, be removed to receive just $\frac{2}{3}$ as much light? Interpret the two results.

CHAPTER XXIV.

ARITHMETICAL PROGRESSION.

303. A Series is a succession of terms formed according to some law.

Exs. 1, 4, 9, 16, 25, ...

$1 - x + x^2 - x^3 + x^4 -$, ...

2, 4, 8, 16, 32, ...

304. An **Arithmetical Progression** is a series each term of which is formed by adding a constant quantity, called the difference, to the preceding term.

Thus, 1, 4, 7, 10, 13, ... is an arithmetical progression in which the difference is 3.

Given an Arithmetical Progression (often denoted by A. P.), to determine the difference, *from any term subtract the preceding term.*

Thus, in the A. P., $\frac{3}{2}, -\frac{3}{4}, -3,$

the difference $= -\frac{3}{4} - \frac{3}{2} = -\frac{9}{4}.$

305. Principal Quantities and Symbols Used. In an A. P. we are concerned with five quantities:

1. The *first term*, denoted by a.
2. The *common difference*, denoted by d.
3. The *last term*, denoted by l.
4. The *number of terms*, denoted by n.
5. The *sum of the terms*, denoted by s.

306. Two Fundamental Formulas. Since in an A. P. each term is formed by adding the common difference, d, to the preceding term, the general form of an A. P. is—

$$a, \quad a + d, \quad a + 2d, \quad a + 3d, + \ldots\ldots$$

Hence, the coefficient of d in each term is one less than the number of the term.

Thus, the 7th term is $a + 6d$,

12th term is $a + 11d$,

nth term is $a + (n-1)d$.

Hence, $l = a + (n-1)d$ . . . . (1)

Also,

$s = a + (a+d) + (a+2d) + \ldots\ldots + (l-d) + l$ . . (2)

Writing the terms of this series in reverse order,

$s = l + (l-d) + (l-2d) + \ldots\ldots + (a+d) + a$ . . (3)

Adding (2) and (3),

$2s = (a+l) + (a+l) + (a+l) + \ldots\ldots + (a+l) + (a+l)$

$= n(a+l)$.

$\therefore s = \frac{n}{2}(a+l)$ . . . . (4)

If we substitute for l in (4) from (1),

$s = \frac{n}{2}[2a + (n-1)d]$ . . . . (5)

Hence, combining results, we have the two fundamental formulas for l and s,

I. $l = a + (n-1)d$.

II. $s = \frac{n}{2}(a+l)$

$s = \frac{n}{2}[2a + (n-1)d]$.

Ex. 1. Find the 12th term and the sum of 12 terms of the A. P., 5, 3, 1, -1, -3, ...

In this series $a = 5, \quad d = -2, \quad n = 12.$

From I., $l = 5 + (12-1)(-2) = 5 - 22 = -17.$

From II., $s = \frac{12}{2}(5 - 17) = -72$, *Sum.*

Ex. 2. Find the sum of n terms of the A. P.,

$$\frac{a+b}{2}, \quad \frac{a-b}{2}, \quad \frac{a-3b}{2}, \; \ldots\ldots$$

Here $$a = \frac{a+b}{2}, \quad d = -b, \quad n = n.$$

Substituting in the fundamental formula, $s = \frac{n}{2}[2a + (n-1)d]$,

$$s = \frac{n}{2}[a + b + (n-1)(-b)]$$
$$= \frac{n}{2}[a + (2-n)b], \text{ } \textit{Sum.}$$

EXERCISE 111.

1. Find the 8th term in the series 3, 7, 11, ...

2. Find the 9th term and the sum of 9 terms in 7, 3, -1, ...

3. Find the 13th term and the sum of 13 terms in -10, -13, -16, ...

4. Find the 20th and 28th terms in 5, $\frac{18}{8}$, $\frac{11}{8}$, ...

5. Find the 16th and 25th terms in $-13\frac{1}{2}$, -9, $-4\frac{1}{2}$, ...

6. Find the 7th and 10th terms and the sum of 10 terms in the series $\frac{3}{4}$, $\frac{2}{3}$, $\frac{7}{12}$, ...

7. Find the 18th term and the sum of 18 terms in the series 3, 2.4, 1.8, ...

Find the sum of the series—

8. 3, 8, 13, ... to 8 terms.
9. -4, -7, -10, ... to 6 terms.
10. 3, -3, -9, ... to 9 terms.
11. $2\frac{1}{2}$, $3\frac{3}{4}$, 5, ... to 14 terms.
12. $\frac{5}{6}$, $\frac{3}{4}$, $\frac{2}{3}$, ... to 96 terms.
13. $-\frac{1}{3}$, $\frac{1}{2}$, $\frac{4}{3}$, ... to 38 terms.
14. $-\frac{3}{4}$, $-\frac{5}{6}$, $-\frac{11}{12}$, ... to 55 terms.
15. $3c$, $\frac{1}{2}c$, $-2c$, ... to b terms.
16. $2x-y$, $x+y$, $3y$, ... to r terms.
17. $5\sqrt{2}-2\sqrt{3}$, $4\sqrt{2}-3\sqrt{3}$, ... to 11 terms.
18. $3a-\frac{1}{a}$, $2a$, $a+\frac{1}{a}$, ... to 12 terms.

307. Problem I. *Given any three of the five quantities, a, d, l, n, s, to find the other two.*

If we substitute for the three given quantities their values in the two fundamental formulas (I. and II., Art. 306), we shall have as a result two equations with two unknown quantities. The values of these unknown quantities may then be found by solving the two equations. Hence, by the use of these fundamental formulas, problems relating to A. P. are converted into problems relating to the solution of equations, processes already mastered.

Ex. Given $d=2$, $l=21$, $s=121$, find a, n.

Substitute for d, l, s in Formulas I. and II.,

$$21 = a + (n-1)2 \quad \ldots\ldots (1)$$

$$121 = \frac{n(a+21)}{2} \quad \ldots\ldots (2)$$

$$\therefore a + 2n = 23 \quad \ldots\ldots (3)$$

$$an + 21n = 242 \quad \ldots\ldots (4)$$

Substitute for a in (4) from (3),

$$n(23-2n) + 21n = 242$$

Whence $\qquad n = 11$

Hence, from (3), $\qquad a = 1.$

308. Problem II. *Given three of the five quantities, a, d, l, n, s, to obtain a formula for one or both of the other two in terms of the three given quantities.*

Ex. Given d, l, s, obtain a formula for n.

Since $\qquad l = a + (n-1)d \quad \ldots\ldots (1)$

$$s = \frac{n}{2}[2a + (n-1)d] \quad \ldots\ldots (2)$$

Substitute for a from (1) in (2),

$$s = \frac{n}{2}[2l - 2(n-1)d + (n-1)d]$$

$$\therefore 2s = 2ln - n(n-1)d$$

Whence $dn^2 - (d+2l)n = -2s$

Solving for n, $\qquad n = \dfrac{d + 2l \pm \sqrt{(d+2l)^2 - 8ds}}{2d}.$

No.	Given.	Required.	Formulas.
1	a, d, n	l	$l = a + (n - 1)d.$
2	a, d, s		$l = -\frac{1}{2}d \pm \sqrt{2ds + (a - \frac{1}{2}d)^2}.$
3	a, n, s		$l = \frac{2s}{n} - a.$
4	d, n, s		$l = \frac{s}{n} + \frac{(n-1)d}{2}.$
5	d, n, l	a	$a = l - (n - 1)d.$
6	d, l, s		$a = \frac{1}{2}d \pm \sqrt{(\frac{1}{2}d + l)^2 - 2ds}.$
7	n, l, s		$a = \frac{2s}{n} - l.$
8	d, n, s		$a = \frac{s}{n} - \frac{(n-1)d}{2}.$
9	a, d, n	s	$s = \frac{1}{2}n[2a + (n - 1)d].$
10	a, d, l		$s = \frac{l + a}{2} + \frac{l^2 - a^2}{2d}.$
11	a, n, l		$s = \frac{n}{2}(a + l).$
12	d, n, l		$s = \frac{1}{2}n[2l - (n - 1)d].$
13	a, n, l	d	$d = \frac{l - a}{n - 1}.$
14	a, n, s		$d = \frac{2(s - an)}{n(n - 1)}.$
15	a, l, s		$d = \frac{l^2 - a^2}{2s - l - a}.$
16	n, l, s		$d = \frac{2(nl - s)}{n(n - 1)}.$
17	a, d, l	n	$n = \frac{l - a}{d} + 1.$
18	a, d, s		$n = \frac{d - 2a \pm \sqrt{(d - 2a)^2 + 8ds}}{2d}.$
19	a, l, s		$n = \frac{2s}{l + a}.$
20	d, l, s		$n = \frac{2l + d \pm \sqrt{(2l + d)^2 - 8ds}}{2d}.$

EXERCISE 112.

Find the first term and the sum of the series when—

1. $d=3,\ l=40,\ n=13.$
2. $d=\frac{3}{4},\ l=18\frac{1}{3},\ n=33.$

Find the first term and the common difference when—

3. $s=275,\ l=45,\ n=11.$
4. $s=4,\ l=-10,\ n=8.$
5. $s=-246\frac{1}{2},\ l=-34\frac{1}{2},\ n=17.$
6. $s=-38\frac{1}{2},\ l=-5\frac{1}{6},\ n=21.$
7. $s=9,\ l=2\frac{2}{3},\ n=9.$
8. $s=-\frac{47}{6},\ l=-4,\ n=47.$

Find n and d when—

9. $a=-5, l=15, s=105.$
10. $a=19, l=-21, s=-21.$
11. $a=\frac{7}{12}, l=-\frac{5}{6}, s=-2\frac{1}{4}.$
12. $a=-3\frac{1}{4}, l=9\frac{1}{4}, s=48,$

Find a and n when—

13. $l=-8, d=-3, s=-3.$
14. $l=1, d=\frac{2}{3}, s=-20.$
15. $l=2, d=-\frac{1}{4}, s=19\frac{1}{4}.$
16. $l=-\frac{4}{3}, d=-\frac{1}{15}, s=-\frac{40}{3}.$

How many consecutive terms must be taken from—

17. $1,\ 1\frac{1}{2},\ 2$ . . . . to make 45?
18. $\frac{3}{4},\ \frac{1}{2},\ \frac{1}{4}$ . . . . to make -1?
19. $\frac{19}{6},\ 2,\ \frac{5}{6}$ . . . . to make $-20\frac{5}{6}$?
20. $\frac{3}{2},\ \frac{5}{4},\ 1$ . . . . to make 4.5?

309. Arithmetical Means. If it be required to insert a given number of arithmetical means between two given numbers, the solution of the problem is readily effected by means of Problem I. (Art. 307). The quantities in the A. P. which are given are seen to be a, l, n, and it is required to find d.

Ex. Insert 9 arithmetical means between 1 and 5.

We have given $a=1,\ \ l=5,\ \ n=11.$

Hence, we find $d=\frac{2}{5}.$

The required means are therefore $1\frac{2}{5}$, $1\frac{4}{5}$, $2\frac{1}{5}$, ...

In case but a single arithmetical mean is to be inserted between two quantities, a and b, this one mean is found most readily by use of the formula $\frac{a+b}{2}$.

For if x denote the required mean, the A. P. is a, x, b.

$$\text{Hence,}\quad x-a=b-x$$
$$2x=a+b$$
$$x=\frac{a+b}{2}.$$

EXERCISE 113.

Insert—

1. Four arithmetical means between 7 and -3.
2. Seven arithmetical means between 4 and 6.
3. Eight arithmetical means between $\frac{3}{2}$ and 3.
4. Thirteen arithmetical means between $\frac{1}{3}$ and $-\frac{5}{6}$.
5. Fifteen arithmetical means between $-4\frac{1}{3}$ and 9.
6. The arithmetical mean between $2\frac{1}{5}$ and $-5\frac{2}{3}$.
7. The arithmetical mean between $x+1$ and $x-1$.
8. The A. M. between $\frac{a}{b}$ and $\frac{b}{a}$. Between $\frac{1}{x+y}$ and $\frac{1}{x-y}$.

310. Miscellaneous Examples.

Ex. 1. The 7th term of an A. P. is 5, and the 14th term is -9. Find the first term.

By the use of Formula I. (Art. 306),

the 7th term is $a + 6d$, the 14th term is $a + 13d$.

$$\therefore a + 6d = 5 \quad \ldots \ldots \quad (1)$$

$$a + 13d = -9 \quad \ldots \ldots \quad (2)$$

Subtracting (1) from (2), $7d = -14$

$$d = -2$$

Substitute for d in (1), $a - 12 = 5$

$$a = 17, \textit{ Result.}$$

Ex. 2. The sum of five numbers in A. P. is 15, and the sum of the first and fourth numbers is 9. Find the numbers.

Denote the numbers by

$$x + 2y, \quad x + y, \quad x, \quad x - y, \quad x - 2y.$$

Add, $5x = 15 \quad \ldots \ldots \quad (1)$

Also, $(x + 2y) + (x - y) = 9$

$$\therefore 2x + y = 9 \quad \ldots \ldots \quad (2)$$

From (1) $x = 3$; hence, from (2), $y = 3$.

Hence, the numbers are 9, 6, 3, 0, -3, *Result.*

Similarly, in dealing with four unknown quantities in A. P., we denote them by

$$x + 3y, \quad x + y, \quad x - y, \quad x - 3y.$$

EXERCISE 114.

Find the first two terms of the series wherein—

1. The 4th term is 11 and the 10th is 23.
2. The 6th term is -3 and the 12th is -12.
3. The 7th term is $-\frac{1}{6}$ and the 16th is $2\frac{1}{3}$.
4. The fifth term is $c - 3b$ and the 11th is $3b - 5c$.
5. Find the sum of the first n odd numbers.
6. Find the sum of the first n numbers divisible by 7.
7. Which term in the series $1\frac{1}{6}$, $1\frac{1}{3}$, $1\frac{1}{2}$, ... is 18?
8. The first term of an arithmetical progression is 8; the 3d term is to the 7th as the 8th is to the 10th. Find the series.

9. Find four numbers in A. P., such that the sum of the first two is 1, and the sum of the last two is -19.

10. Find four numbers in A. P. whose sum is 16 and product is 105.

11. A man travels $2\frac{1}{3}$ miles the first day, $2\frac{2}{3}$ the second, 3 the third, and so on; at the end of his journey he finds that if he had traveled $6\frac{1}{3}$ miles every day he would have required the same time. How many days was he walking?

12. The sum of 10 numbers in an A. P. is 145, and the sum of the fourth and ninth terms is 5 times the third term. Find the series.

13. If the 11th term is 7 and the 21st term is $8\frac{2}{3}$, find the 41st term of the same A. P.

14. In an A. P. of 21 terms the sum of the last three terms is 23, and the sum of the middle three is 5. Find the series.

15. Required five numbers in A. P., such that the sum of the first, third, and fourth terms shall be 8, and the product of the second and fifth shall be -54.

16. The sum of five numbers in A. P. is 40, and the sum of their squares is 410. Find them.

17. The 14th term of an A. P. is 38; the 90th term is 152, and the last term is 218. Find the number of terms.

18. How many numbers of two figures are there divisible by 3? By 7? How many numbers of three figures are divisible by 6? By 9?

19. How many numbers of four figures are there divisible by 11? Find the sum of all the numbers of three figures divisible by 7.

20. If a body falls $16\frac{1}{12}$ ft. the first second of its fall; three times this distance the second; five times the third, and so on, how far will it fall the 30th second? How far will it have fallen during the 30 seconds?

21. If a, b, c, d are in A. P. prove: (1) that $a+d=b+c$: (2) that ak, bk, ck, dk are also in A. P.; and (3) that $a+k$, $b+k$, $c+k$, $d+k$ are in A. P. State this problem without the use of the symbols, a, b, c, d, k.

CHAPTER XXV.

GEOMETRICAL AND HARMONICAL PROGRESSIONS.

311. A Geometrical Progression is a series each term of which is formed by multiplying the preceding term by a constant quantity called the ratio.

Thus, 1, 3, 9, 27, 81, ... is a geometrical progression (or G. P.) in which the ratio is 3.

Given a geometrical progression, to determine the ratio: *divide any term by the preceding term.*

Thus, in the G. P., $-3, \frac{3}{2}, -\frac{3}{4}, \ldots\ldots$

the ratio $= \frac{\frac{3}{2}}{-3} = -\frac{1}{2}.$

312. Quantities and Symbols Used. a, l, n, s are used, as in A. P. Besides these, r is used to denote the ratio.

313. Two Fundamental Formulas. Since in a G. P. each term is formed by multiplying the preceding term by the common ratio, r, the general form of a G. P. is—

$$a, \quad ar, \quad ar^2, \quad ar^3, \quad ar^4, \ldots\ldots$$

Hence, the exponent of r in each term is one less than the number of the term. Thus,

The 10th term is ar^9.

The 15th term is ar^{14}.

The nth term, or $l = ar^{n-1}$ . . . . (1)

In deriving a formula for the sum, we know, also,

$$s = a + ar + ar^2 + \ldots\ldots + ar^{n-1} \quad . \quad . \quad . \quad (2)$$

Multiply (2) by r,

$$rs = ar + ar^2 + ar^3 + \ldots\ldots + ar^{n-1} + ar^n \quad . \quad (3)$$

Subtract (2) from (3),

$$rs - s = ar^n - a.$$

$$\therefore s = \frac{ar^n - a}{r-1} \quad . \quad . \quad . \quad . \quad . \quad . \quad . \quad . \quad . \quad . \quad . \quad . \quad (4)$$

Multiply (1) by r,

$$rl = ar^n.$$

Substitute rl for ar^n in (4),

$$s = \frac{rl - a}{r-1} \quad . \quad . \quad . \quad . \quad . \quad . \quad . \quad . \quad . \quad . \quad . \quad . \quad (5)$$

Hence, collecting the results obtained in (1), (4), (5), we have the two fundamental formulas for l and s:

$$\text{I.} \quad l = ar^{n-1}.$$

$$\text{II.} \quad s = \frac{ar^n - a}{r-1}$$

$$s = \frac{rl - a}{r-1}.$$

Ex. 1. Find the 8th term and sum of 8 terms of the G. P., 1, 3, 9, 27, ...

In this case, $a = 1, \quad r = 3, \quad n = 8.$

From I., $l = 1 \times 3^7 = 2187.$

From II., $s = \dfrac{3 \times 2187 - 1}{3 - 1} = 3280.$

Ex. 2. Find the 10th term and the sum of 10 terms of the G. P., 4, -2, 1, $-\frac{1}{2}$, ...

Here $a = 4, \quad r = -\frac{1}{2}, \quad n = 10.$

Hence, $l = 4(-\frac{1}{2})^9 = -\frac{4}{512} = -\frac{1}{128}.$

$$s = \frac{(-\frac{1}{2})(-\frac{1}{128}) - 4}{-\frac{1}{2} - 1} = \frac{341}{128}.$$

EXERCISE 115.

1. Find the sixth term in the series 2, 6, 18, ...
2. Find the 7th term in 3, 6, 12, ...
3. Find the 6th and the sum of 6 terms in 45, -15, 5, ...
4. Find the 5th and the sum of 5 terms in 81, -54, ...
5. Find the 7th and the sum of 7 terms in $1\frac{1}{8}$, $-\frac{3}{4}$, ...
6. Find the 9th term in the series 2, $2\sqrt{2}$, 4, ...

Find the sum of the series—

7. 3, -6, 12, ... to 6 terms.
8. 27, -18, 12, ... to 7 terms.
9. $-\frac{8}{9}$, $1\frac{1}{3}$, -2, ... to 9 terms.
10. $\frac{2}{3}$, $-\frac{1}{6}$, $\frac{1}{24}$, ... to 8 terms.
11. $\frac{1}{\sqrt{3}}$, 1, $\sqrt{3}$, ... to 8 terms.
12. $\sqrt{3}$, $\sqrt{6}$, $2\sqrt{3}$, ... to 10 terms.
13. $\sqrt{2}-1$, 1, $\sqrt{2}+1$, ... to 6 terms.

314. Problem I. *Given three of the five quantities, a, l, n, s, r, to determine the other two.*

As in A. P., in the two fundamental formulas (I. and II., Art. 313) substitute for the three known quantities, and determine the other two quantities by solving the resulting equations.

Ex. 1. Given $a=-2$, $n=7$, $l=-128$; find r, s.

From I., $\qquad -128=-2r^6.$

Hence, $\qquad r^6=64, \qquad r=\pm 2.$

From II., if $r=+2$, $s=\frac{2(-128)-(-2)}{2-1}=-256+2=-254.$

If $r=-2$, $\qquad s=\frac{(-2)(-128)-(-2)}{-2-1}=\frac{256+2}{-3}=-86.$

Hence, there are two sets of answers; viz., $r=+2$, $\quad s=-254$.

$r=-2$, $\quad s=-86$.

No.	Given.	Required.	Formulas.
1	a, r, n	l	$l = ar^{n-1}$.
2	a, r, s		$l = \frac{a + (r-1)s}{r}$.
3	r, n, s		$l = \frac{(r-1)sr^{n-1}}{r^n - 1}$.
4	a, n, s		$l(s-l)^{n-1} - a(s-a)^{n-1} = 0$.
5	r, n, l	a	$a = \frac{l}{r^{n-1}}$.
6	r, n, s		$a = \frac{(r-1)s}{r^n - 1}$.
7	r, l, s		$a = rl - (r-1)s$.
8	n, l, s		$a(s-a)^{n-1} - l(s-l)^{n-1} = 0$.
9	a, r, n	s	$s = \frac{a(r^n - 1)}{r - 1}$.
10	a, r, l		$s = \frac{lr - a}{r - 1}$.
11	a, n, l		$s = \frac{\sqrt[n-1]{l^n} - \sqrt[n-1]{a^n}}{\sqrt[n-1]{l} - \sqrt[n-1]{a}}$.
12	r, n, l		$s = \frac{(r^n - 1)l}{(r-1)r^{n-1}}$.
13	a, n, l	r	$r = \sqrt[n-1]{\frac{l}{a}}$.
14	a, n, s		$r^n - \frac{s}{a}r + \frac{s-a}{a} = 0$.
15	a, l, s		$r = \frac{s-a}{s-l}$.
16	n, l, s		$r^n - \frac{s}{s-l}r^{n-1} + \frac{l}{s-l} = 0$.
17	a, r, l	n	$n = \frac{\log l - \log a}{\log r} + 1$.
18	a, r, s		$n = \frac{\log[a + (r-1)s] - \log a}{\log r}$.
19	a, l, s		$n = \frac{\log l - \log a}{\log(s-a) - \log(s-l)} + 1$.
20	r, l, s		$n = \frac{\log l - \log[lr - (r-1)s]}{\log r} + 1$.

Ex. 2. Given $a=\frac{3}{4}$, $r=-\frac{1}{3}$, $s=\frac{91}{162}$; find l, n.

From I., $$l=\tfrac{3}{4}(-\tfrac{1}{3})^{n-1} \quad \ldots\ldots (1)$$

From II., $$\tfrac{91}{162}=\frac{\frac{3}{4}(-\frac{1}{3})^{n}-\frac{3}{4}}{-\frac{1}{3}-1} \quad \ldots\ldots (2)$$

Whence $$\tfrac{182}{243}=\frac{(-\frac{1}{3})^{n}-1}{-\frac{4}{3}}$$

$$-\tfrac{728}{729}=(-\tfrac{1}{3})^{n}-1$$

$$(-\tfrac{1}{3})^{n}=\tfrac{1}{729}=(-\tfrac{1}{3})^{6}.$$

Whence $n=6$.

Substitute for n in (1), $l=-\frac{1}{324}$.

315. Problem II. *Given three of the five quantities,* ***a, l, n, s, r,*** *to obtain a formula for one or both of the other two in terms of the three given quantities.*

Ex. Given n, r, s, obtain a formula for l.

Using $$l=ar^{n-1} \quad \ldots\ldots (1)$$

$$s=\frac{rl-a}{r-1} \quad \ldots\ldots (2)$$

Solve (2) for a, $$a=rl-s(r-1) \quad \ldots\ldots (3)$$

Substitute in (1) for a from (3),

$$l=r^{n}l-sr^{n-1}(r-1).$$

Hence, $$(r^{n}-1)l=sr^{n-1}(r-1)$$

$$l=\frac{sr^{n-1}(r-1)}{r^{n}-1}.$$

A complete table of all possible formulas for G. P. is given on the opposite page. These, the student should be required to derive for himself (except those for n).

EXERCISE 116.

Find the first term and the sum when—

1. $n=6$, $r=3$, $l=486$.
2. $n=8$, $r=-2$, $l=-640$.
3. $n=7$, $r=\frac{3}{2}$, $l=\frac{729}{4}$.
4. $n=8$, $r=-\frac{2}{3}$, $l=-\frac{640}{2187}$.
5. $n=9$, $r=-3$, $l=-1215$.
6. $n=7$, $r=\frac{1}{2}\sqrt{6}$, $l=3$.

Find the ratio when—

7. $a=-2$, $l=2048$, $n=6$.
8. $a=9$, $l=\frac{16}{9}$, $s=23\frac{4}{9}$.
9. $a=2\frac{26}{27}$, $l=-\frac{45}{64}$, $n=6$.
10. $a=-16\frac{1}{5}$, $l=\frac{1}{15}$, $s=-12\frac{2}{15}$.

Find the number of terms when—

11. $a=\frac{4}{3}, l=\frac{1}{12}, r=\frac{1}{2}$.

12. $a=3, l=-96, s=-63$.

13. $a=18, r=-\frac{2}{3}, s=12\frac{2}{9}$.

14. $l=-8, r=-2, s=-5\frac{1}{4}$.

How many consecutive terms must be taken from the series—

15. $\frac{3}{4}, \frac{1}{2}, \frac{1}{3}, \ldots .$ to make $\frac{211}{108}$?

16. $15\frac{5}{8}, -6\frac{1}{4}, 2\frac{1}{2}, \ldots .$ to make $10\frac{7}{8}$?

17. $5\frac{1}{3}, -8, 12, \ldots .$ to make $-22\frac{1}{6}$?

316. Geometrical Means. If it be required to insert a given number of geometrical means between two given numbers, the solution of the problem is readily effected by means of Problem I. (Art. 314). The quantities which are given are seen to be a, l, n, and it is required to find r.

Ex. Insert 5 geometrical means between 3 and $\frac{1}{243}$.

We have given $a=3$, $l=\frac{1}{243}$, $n=7$, to find r.

Solving by Problem I., $r=\frac{1}{3}$.

Hence, the required geometrical means are,

$$1, \frac{1}{3}, \frac{1}{9}, \frac{1}{27}, \frac{1}{81}.$$

In case but one geometrical mean is to be inserted between two given quantities, a and b, this one mean is found most readily by using the formula $\sqrt{ab}$. For if x represent the geometrical mean between a and b, the series will be

$$a, \quad x, \quad b.$$

$$\text{Hence, } \frac{x}{a}=\frac{b}{x}.$$

$$\therefore x^2=ab.$$

$$x=\sqrt{ab}.$$

EXERCISE 117.

Insert—

1. Three geometrical means between 8 and $\frac{1}{2}$.
2. Three geometrical means between $\frac{9}{4}$ and $\frac{4}{9}$.
3. Six geometrical means between $\frac{1}{24}$ and $-\frac{16}{3}$.
4. Four geometrical means between $-\frac{7}{2}$ and 3584.
5. Six geometrical means between 56 and $-\frac{7}{16}$.
6. Five geometrical means between $\frac{3}{2}$ and 12.

Find the geometrical mean between—

7. $4\frac{1}{6}$ and $\frac{3}{8}$.
8. $3\frac{8}{9}$ and $6\frac{3}{7}$.
9. $5\sqrt{2}+1$ and $5\sqrt{2}-1$.
10. $3\sqrt{5}+2\sqrt{3}$ and $3\sqrt{5}-2\sqrt{3}$.
11. $28a^3x$ and $63axy^4$.
12. $\dfrac{a\sqrt{x}}{c^2\sqrt[3]{y}}$ and $\dfrac{y\sqrt[3]{a^2}}{x\sqrt{c^3}}$.
13. Insert 6 geometrical means between $\dfrac{a^3}{16}$ and $\dfrac{8}{\sqrt{a}}$.
14. Insert 7 geometrical means between $\dfrac{8}{n^2}$ and $\dfrac{n^2}{2}$.

317. Problem III. *To find the limit of the sum of an infinite decreasing geometrical progression.*

If a line AB

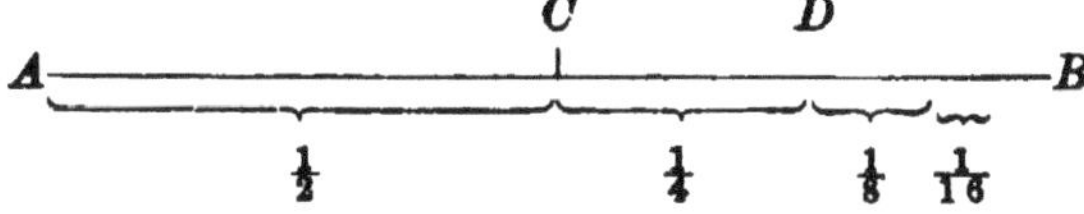

be given of unit length, and one-half of it (AC) be taken, and then one-half of the remainder (CD), and one-half of the remainder, and so on, the sum of the parts taken will be

$$\tfrac{1}{2}+\tfrac{1}{4}+\tfrac{1}{8}+\tfrac{1}{16}+\tfrac{1}{32}+\ldots\ldots$$

This is an infinite decreasing G. P. in which $r = \frac{1}{2}$. But the sum of all these parts must be less than 1, but approach closer and closer to 1 as a limit, the greater the number of parts taken. This is an illustration of the meaning of the limit of an infinite decreasing G. P.

In general, to find the limit of an infinite decreasing G. P. we have the formula

$$s = \frac{a}{1-r} \quad . \; . \; . \; . \; . \; . \; . \; . \; . \; . \; \text{III.}$$

Formula II. of Art. 313 may be written, $s = \frac{a - rl}{1-r}$. Then, as the number of terms increases,

l approaches indefinitely to 0.

$\therefore \; rl$ " " 0.

$\therefore \; a - rl$ " " $a - 0 = a$.

$\therefore \; \frac{a-rl}{1-r}$ " " $\frac{a}{1-r}$.

$$\therefore \; s = \frac{a}{1-r}.$$

Ex. Find the sum of 9, -3, 1, $-\frac{1}{3}$, ... to infinity.

Here $a = 9$, $r = -\frac{1}{3}$.

$$\therefore \; s = \frac{9}{1-(-\frac{1}{3})} = \frac{9}{1+\frac{1}{3}} = \frac{27}{4}, \textit{ Sum.}$$

318. Repeating Decimals. By the use of the Formula III. of Art. 317, the value of repeating decimals may be determined.

Ex. 1. Find the value of 0.373737 . . . . .

$$0.373737 \; . \; . \; . \; . \; . = .37 + .0037 + .000037 + \; . \; . \; . \; . \; .$$

Here $a = .37$, $r = .01$.

$$\therefore \; s = \frac{.37}{1-.01} = \frac{.37}{.99} = \frac{37}{99}.$$

Ex. 2. Find the value of 3.1186186 . . . . .

$$3.1186186 = 3.1 + .0186 + .0000186 + \ldots\ldots$$

Setting aside 3.1, and treating the remaining terms as a G. P.,

$$a = .0186, \qquad r = .001.$$

$$\therefore s = \frac{.0186}{1 - .001} = \frac{.0186}{.999} = \tfrac{186}{9990} = \tfrac{62}{3330}.$$

$$\therefore 3.1186186 \ldots\ldots = 3\tfrac{1}{10} + \tfrac{62}{3330} = 3\tfrac{79}{666}.$$

EXERCISE 118.

Find the sum to infinity of the series—

1. $2, \frac{4}{3}, \frac{8}{9}, \ldots\ldots$
2. $2, -1, \frac{1}{2}, \ldots\ldots$
3. $-9, 6, -4, \ldots\ldots$
4. $1\frac{7}{8}, 1\frac{1}{4}, \frac{5}{6}, \ldots\ldots$
5. $4\frac{1}{6}, -2\frac{1}{2}, 1\frac{1}{2}, \ldots\ldots$
6. $\frac{1}{12}, \frac{1}{18}, \frac{1}{27}, \ldots\ldots$
7. $2\frac{13}{16}, -1\frac{7}{8}, 1\frac{1}{4}, \ldots\ldots$
8. $6, 3\sqrt{2}, 3, \ldots\ldots$
9. $\dfrac{1}{\sqrt{2}-1}, 1, \dfrac{1}{\sqrt{2}+1}, \ldots\ldots$
10. $\frac{1}{2}\sqrt{2} + \frac{1}{3}\sqrt{3} + \frac{1}{3}\sqrt{2} \ldots\ldots$
11. $1 + \left(1 - \dfrac{1}{a}\right) + \left(1 - \dfrac{1}{a}\right)^2 + \ldots\ldots$

Find the values of—

12. $0.\dot{6}\dot{3}$.
13. $0.\dot{4}1\dot{7}$.
14. $5.\dot{8}4\dot{6}$.
15. 3.52424 . . . . .
16. 1.4037037 . . . . .
17. 3.215454 . . . . .
18. 1.02727 . . . . .
19. 1.027027 . . . . .
20. 0.30102102 . . . . .

319. Miscellaneous Problems.

Ex. Find four members in G. P., such that the sum

of the first and fourth is 56, and of the second and third is 24.

Denote the required numbers by a, ar, ar^2, ar^3.

Then $$a + ar^3 = 56$$

$$ar + ar^2 = 24.$$

Or, $$a(1 + r^3) = 56 \quad \ldots\ldots\ldots\ldots (1)$$

$$ar(1 + r) = 24 \quad \ldots\ldots\ldots\ldots (2)$$

Divide (1) by (2), $$\frac{1 - r + r^2}{r} = \tfrac{7}{3}.$$

Hence, $$3 - 3r + 3r^2 = 7r.$$

$$3r^2 - 10r = -3$$

$$r = 3, \quad \text{or } \tfrac{1}{3}.$$

And $$a = 2, \quad \text{or } 54.$$

Hence, the numbers are, 2, 6, 18, 54.

Or, 54, 18, 6, 2.

EXERCISE 119.

Find the first two terms of the series wherein—

1. The 3d term is 2, and the 5th is 18.
2. The 4th term is $\frac{3}{2}$ and the 9th is 48.
3. The 3d term is 5 and the 8th is $\frac{1215}{32}$.
4. The 5th term is 6 and the 11th is $\frac{3}{32}$.

Determine the nature, whether Ar. or Geom., of each—

5. $\frac{1}{4}, \frac{1}{3}, \frac{4}{9} \ldots\ldots$
6. $\frac{1}{4}, \frac{1}{6}, \frac{1}{9} \ldots\ldots$
7. $\frac{1}{4}, \frac{1}{6}, \frac{1}{12} \ldots\ldots$
8. $\frac{2}{3}, \frac{3}{4}, \frac{5}{6} \ldots\ldots$
9. $3\frac{1}{5}, 4\frac{4}{5}, 7\frac{1}{5} \ldots\ldots$
10. $7\frac{1}{3}, 5\frac{7}{8}, 4\frac{5}{12} \ldots\ldots$

11. Divide 65 into 3 parts in geometrical progression, such that the sum of the first and third is $3\frac{1}{3}$ times the second part.

12. There are 3 numbers in G. P. whose sum is 49, and the sum of the first and second is to the sum of the first and third as 3 to 5. Find them.

13. The sum of three numbers in G. P. is 21, and the sum of their reciprocals is $\frac{7}{12}$. Find them.

14. Find four numbers in G. P., such that the sum of the first and third is 10, and of the second and fourth is 30.

15. Three numbers whose sum is 24 are in A. P., but if 3, 4, and 7 be added to them respectively, these sums will be in G. P. Find the numbers.

16. The sum of $225 was divided among four persons in such a manner that the shares were in G. P., and the difference between the greatest and least was to the difference between the means as 7 is to 2. Find each share.

17. Find the sum of $\frac{1}{\sqrt{2}-1}$, $\sqrt{2}$, $\frac{2}{\sqrt{2}+1}$ . . . . *ad infinitum.*

18. There are 4 numbers the first 3 of which are in G. P., and the last 3 are in A. P.; the sum of the first and last is 14, and of the means is 12. Find them.

19. If the series $\frac{3}{4}$, $\frac{1}{2}$ . . . . be arithmetical, find the 102d term; if geometrical, find the sum to infinity.

20. Divide $369 among A, B, C, and D so that their shares may be in G. P., and the sum of A's and B's shares shall be $144.

21. The ages of three men, A, B, and C, are in G. P., and the sum of their years is 111; but $\frac{3}{8}$ of A's age, $\frac{7}{12}$ of B's, and $\frac{1}{2}$ of C's form an A. P. Find their ages.

22. Insert between 2 and 9 two numbers, such that the first three of the four may be in A. P. and the last three in G. P.

23. Prove that the series $\sqrt{2}-1$, $3\sqrt{2}-4$, $2(5\sqrt{2}-7)$ . . . . is geometrical; that its ratio is $2-\sqrt{2}$; and that its sum to infinity is unity.

HARMONICAL PROGRESSION.

320. An **Harmonical Progression** is a series the reciprocals of whose terms form an arithmetical progression.

Thus, 1, $\frac{1}{4}$, $\frac{1}{7}$, $\frac{1}{10}$, ... is an harmonical progression, since the reciprocals of its terms,

$$1, 4, 7, 10, \ldots\ldots$$

form an arithmetical progression.

The general form of an harmonical progression is—

$$\frac{1}{a}, \frac{1}{a+d}, \frac{1}{a+2d}, \frac{1}{a+3d}, \ldots\ldots \frac{1}{a+(n-1)d}.$$

321. General Principle. Problems relating to harmonical progression are solved to best advantage by taking the reciprocals of the terms of the given progression, and solving the A. P. thus formed.

Ex. 1. Find the 15th term of the H. P., $\frac{1}{2}, \frac{1}{5}, \frac{1}{8}, \frac{1}{11} \ldots\ldots$

Taking the reciprocals of the terms of the given series, we obtain the A. P.,

$$2, 5, 8, 11 \ldots\ldots$$

In this $a=2, \quad d=3, \quad n=15.$

$$\therefore l = 2 + 14 \times 3 = 44.$$

Hence, the 15th term of the given H. P. is the reciprocal of 44; that is, $\frac{1}{44}$.

Ex. 2. Insert 7 harmonic means between -3 and 4.

To solve this problem it is necessary to insert 7 arithmetical means between $-\frac{1}{3}$ and $\frac{1}{4}$.

Hence, $a = -\frac{1}{3}, \quad l = \frac{1}{4}, \quad n = 9.$

$$\therefore \frac{1}{4} = -\frac{1}{3} + 8d, \quad d = \frac{7}{96}.$$

Inserting arithmetical means, the A. P. is

$$-\frac{1}{3}, -\frac{25}{96}, -\frac{3}{16}, -\frac{11}{96}, -\frac{1}{24}, \frac{1}{32}, \frac{5}{48}, \frac{17}{96}, \frac{1}{4}.$$

Hence, the H. P. is

$$-3, -\frac{96}{25}, -\frac{16}{3}, -\frac{96}{11}, -24, 32, \frac{48}{5}, \frac{96}{17}, 4.$$

322. Harmonic Mean. If but one harmonic mean is to be inserted between a and b, and this be denoted by H, then $\frac{1}{H}$ is the arithmetical mean between $\frac{1}{a}$ and $\frac{1}{b}$; and by Art. 309,

$$\frac{1}{H} = \frac{\frac{1}{a}+\frac{1}{b}}{2} = \frac{a+b}{2ab}. \qquad \therefore H = \frac{2ab}{a+b}.$$

If the arithmetical, geometrical, and harmonic means between a and b be denoted by A, G, H, respectively,

we have $\quad A = \frac{a+b}{2}, \quad G = \sqrt{ab}, \quad H = \frac{2ab}{a+b}.$

$$\therefore H = ab \times \frac{2}{a+b} = G^2 \times \frac{1}{A}.$$

$$\therefore G^2 = AH \text{ and } G = \sqrt{AH}.$$

Hence, *the geometrical mean between two quantities is also the geometrical mean between their arithmetical and harmonic means.*

EXERCISE 120.

Find the last term in the series—

1. $\frac{4}{5}, \frac{4}{7}, \frac{4}{9}, \ldots$ to 20 terms.
2. $\frac{6}{5}, 2, 6, \ldots$ to 18 terms.
3. $-\frac{3}{13}, -\frac{3}{8}, -1, \ldots$ to 27 terms.

Insert—

4. Four harmonic means between $\frac{4}{3}$ and $\frac{1}{2}$.
5. Five harmonic means between $\frac{6}{5}$ and $\frac{6}{17}$.
6. Seven harmonic means between $-3\frac{1}{3}$ and $-\frac{2}{7}$.

Find the harmonic mean between—

7. $2\frac{2}{3}$ and 3. 8. $4\frac{1}{2}$ and $-3\frac{3}{5}$. 9. $\frac{x-1}{x+1}$ and $\frac{x+1}{x-1}$.

10. The first term of a H. P. is x, and the second term is y. Find the next two terms.

11. The arithmetical mean between two numbers is 15, and the H. M. is $14\frac{2}{5}$. Find the numbers.

12. The fifth term of a H. P. is -1, and the 14th, is $\frac{2}{7}$. Find the series.

13. The G. M. between two numbers is 16, and the H. M. is $12\frac{4}{5}$. Find the numbers.

CHAPTER XXVI.

PERMUTATIONS AND COMBINATIONS.

323. The **Permutations** of a group of objects are the different *arrangements* which can be made of the objects *with respect to their order.*

Thus, the permutations of the three letters *a*, *b*, *c* are, *abc*, *acb*, *bac*, *bca*, *cab*, *cba*. Taken two letters at a time, the permutations of *a*, *b*, *c* are, *ab*, *ba*, *ac*, *ca*, *bc*, *cb*.

324. The **Combinations** of a group of objects are the different *collections* which can be made from them *without respect to order.*

Thus, the combinations which can be made from three letters, *a*, *b*, *c*, taken two at a time, are, *ab*, *ac*, *bc*. It is seen that *ba* and *ab*, for instance, are different permutations, but are the same combination.

325. The Number of Permutations in a Group of *n* Elements. To determine the number of permutations that can be made of a group of objects by actually writing out the permutations and counting them usually involves so much labor as to make the method impracticable. Instead of this, it is possible to establish a formula by which the number of permutations in a group of objects can be determined by a simple process of multiplication from the number of objects or elements in the group.

Thus, to determine the number of permutations which can be made with a group of four objects, as *a*, *b*, *c*, *d*, we conceive the permutations to be formed by writing each of the four elements in the first place in turn; after each of which the remaining three elements may be placed; after each of which the remaining two, and so on.

Thus, we obtain

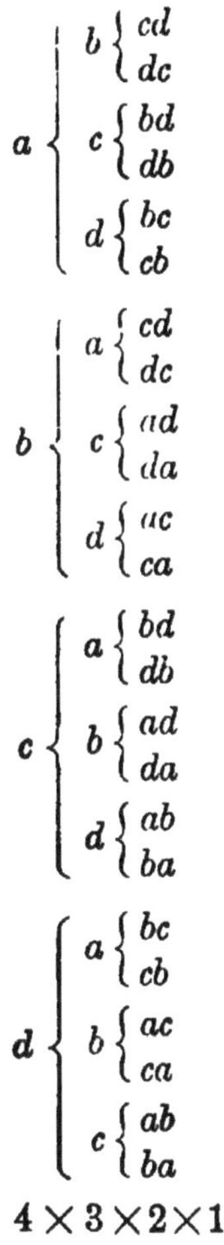

Whence we have all the permutations, *abcd*, *abdc*, etc.; as, $4 \times 3 \times 2 \times 1$, or 24 in number.

Similarly, to form the number of permutations which can be made with a group of n elements, we may write each of the n elements in the first place; after each of which the remaining $n-1$ elements may be written; after each of which the remaining $n-2$ elements, and so on.

Hence, the number of permutations formed from a group of n elements is

$$n(n-1)(n-2) \ldots . 3 \times 2 \times 1.$$

If the permutations of four objects be formed, taking two objects at a time, in the first place each of the four objects may be written, after each of which the remaining three, which exhausts the number of permutations so formed. Hence, the permutations of four objects, taken two at a time, is 4×3.

Similarly, the permutations of n objects taken 2 at a time is $n(n-1)$; of n objects taken r at a time, is

$$n(n-1)(n-2) \ldots . (n-r+1);$$

[r factors in all].

326. Symbols Used. For the number of permutations of a group of n objects, taken r at a time, the symbol, ${}_nP_r$, is used. The product $n(n-1) \ldots . 3 \times 2 \times 1$ is called *factorial* n, and is abbreviated into the form, $\underline{|n}$, or $n!$.

Hence, from the preceding Article,

$${}_nP_n = n(n-1)(n-2) \ldots . 3 \times 2 \times 1 = \underline{|n}, \text{ or } n!.$$

$${}_nP_r = n(n-1)(n-2) \ldots . (n-r+1).$$

Ex. 1. In how many ways may the letters which form the word *Baltimore* be arranged?

Since there are 9 *different* letters in the word Baltimore, we have

$_9P_9 = 9 \times 8 \times 7 \times 6 \times 5 \times 4 \times 3 \times 2 \times 1 = 362,880$ *Permutations.*

Ex. 2. How many of the permutations in Ex. 1 will begin with the letter *l*?

Since the letter *l* remains fixed in the first place, 8 letters are arranged in different orders. Hence, we have

$$_8P_8 = \underline{|8} = 40,320 \text{ Permutations.}$$

327. The Number of Combinations in a Group of *n* Elements. We denote the number of combinations which can be made with a group of *n* elements, taken *r* at a time, by $_nC_r$.

The permutations of a group of *n* objects, taken *r* at a time, might be formed by writing all the different combinations of the *n* objects taken *r* at a time, and then making all possible permutations (that is, arrangements) of the elements of each combination of *r* objects; viz. $\underline{|r}$ arrangements.

Thus, for example,

$$_9P_3 = {_9C_3} \times \underline{|3}$$

Or, in general,

$$_nP_r = {_nC_r} \times \underline{|r}$$

$$\therefore {_nC_r} = \frac{_nP_r}{\underline{|r}}$$

$$_nC_r = \frac{n(n-1)(n-2)\ldots\ldots(n-r+1)}{r(r-1)(r-2)\ldots\ldots 1}.$$

Ex. How many combinations can be formed with the letters of the word *Baltimore*, taken four at a time?

We have

$$_9C_4 = \frac{9 \times 8 \times 7 \times 6}{4 \times 3 \times 2 \times 1} = 126 \text{ Combinations.}$$

328. Other Formulas for $_nC_r$. The formula for $_nC_r$ may be put into another and often more convenient form by multiplying both numerator and denominator by $\lfloor n-r$.

Thus, $$_nC_r = \frac{n(n-1)\ldots.(n-r+1)\lfloor n-r}{\lfloor r\lfloor n-r} = \frac{\lfloor n}{\lfloor r\lfloor n-r}.$$

It is also to be observed that for each different selection (or combination) taken from a group of objects a different group is left. Thus, if from a, b, c, d, e we take a, b, d, we have left c, e; if we take b, c, e, we have left a, d, and so on.

Hence, $\quad {_5C_3} = {_5C_2}$

Or, in general, $\quad {_nC_r} = {_nC_{n-r}}$.

Ex. Determine the number of combinations of 20 letters taken 18 at a time.

$$_{20}C_{18} = {_{20}C_2}$$
$$= \frac{20\times19}{1\times2} = 190 \textit{ Combinations.}$$

EXERCISE 121.

Find the values of—

1. $_6P_4$.
3. $_7P_5$.
5. $_{19}C_{16}$.
7. $_{21}P_5$.
2. $_{20}C_9$.
4. $_{11}C_5$.
6. $_{14}P_4$.
8. $_{18}C_9$.

9. How many different numbers of 5 different figures each can be formed from the nine significant digits?

10. From a company of 24 men, in how many ways can a committee of 10 be selected?

11. How many words of 5 different letters can be formed from our alphabet?

12. There are 13 points in a plane, no three of which are in the same straight line. How many different triangles can be formed having three of the points for vertices?

13. How many different words of eight letters each can be formed from the letters in the word *republic?*

14. How many different numbers of 4 different figures can be formed from the ten digits 0, 1, 2, ... ?

15. From the letters in the word *universal* how many words can be formed, taking 6 at a time? Taking 7 at a time?

16. Five persons enter a car in which there are seven seats. In how many ways may they take their places?

17. How many different throws can be made with 2 dice?

18. How many different throws can be made with 3 dice?

19. How many numbers is it possible to write by use of our 10 digits without repeating any digit?

SPECIAL CASES.

329. I. Elements of a Group repeated. It may happen that some of the elements in a group of objects are the same. Thus, if it be required to determine the number of permutations which may be made of the letters in the word *Chicago*, it is observed that in this word the letter c occurs twice; hence, all the permutations formed by interchanging the c's are identical. Hence, for the total number of permutations we have

$$\frac{7!}{2!} = 2520 \text{ Permutations.}$$

In general, if in a group of n elements, one set of elements, i in number, are identical, and another set, k in number, are identical, the total number of permutations in the entire group, taken all together, is

$$\frac{n!}{i!\,k!}.$$

Ex. How many permutations can be formed from the letters of *United States*, taken all together?

We have 12 letters in all, but t used three times, s used twice, and e twice.

Hence, $\frac{|\underline{12}}{|\underline{3}\,|\underline{2}\,|\underline{2}} = 19{,}958{,}400$ *Permutations.*

330. II. **Different Groups taken Together.** If combinations be formed from one group of elements, and other combinations from another group of elements, and these combinations be taken together in pairs, one from each group, it is evident that the total number of combinations will be the product of the number of combinations in one group by the number in the other group. For each combination in the first group may be joined separately to each combination of the second group.

Ex. Out of 15 Republicans and 10 Democrats, how many different committees may be formed of 4 Republicans and 3 Democrats?

We have for the number of partial committees from the Republicans,

$$_{15}C_4 = \frac{15 \times 14 \times 13 \times 12}{4 \times 3 \times 2 \times 1} = 1365.$$

From Democrats,

$$_{10}C_3 = \frac{10 \times 9 \times 8}{3 \times 2 \times 1} = 120.$$

Hence, the number of entire committees will be

$1365 \times 120 = 163{,}800$, *Result.*

EXERCISE 122.

Find the number of different permutations that can be made from each of the following words, taking all the letters:

1. Inning.
2. Successes.
3. Upper House.
4. Mississippi.
5. Independence.
6. Unconsciousness.

7. How many different arrangements can be made from $a^5b^3c^6$, when written in the expanded form?

8. How many committees of 2 teachers and 3 boys can be selected from a school of 5 teachers and 25 boys?

9. From 9 red balls, 5 white balls, and 4 black balls, how many different combinations can be formed, each consisting of 5 red, 3 white, and 2 black balls?

10. From 9 merchants, 14 lawyers, and 7 teachers, how many different companies can be formed, each consisting of 3 merchants, 5 lawyers, and 2 teachers?

11. How many different words, each consisting of 3 consonants and 1 vowel, can be formed from 12 consonants and 3 vowels?

HINT. The number of possible *combinations* is ${}_{12}C_3 \times {}_3C_1$. But each combination contains 4 letters, and may be arranged into $\lfloor 4$ different *permutations.*

12. How many different words, each consisting of 4 consonants and 2 vowels, can be formed from 8 consonants and 4 vowels?

13. Find the number of different words of 3 consonants and 2 vowels each, that can be formed from an alphabet of 5 vowels and 21 consonants.

14. In how many ways can 2 ladies and 2 gentlemen be chosen to make a set at lawn tennis from a company of 6 ladies and 8 gentlemen?

How many different words can be formed from the letters in the following words, using all the letters in each instance:

15. *Volume,* the second, fourth, and sixth letters being vowels?

16. *Numerical,* the even places always to be occupied by vowels?

17. *Absolute,* the first and last letters to be vowels?

18. *Parallel,* the first and last letters to be consonants?

19. How many different quadrilaterals can be formed from 20 points in a plane, as the vertices, if no three are in the same straight line?

20. A man has 7 pairs of trousers, 5 vests, and 6 coats. In how many different costumes may he appear?

21. In how many different ways can a base-ball nine be arranged, the pitcher and catcher being always the same, but the others playing in any other position?

22. How many different sums of money can be obtained from a cent, a five-cent piece, a dime, a quarter-dollar, and a half-dollar?

23. From 5 labials, 4 palatals, and 6 vowels how many words can be formed, each containing 2 vowels, 2 labials, and a palatal?

24. There are 12 persons in a coach at the time of an accident at which 3 are killed, 4 are wounded, and 5 escape uninjured. In how many ways might all this happen?

25. The number of committees of three members each which can be formed from a certain number of boys, is to the number of similar committees possible if there were one more boy, as 10 to 11. Required the number of boys.

26. The number of words of four letters each which can be made from a certain number of letters, is $\frac{1}{21}$ of the number of words of five letters each which could be made if there were two more letters. Required the number of letters. Verify both results.

27. If ${}_nC_3 : {}_{2n}C_2 = 7 : 15$, find n.

28. If ${}_{20}C_r = {}_{20}C_{r+2}$, find ${}_rP_4$.

29. Show that ${}_{n+1}C_r = {}_nC_r + {}_nC_{r-1}$.

30. In how many ways can nine children form a ring, all facing the centre?

CHAPTER XXVII.

UNDETERMINED COEFFICIENTS.

331. Convergency of Series. Using the result obtained in Art. 317, viz.,

$$s = \frac{a}{1-r},$$

it is found that the infinite series

$$1 + \tfrac{1}{2} + \tfrac{1}{4} + \tfrac{1}{8} + \ldots\ldots\ldots \quad (1)$$

approaches a limit, $\dfrac{1}{1-\frac{1}{2}} = 2.$

Similarly, if we take the infinite series

$$1 + x + x^2 + x^3 + \ldots\ldots\ldots \quad (2),$$

we find that for any value of x between 0 and 1, as $\dfrac{b}{a}$, where $b < a$, the series has a definite limit; viz.,

$$\frac{1}{1-\dfrac{b}{a}} = \frac{a}{a-b}.$$

But if in series (2) we let $x = 1$, the value of the series is

$$1 + 1 + 1 + 1 + \ldots,$$

which becomes greater without limit, the larger the number of terms which is taken.

Similarly, if x have any value greater than 1 in series (2), the series has no limit to its value.

Hence, in one kind of infinite series the sum approaches a fixed quantity as limit, while in the other kind the sum exceeds any assigned limit.

A **Convergent Series** is one in which the sum of the first n terms approaches a certain fixed limit, no matter how great n may be.

A **Divergent Series** is one in which the sum of the first n terms can be made to exceed any assigned quantity by making n sufficiently large.

332. Use of Infinite Series. Infinite series are useful in representing complex algebraic expressions, since such series ordinarily consist of simple combinations of the functions, sum, difference, product, and power of quantities used. It is evident, however, that infinite series can be used in a valid way only when the series used is convergent.

333. Identities. An identity is a statement of equality between two algebraic expressions, which is satisfied by all values whatever of the unknown quantity or quantities used in the expressions.

$$\text{Ex. } x^2 - 4 \equiv (x-2)(x+2)$$

is an identity, since it is satisfied by any value of x, as 1, 2, 3, -1, etc.; whereas, for example, in

$$7x - 6 = 10 - x,$$

x has but a single value; viz., $x = 2$.

The correct *sign* for an identity is $\equiv$, but in the case of such identities as occur in the early part of the algebra, the custom prevails of using the equality mark for these identities as well as for equations. The student, however, should form the habit of carefully discriminating between identities and equalities when the equality mark is thus used for both.

In this chapter the identity mark will be used for identities, and the equality mark for equalities.

334. Fundamental Property of Identities. *In the identity* $A + Bx + Cx^2 + Dx^3 + \dots \equiv A' + B'x + C'x^2 + D'x^3 + \dots$ *(each member being either finite or convergent), the coefficients of like powers of x are equal; that is,* $A = A'$, $B = B'$, $C = C'$, etc.

For, since the identity

$$A + Bx + Cx^2 + Dx^3 + \dots \equiv A' + B'x + C'x^2 + D'x^3 + \dots (1)$$

is true for all values of x, it is true when $x = 0$.

Hence, from (1) $\quad A = A' \dots \dots (2)$

Subtracting (2) from (1), and dividing the result by x,

$$B + Cx + Dx^2 + \dots \equiv B' + C'x + D'x^2 + \dots \dots (3)$$

This identity is true for all values of x.

$\therefore$ letting $x = 0$,

$$B = B' \dots \dots (4)$$

Subtracting (4) from (3), and proceeding as before,

$$C = C', \quad D = D', \text{ etc., etc.}$$

Ex. Find the condition that $x^2 + px + q$ is divisible by $x + a$.

If $x^2 + px + q$ is divisible by $x + a$, the quotient must be of the form $x + k$.

Hence, $\quad x^2 + px + q \equiv (x + a)(x + k)$,

$$x^2 + px + q \equiv x^2 + (a + k)x + ak.$$

Hence, by the fundamental properties of identities,

$$p = a + k \dots \dots (1)$$

$$q = ak \dots \dots (2)$$

Eliminating k from these equalities,

$$q = a(p - a),$$

which expresses the required condition.

EXERCISE 123.

Find the value of k that will make—

1. $8x^2 + kx - 91$ exactly divisible by $2x + 7$.
2. $6x^3 + x^2 + kx + 4$ exactly divisible by $3x - 4$.
3. $3x^3 + kx^2 + 25$ exactly divisible by $x^2 - 3x + 5$.
4. $2x^4 - x^3 - 3x^2 + 10x + k$ divisible by $x^2 + x - 2$.

Determine the conditions necessary that—

5. $x^2 + ax - b$ may be exactly divisible by $x - c$.
6. $ax^2 + bx + c$ may be exactly divisible by $x + d$.
7. $x^3 + ax + b$ may be exactly divisible by $x^2 - x + d$.
8. $x^4 + ax^3 + bx^2 + cx + d$ may be divisible by $x^2 + mx + n$.

Prove that if—

9. $ax^3 + bx^2 + cx + d$ is divisible by $x^2 - l^2$, then $ad = bc$.

10. $ax^2 + 2hxy + by^2 + 2gx + 2fy + c$ is the square of $a'x + b'y + c'$, then $af = gh$, $bg = fh$, and $ch = fg$.

APPLICATIONS.

I. Expansion of a Fraction into a Series.

335. General Method by Undetermined Coefficients. A fraction may be expanded into a series by dividing the numerator by the denominator, but it is often more convenient to make the transformation by the use of undetermined coefficients.

Ex. Expand $\dfrac{1 - 2x^2}{1 + 2x - 3x^2}$ into a series by the use of undetermined coefficients.

Let $$\frac{1 - 2x^2}{1 + 2x - 3x^2} \equiv A + Bx + Cx^2 + Dx^3 + \ldots\ldots\ldots\ldots \quad (1)$$

where $A, B, C, D \ldots\ldots$ are unknown numbers to be determined.

Clear (1) of fractions, collecting the coefficients of x, x^2, etc. by use of the vinculum in the vertical position instead of by use of parenthesis, thus, for $(B + 2A)x$ use $\left.\begin{array}{r} +\ B \\ +\ 2A \end{array}\right| x.$

$$1 - 2x^2 \equiv A + \left.\begin{array}{r} B \\ +\,2A \\ \end{array}\right| x + \left.\begin{array}{r} C \\ +\,2B \\ -\,3A \end{array}\right| x^2 + \left.\begin{array}{r} D \\ +\,2C \\ -\,3B \end{array}\right| x^3 + \ldots\ldots \quad (2)$$

Since (2) is an identity, the coefficients of like powers of x are equal.

Hence,

$$\left.\begin{aligned} A &= 1, \\ B + 2A &= 0, \\ C + 2B - 3A &= -2, \\ D + 2C - 3B &= 0, \end{aligned}\right\} \text{ whence } \left\{\begin{aligned} A &= 1, \\ B &= -2, \\ C &= 5, \\ D &= -16, \end{aligned}\right.$$

etc. etc.

Substitute for A, B, C, D, ... in (1),

$$\frac{1 - 2x^2}{1 + 2x - 3x^2} \equiv 1 - 2x + 5x^2 - 16x^3 + \ldots, \quad \textit{Series.}$$

This result may also be obtained by division.

336. Special Cases. When the lowest power of x in the denominator is higher than the lowest power in the numerator, we may proceed in either of two ways, as illustrated by the following example:

Ex. Expand $\dfrac{1+2x}{x^2-3x^3}$ into a series.

$$\frac{1 + 2x}{x^2 - 3x^3} \equiv \frac{1}{x^2}\left(\frac{1 + 2x}{1 - 3x}\right).$$

Let
$$\frac{1 + 2x}{1 - 3x} \equiv A + Bx + Cx^2 + Dx^3 + \ldots.$$

Solving this identity,

$$\frac{1 + 2x}{1 - 3x} \equiv 1 + 5x + 15x^2 + 45x^3 + \ldots.$$

Multiplying this result by $\dfrac{1}{x^2}$, that is, by x^{-2}, we obtain

$$\frac{1 + 2x}{x^2 - 3x^3} \equiv x^{-2} + 5x^{-1} + 15 + 45x + \ldots, \quad \textit{The required series.}$$

Or, at the outset we might have divided the first term of the numerator, 1, by x^2, the first term of the denominator, and determined x^{-2} as the power of x occurring first in the required series, and let

$$\frac{1 + 2x}{x^2 - 3x^3} \equiv Ax^{-2} + Bx^{-1} + C + Dx + \ldots,$$

and determined A, B, C, D, ... as usual.

Again, if the numerator and denominator of the given fraction contain only *even* powers of x, the process of expansion may be shortened by using only even powers in the assumed series.

Thus, let $\dfrac{3+x^2-2x^4}{1-2x^2+3x^4} \equiv A + Bx^2 + Cx^4 + Dx^6 + \ldots\ldots$

EXERCISE 124.

Expand into a series in ascending powers of x—

1. $\dfrac{1+2x}{1+x}$.
2. $\dfrac{2-3x}{1-2x}$.
3. $\dfrac{1-5x}{1+x-x^2}$.
4. $\dfrac{3-x^2}{1+2x^2}$.
5. $\dfrac{1-x-x^2}{1+x+x^2}$.
6. $\dfrac{x+3x^2}{1-2x+3x^2}$.
7. $\dfrac{2-x-4x^2}{2+x-x^2}$.
8. $\dfrac{3+x-2x^2}{3-x^2+x^3}$.
9. $\dfrac{1-x}{2+6x-x^2}$.
10. $\dfrac{2-3x+2x^2}{3+x-2x^2}$.
11. $\dfrac{3-x^2+4x^3}{2+x^2-x^3}$.
12. $\dfrac{1-2x}{x^2+x^3}$.
13. $\dfrac{2-x^2}{x-x^2+x^3}$.
14. $\dfrac{4+7x}{2x+3x^2}$.
15. $\dfrac{2+x-4x^2}{2x^3+x^4}$.
16. $\dfrac{1-2x^2-x^3}{x^2+x^3-x^4}$.
17. $\dfrac{1-x+x^2}{x^2+x^4}$.

II. Expansion of a Radical into a Series.

337. Illustrative Example.

Ex. Expand $\sqrt{1+2x}$ into a series by use of undetermined coefficients.

Let $\qquad \sqrt{1+2x} \equiv A + Bx + Cx^2 + Dx^3 + \ldots\ldots$

Squaring both sides, using method of Art. 89 for the right member,

$$1 + 2x \equiv A^2 + 2AB \Big| \, x + \begin{array}{c} B^2 \\ 2AC \end{array} \Big| \, x^2 + \begin{array}{c} 2AD \\ +\,2BC \end{array} \Big| \, x^3 + \ldots\ldots$$

In this identity equate the coefficients of like powers of x.

$$\begin{array}{rll} A^2 = 1, & \text{hence,} & A = 1. \\ 2AB = 2, & \text{"} & B = 1. \\ B^2 + 2AC = 0, & \text{"} & C = -\tfrac{1}{2}. \\ 2AD + 2BC = 0, & \text{"} & D = \tfrac{1}{2}. \\ \text{etc.} & & \text{etc.} \end{array}$$

$$\therefore \sqrt{1+2x} \equiv 1 + x - \tfrac{1}{2}x^2 + \tfrac{1}{2}x^3 + \ldots, \quad \textit{Series.}$$

Another series may also be obtained by taking the negative value of A obtained from $A^2 = 1$,—viz. $A = -1$,—and determining corresponding values for B, C, D, etc.

EXERCISE 125.

Expand into a series in ascending powers of x—

1. $\sqrt{1-2x}$. 3. $\sqrt{4-3x+x^2}$. 5. $\sqrt[3]{1+x}$.

2. $\sqrt{1+4x+2x^2}$. 4. $\sqrt{a^2-x^2}$. 6. $\sqrt[3]{1-3x-6x^2}$.

III. Separating a Fraction into Partial Fractions.

338. General Method. If the degree of the numerator of a fraction be greater than the degree of the denominator, the fraction may be converted into a mixed number, the fraction obtained having the degree of the numerator less than the degree of the denominator.

If we consider proper fractions only, such fractions, if their denominators can be factored, may be separated into partial fractions by the use of the properties of identities.

It is evident that if in the original fraction the degree of the numerator is less than the degree of the denominator, the same must be true in each partial or component fraction.

The problem before us is the inverse of that treated in Arts. 143 and 144. There, the several fractions were given to find their sum, but here, the sum is given to determine the constituent fractions. The importance and usefulness of the methods here presented will be more fully appreciated by the pupil when he has advanced to the Integral Calculus.

CASE I.

339. Factors of the Denominator are of the First Degree and Unequal.

Ex. 1. Separate $\dfrac{5x-14}{x^2-6x+8}$ into partial fractions.

Let $$\frac{5x-14}{x^2-6x+8} \equiv \frac{A}{x-2} + \frac{B}{x-4} \quad \ldots \ldots \ldots \ldots (1)$$

Clearing of fractions,

$$5x - 14 \equiv A(x-4) + B(x-2) \quad \ldots \ldots \ldots (2)$$

Hence, $$5x - 14 \equiv (A+B)x - 4A - 2B \quad \ldots \ldots \ldots (3)$$

Equating coefficients of like powers of x,

$$\left.\begin{aligned} A + B &= 5, \\ -4A - 2B &= -14, \end{aligned}\right\} \quad \begin{aligned} &\text{whence } A = 2, \\ &\text{and } \quad B = 3. \end{aligned}$$

Substituting for A and B in (1),

$$\frac{5x-14}{x^2-6x+8} \equiv \frac{2}{x-2} + \frac{3}{x-4}.$$

The values of A and B may also be obtained from (2) in another way, which usually involves less labor in this case. Thus, since (2) is an identity, and therefore true for any value of x whatever, let $x = 2$,

then from (2) $\qquad 10 - 14 = A(2-4).$

Hence $\qquad A = 2.$

Again in (2) let $\qquad x = 4, \qquad$ then $\quad B = 3.$

Ex. 2. Separate $\dfrac{x^2-x-3}{x^3-4x}$ into partial fractions.

$$\frac{x^2-x-3}{x^3-4x} \equiv \frac{A}{x} + \frac{B}{x-2} + \frac{C}{x+2}.$$

Hence, $x^2 - x - 3 \equiv A(x-2)(x+2) + Bx(x+2) + Cx(x-2) \quad .(1)$

In (1) let $\quad x = 0, \qquad$ then $\quad -3 = A(-2)(+2), \qquad$ hence, $\quad A = \frac{3}{4}.$

In (1) let $\quad x = 2, \qquad$ " $\quad B = -\frac{1}{8}.$

In (1) let $\quad x = -2, \qquad$ " $\quad C = \frac{3}{8}.$

Hence, $$\frac{x^2-x-3}{x^3-4x} \equiv \frac{3}{4x} - \frac{1}{8(x-2)} + \frac{3}{8(x+2)}.$$

EXERCISE 126.

Separate into partial fractions—

1. $\dfrac{x-12}{x(x-3)}$. 3. $\dfrac{2x+24}{x^2-9}$. 5.* $\dfrac{2x^3-13x+9}{2x^2-6x}$.

2. $\dfrac{3x}{x^2-x-2}$. 4. $\dfrac{2x^2+7x-1}{x(4x^2-1)}$. 6. $\dfrac{x^2-6x-1}{2x^3-x^2-x}$.

7. $\dfrac{12x^3-73x^2+53}{12x^2-x-6}$. 10. $\dfrac{10(4x^2-x-6)}{(x^2-1)(4x^2-9)}$.

8. $\dfrac{x^4-2x^3+4x+5}{x^2+x-2}$. 11. $\dfrac{ax-14a^2}{x^2-3ax-4a^2}$.

9. $\dfrac{2-11x-3x^2}{18x^3-18x^2+4x}$. 12. $\dfrac{2x^3+5ax^2}{2x^2+ax-a^2}$.

13. $\dfrac{x^6-2x^3+16}{x^3-8}$. 14. $\dfrac{12(x-1)}{x^3-4x^2+2x}$. 15. $\dfrac{4x-14}{4x^2-20x+23}$.

CASE II.

340. Factors of the Denominator are of the First Degree and some of them Repeated.

In this case some factors of the denominator will be of the form x^n or $(ax+b)^n$.

If we consider the fraction $\dfrac{3x-1}{x^3(x+2)}$, it is evidently separable into $\dfrac{Ax^2+Bx+C}{x^3}+\dfrac{D}{x+2}$, the condition being that the degree of the numerator in each partial fraction must be one less than the degree of the denominator.

But
$$\frac{Ax^2+Bx+C}{x^3}\equiv\frac{Ax^2}{x^3}+\frac{Bx}{x^3}+\frac{C}{x^3}$$
$$\equiv\frac{A}{x}+\frac{B}{x^2}+\frac{C}{x^3},$$
the latter being the more convenient form.

---

* First reduce to a mixed number.

Hence, we write

$$\frac{3x-1}{x^3(x+2)} \equiv \frac{A}{x} + \frac{B}{x^2} + \frac{C}{x^3} + \frac{D}{x+2},$$

and find A, B, C, D as usual.

Again, if the factor of the denominator which is a power be a binomial, it is similarly separable.

Thus,
$$\frac{3x+5}{x(x-2)^3} \equiv \frac{A}{x} + \frac{Bx^2+Cx+D}{(x-2)^3}$$

$$\equiv \frac{A}{x} + \frac{Bx^2-4Bx+4B+(C+4B)x+D-4B}{(x-2)^3}$$

$$\equiv \frac{A}{x} + \frac{B}{x-2} + \frac{C'x+D'}{(x-2)^3}$$

$$\equiv \frac{A}{x} + \frac{B}{x-2} + \frac{C'x-2C'+D'+2C'}{(x-2)^3}$$

$$\equiv \frac{A}{x} + \frac{B}{x-2} + \frac{C'}{(x-2)^2} + \frac{D''}{(x-2)^3},$$

where A, B, C', D'' are numbers to be determined.

Ex. Separate $\dfrac{1}{(x-1)^2(x+1)}$ into partial fractions.

$$\frac{1}{(x-1)^2(x+1)} \equiv \frac{A}{x-1} + \frac{B}{(x-1)^2} + \frac{C}{x+1}.$$

Clearing of fractions,

$$1 \equiv A(x^2-1) + B(x+1) + C(x-1)^2$$

Hence,
$$1 \equiv (A+C)x^2 + (B-2C)x - A + B + C.$$

Hence,
$$\left.\begin{aligned} A+C&=0,\\ B-2C&=0,\\ -A+B+C&=1,\end{aligned}\right\}\quad \text{whence}\quad \left\{\begin{aligned} A&=-\tfrac{1}{4}.\\ B&=\tfrac{1}{2}.\\ C&=\tfrac{1}{4}.\end{aligned}\right.$$

$$\therefore \frac{1}{(x-1)^2(x+1)} \equiv -\frac{1}{4(x-1)} + \frac{1}{2(x-1)^2} + \frac{1}{4(x+1)}.$$

EXERCISE 127.

Separate into partial fractions—

1. $\dfrac{4x^2-1}{x^2(x+1)}$.

2. $\dfrac{x^2}{(x+1)^3}$.

3. $\dfrac{2-3x^2}{(x+2)^3}$.

4. $\dfrac{3x-20}{3x^3-4x^2}$.

5. $\dfrac{3x^2+x-2}{(x-2)^2(1-2x)}$.

6. $\dfrac{28x^2-4x}{(2x-1)^3}$.

7. $\dfrac{1-2x-17x^2}{x^3(5x-1)}$.

8. $\dfrac{8x^2+9x-35}{(x-1)^2(x+2)^2}$.

9. $\dfrac{x^3-5}{(x-2)^4}$.

10. $\dfrac{8x^4-81x-54}{x^2(2x+3)^3}$.

11. $\dfrac{5(9+11x)}{(2x^2+x-3)^2}$.

12. $\dfrac{16x^2+15x-50}{x^2(x+1)(2x-5)^2}$.

CASE III.

341. One or More Factors of the Denominator is of the Second Degree.

It is necessary in each case to let the degree of the numerator of each partial fraction be one less than the degree of the denominator.

Ex. 1. Separate $\dfrac{x}{(x+1)(x^2+1)}$ into partial fractions.

Let $\dfrac{x}{(x+1)(x^2+1)} \equiv \dfrac{A}{x+1} + \dfrac{Bx+C}{x^2+1}$.

Hence, $x \equiv Ax^2 + A + Bx^2 + Bx + Cx + C$.

$x \equiv (A+B)x^2 + (B+C)x + A + C$.

Hence, $\left.\begin{array}{l} A+B=0, \\ B+C=1, \\ A+C=0, \end{array}\right\}$ whence $\left\{\begin{array}{l} A=-\frac{1}{2}. \\ B=\frac{1}{2}. \\ C=\frac{1}{2}. \end{array}\right.$

$\therefore \dfrac{x}{(x+1)(x^2+1)} \equiv \dfrac{-1}{2(x+1)} + \dfrac{x+1}{2(x^2+1)}$.

Ex. 2. Separate $\dfrac{3}{x^3+1}$ into partial fractions.

$$\frac{3}{x^3+1} \equiv \frac{A}{x+1} + \frac{Bx+C}{x^2-x+1}$$

$$3 \equiv A(x^2-x+1) + (Bx+C)(x+1)$$

$$3 \equiv (A+B)x^2 + (B+C-A)x + A + C$$

Hence,

$$\left.\begin{aligned} A+B&=0,\\ B+C-A&=0,\\ A+C&=3, \end{aligned}\right\} \quad \text{whence} \quad \left\{\begin{aligned} A&=1,\\ B&=-1,\\ C&=2. \end{aligned}\right.$$

$$\therefore \frac{3}{x^3+1} \equiv \frac{1}{x+1} - \frac{x-2}{x^2-x+1}.$$

The following are other possible examples, with their methods of separation:

$$\frac{3x-2}{x(x^4+2)} \equiv \frac{A}{x} + \frac{Bx^3+Cx^2+Dx+E}{x^4+2}.$$

$$\frac{5x+7}{(x+1)(x^2+2)^2} \equiv \frac{A}{x+1} + \frac{Bx+C}{x^2+2} + \frac{Dx+E}{(x^2+2)^2}.$$

It will be observed that the nnmber of known letters used in the solution of all examples in Partial Fractions is the same as the degree of the Denominator.

EXERCISE 128.

Separate into partial fractions—

1. $\dfrac{5x+4}{x(x^2+2)}$.

2. $\dfrac{16x+4}{x^3-8}$.

3. $\dfrac{2x^2-7x+1}{x^3-1}$.

4. $\dfrac{-5-3x}{x^2(x^2+1)}$.

5. $\dfrac{10x+2}{x^4+x^2+1}$.

6. $\dfrac{x^3+4x^2}{(x^2-x+3)^2}$.

7. $\dfrac{2x^3-x-3}{x(2x^2-3)^2}$.

8. $\dfrac{3+2x-2x^2-4x^3-3x^4}{2x(x^2+1)^2(x-1)}$.

IV. Reversion of Series.

342. General Method. If, for example, the series

$$y = x + 2x^2 + 3x^3 + 4x^4 + \ldots\ldots$$

be given, it is impossible by direct substitution to find the value of x for a given value of y; as, $\frac{1}{10}$.

If, however, we convert the series into the form

$$x = ay + by^2 + cy^3 + dy^4 + \dots,$$

where the coefficients $a, b, c, d \dots$ are known, the value of x may be determined by direct substitution of the value of y.

Reversion of Series is the process of converting a series in which y (or any unknown) is given in terms of x (another unknown) into a series in which x is given in terms of y.

Ex. Revert the series

$$y = x - 2x^2 + 3x^3 - 4x^4 + \dots \dots \quad (1)$$

Let $$x = Ay + By^2 + Cy^3 + Dy^4 + \dots \dots \quad (2)$$

In (2) substitute for y its value from (1),

$$\begin{aligned} x = & A(x - 2x^2 + 3x^3 - 4x^4 + \dots) \\ & + B(x^2 - 4x^3 + 10x^4 + \dots) \\ & + C(x^3 - 6x^4 + \dots) \\ & + D(x^4 + \dots) \\ & + \text{ etc.} \end{aligned}$$

Hence, $$x = Ax \begin{array}{c|} -2A \\ +B \end{array}\; x^2 \begin{array}{c|} +3A \\ -4B \\ +C \end{array}\; x^3 \begin{array}{c|} -4A \\ +10B \\ -6C \\ +D \end{array}\; x^4 + \dots$$

Equating coefficients of like powers of x,

$$\left.\begin{aligned} A &= 1, \\ -2A + B &= 0, \\ 3A - 4B + C &= 0, \\ -4A + 10B - 6C + D &= 0, \\ \text{etc.} \end{aligned}\right\} \text{ whence } \left\{\begin{aligned} A &= 1, \\ B &= 2, \\ C &= 5, \\ D &= 14, \\ \text{etc.} \end{aligned}\right.$$

Substituting for A, B, C, D in (2),

$$x = y + 2y^2 + 5y^3 + 14y^4 + \dots$$

In case a given series contain only odd powers of the unknown number, the process of reverting the series may be abbreviated by assuming only odd powers in the second series used.

Thus, given

$$y = x - 3x^3 + 5x^5 - 7x^7 + \ldots \ldots \quad (1)$$

let

$$x = Ay + By^3 + Cy^5 + Dy^7 + \ldots \ldots \quad (2)$$

For if we assume even powers also of y in the second series, it will be found that all even powers of the value of y contain even powers only of x, and since the left-hand member of (2) contains no even powers of x, all the coefficients of the assumed even powers equal zero.

EXERCISE 129.

Revert to four terms, the series—

1. $y = x + x^2 + x^3 + x^4 \ldots\ldots$
2. $y = x - 3x^2 + 5x^3 - 7x^4 \ldots\ldots$
3. $y = x + 2x^2 + 3x^3 + 4x^4 \ldots\ldots$
4. $y = x - \frac{1}{2}x^2 + \frac{1}{3}x^3 - \frac{1}{4}x^4 \ldots\ldots$
5. $y = x + \dfrac{x^2}{2} + \dfrac{x^3}{4} + \dfrac{x^4}{8} \ldots\ldots$
6. $y = \frac{1}{2}x - \frac{1}{3}x^2 + \frac{1}{4}x^3 - \frac{1}{5}x^4 \ldots\ldots$
7. $y = 2x + 3x^2 + 4x^3 + 5x^4 \ldots\ldots$
8. $y = x + \frac{1}{2}x^3 + \frac{2}{3}x^5 + \frac{3}{4}x^7 \ldots\ldots$
9. $y = x - \dfrac{x^3}{3} + \dfrac{x^5}{5} - \dfrac{x^7}{7} \ldots\ldots$
10. $y = 1 - 2x + 3x^2 - 4x^3 \ldots\ldots$
11. $y = 1 + x + \dfrac{x^2}{\lfloor 2} + \dfrac{x^3}{\lfloor 3} + \dfrac{x^4}{\lfloor 4} \ldots\ldots$

CHAPTER XXVIII.

BINOMIAL THEOREM.

FOR POSITIVE INTEGRAL EXPONENTS.

348. General Formula. By actual multiplication of the first few powers of a binomial, $x + a$, the following results were obtained and stated in Art. 178:

I. The **Number of Terms** equals the exponent of the power of the binomial, plus one.

II. **Exponents.** The exponent of x in the first term equals the index of the required power, and diminishes by 1 in each succeeding term. The exponent of a in the second term is 1, and increases by 1 in each succeeding term.

III. **Coefficients.** The coefficient of the first term is 1; of the second term, it is the index of the required power.

In each succeeding term the coefficient is found by *multiplying the coefficient of the preceding term by the exponent of x in that term, and dividing by the exponent of a, increased by* 1.

IV. **Signs of Terms.** If the binomial is a difference, the signs of the even terms are minus; otherwise, the signs of all the terms are plus.

These results may be expressed in a formula, thus:

$$(x+a)^n = x^n + nx^{n-1}a + \frac{n(n-1)}{\underline{|2}}x^{n-2}a^2 + \frac{n(n-1)(n-2)}{\underline{|3}}x^{n-3}a^3 + \cdots\cdots$$

This formula is called the **Binomial Theorem.**

We shall now prove that these laws hold true for all positive integral values of n.

344. Proof when n is a Positive Integer. The laws stated in the above formula have been shown to be true for the fourth power by actual multiplication. (See Art. 178.) We shall now prove that if this theorem is true for any power, the nth, it is true for the next higher power; viz. the $n+1$st.

Take

$$(x+a)^n = x^n + nx^{n-1}a + \frac{n(n-1)}{1\times 2}x^{n-2}a^2 + \frac{n(n-1)(n-2)}{1\times 2\times 3}x^{n-3}a^3 + \ldots,$$

and multiply both sides by $x+a$.

$$\begin{array}{l}
(x+a)^n = x^n + nx^{n-1}a + \dfrac{n(n-1)}{1\times 2}x^{n-2}a^2 + \dfrac{n(n-1)(n-2)}{1\times 2\times 3}x^{n-3}a^3 + \ldots\ldots \\
\underline{x+a \qquad , \qquad x+a \qquad\qquad\qquad\qquad\qquad\qquad\qquad\qquad\qquad\qquad\qquad} \\
(x+a)^{n+1} = x^{n+1} + nx^n a + \dfrac{n(n-1)}{1\times 2}x^{n-1}a^2 + \dfrac{n(n-1)(n-2)}{1\times 2\times 3}x^{n-2}a^3 + \ldots\ldots \\
\qquad\qquad\qquad\qquad + x^n a \qquad + nx^{n-1}a^2 \qquad + \dfrac{n(n-1)}{1\times 2}x^{n-2}a^3 + \ldots\ldots \\
\hline
(x+a)^{n+1} = x^{n+1} + (n+1)x^n a + \left[\dfrac{n(n-1)}{1\times 2} + n\right]x^{n-1}a^2 \\
\qquad\qquad + \left[\dfrac{n(n-1)(n-2)}{1\times 2\times 3} + \dfrac{n(n-1)}{1\times 2}\right]x^{n-2}a^3 + \ldots\ldots
\end{array}$$

Or,

$$(x+a)^{n+1} = x^{n+1} + (n+1)x^n a + \frac{(n+1)n}{1\times 2}x^{n-1}a^2 + \frac{(n+1)n(n-1)}{1\times 2\times 3}x^{n-2}a^3 + \ldots\ldots$$

This is the same as the original formula, except that $n+1$ is used instead of n.

By actual multiplication the formula is true for the fourth power. Hence, by the general result just proved it must be true for the next higher power, the fifth; hence, again, for the sixth power, and so on indefinitely. This method of proof is called *Mathematical Induction.*

345. Proof by Use of Combinations. Let us consider the products of different binomial factors like $x+a$, $x+b$, $x+c$, etc., in which we afterward make $a=b=c$, etc.

$$(x + a)(x + b) = x^2 + (a + b)x + ab.$$

$$(x + a)(x + b)(x + c) = x^3 + \left.\begin{array}{l} a \\ +b \\ +c \end{array}\right| x^2 + \left.\begin{array}{l} ab \\ +ac \\ +bc \end{array}\right| x + abc.$$

$$(x + a)(x + b)(x + c)(x + d) = x^4 + \left.\begin{array}{l} a \\ +b \\ +c \\ +d \end{array}\right| x^3 + \left.\begin{array}{l} ab \\ +ac \\ +ad \\ +bc \\ +bd \\ +cd \end{array}\right| x^2 + \left.\begin{array}{l} abc \\ +abd \\ +acd \\ +bcd \end{array}\right| x + abcd.$$

In this last product the coefficient of x^4 is 1; of x^3 it is the sum of the combinations of a, b, c, d, taken one at a time; of x^2 it is the sum of the combinations of the same letters taken two at a time; of x, the same taken three at a time; the last term is the product of a, b, c, d, taken all together.

If, now, b, c, d each be made equal to a, the left-hand member will become $(x + a)^4$; the coefficient of x^3 will be a taken as many times as there are combinations in 4 letters taken 1 at a time; that is, $\frac{4}{1}$; the coefficient of x^2 will be a^2 taken ${}_4C_2$, or $\frac{4 \times 3}{1 \times 2}$ times; of x, will be a^3 taken ${}_4C_3$, or $\frac{4 \times 3 \times 2}{1 \times 2 \times 3}$ times; the last term is a^4.

Hence, $(x + a)^4 = x^4 + 4x^3a + \frac{4 \times 3}{1 \times 2}x^2a^2 + \frac{4 \times 3 \times 2}{1 \times 2 \times 3}xa^3 + a^4.$

It will now be a useful exercise for the pupil to prove that if the above law holds for the product of any number of different binomial factors,

$$(x + a)(x + b)(x + c) \ldots . (x + k),$$

it will hold for the product of this number of binomial factors increased by 1, $(x + l)$.

Then, letting $a = b = c = \ldots . = k$ (n letters),

The product of n binomial factors becomes $(x + a)^n$;

The coefficient of x^{n-1} is ${}_nC_1a = na$;

" " x^{n-2} is ${}_nC_2a^2 = \frac{n(n-1)}{1 \times 2}a^2$;

" " x^{n-3} is ${}_nC_3a^3 = \frac{n(n-1)(n-2)}{1 \times 2 \times 3}a^3$;

And so on for all coefficients.

That is, the general theorem holds true when n is a positive integer.

346. Other Forms of the Binomial Formula. When a is negative, a^3, a^5, etc. are negative; hence,

$$(x-a)^n = x^n - nx^{n-1}a + \frac{n(n-1)}{1\times 2}x^{n-2}a^2 - \frac{n(n-1)(n-2)}{1\times 2\times 3}x^{n-3}a^3 + \ldots\ldots \quad (1)$$

If in the original formula x and a be interchanged,

$$(a+x)^n = a^n + na^{n-1}x + \frac{n(n-1)}{1\times 2}a^{n-2}x^2 + \frac{n(n-1)(n-2)}{1\times 2\times 3}a^{n-3}x^3 + \ldots\ldots \quad (2)$$

If a be made equal to 1,

$$(1+x)^n = 1 + nx + \frac{n(n-1)}{1\times 2}x^2 + \frac{n(n-1)(n-2)}{1\times 2\times 3}x^3 + . \quad (3)$$

347. Key Number and rth Term. In committing the binomial formula to memory, it is helpful to observe that a certain number may be regarded as governing the formation of each of its terms. This number is one less than the number of the term. Thus, for the *third* term we have $\frac{n(n-1)}{1\times 2}x^{n-2}a^2$, in which there are *two* factors in the numerator of the coefficient; *two* in the denominator; the exponent of x is $n-2$, and that of a is 2. Hence, we regard 2 as the *key number* of the term.

The number 3 occurs in a similar way in the formation of the fourth term; 4, in the fifth term, and so on.

For the rth term the key number would be $r-1$.

Hence,

$$r\text{th term} = \frac{n(n-1)\ \ldots\ldots\ (n-r+2)}{\underline{|r-1}}x^{n-r+1}a^{r-1}.$$

348. Examples.

Ex. 1. Expand $\left(2a+\dfrac{1}{3\sqrt{y}}\right)^5$.

$$\left(2a+\frac{1}{3\sqrt{y}}\right)^5=\left(2a+\frac{y^{-\frac{1}{2}}}{3}\right)^5$$

$$\left(2a+\frac{y^{-\frac{1}{2}}}{3}\right)^5=(2a)^5+5(2a)^4\left(\frac{y^{-\frac{1}{2}}}{3}\right)+10(2a)^3\left(\frac{y^{-\frac{1}{2}}}{3}\right)^2$$
$$+10(2a)^2\left(\frac{y^{-\frac{1}{2}}}{3}\right)^3+5(2a)\left(\frac{y^{-\frac{1}{2}}}{3}\right)^4+\left(\frac{y^{-\frac{1}{2}}}{3}\right)^5$$
$$=32a^5+\tfrac{80}{3}a^4y^{-\frac{1}{2}}+\tfrac{80}{9}a^3y^{-1}+\tfrac{40}{27}a^2y^{-\frac{3}{2}}+\tfrac{10}{81}ay^{-2}+\tfrac{1}{243}y^{-\frac{5}{2}}.$$

Ex. 2. Expand $\left(\dfrac{x^2}{2}-\dfrac{1}{\sqrt[3]{x^2}}\right)^6$.

$$\left(\frac{x^2}{2}-\frac{1}{\sqrt[3]{x^2}}\right)^6=\left(\frac{x^2}{2}-x^{-\frac{2}{3}}\right)^6$$
$$=\left(\frac{x^2}{2}\right)^6-6\left(\frac{x^2}{2}\right)^5\left(x^{-\frac{2}{3}}\right)+15\left(\frac{x^2}{2}\right)^4\left(x^{-\frac{2}{3}}\right)^2-20\left(\frac{x^2}{2}\right)^3\left(x^{-\frac{2}{3}}\right)^3$$
$$+15\left(\frac{x^2}{2}\right)^2\left(x^{-\frac{2}{3}}\right)^4-6\left(\frac{x^2}{2}\right)\left(x^{-\frac{2}{3}}\right)^5+\left(x^{-\frac{2}{3}}\right)^6$$
$$=\frac{x^{12}}{64}-\frac{3x^{\frac{28}{3}}}{16}+\frac{15x^{\frac{20}{3}}}{16}-\frac{5x^4}{2}+\frac{15x^{\frac{4}{3}}}{4}-3x^{-\frac{4}{3}}+x^{-4}.$$

Ex. 3. Find the sixth term of $\left(\dfrac{x}{2}-\dfrac{2}{3\sqrt{xy}}\right)^9$.

Use the formula of Art. 347.

$$r=6,\quad n=9,\quad x=\frac{x}{2},\quad a=-\tfrac{2}{3}x^{-\frac{1}{2}}y^{-\frac{1}{2}}.$$

$$\text{Sixth term}=\frac{9\times8\times7\times6\times5}{1\times2\times3\times4\times5}\left(\frac{x}{2}\right)^{9-6+1}\left(-\tfrac{2}{3}x^{-\frac{1}{2}}y^{-\frac{1}{2}}\right)^5$$
$$=-\tfrac{126}{1}\cdot\frac{x^4}{2^4}\cdot\frac{2^5x^{-\frac{5}{2}}y^{-\frac{5}{2}}}{3^5}=-\tfrac{28}{27}x^{\frac{3}{2}}y^{-\frac{5}{2}}.$$

Ex. 4. Expand $(1+2x-3x^2)^3$ by use of the binomial formula.

$$
\begin{aligned}
(1+2x-3x^2)^3 &= [(1+2x)-3x^2]^3 \\
&= (1+2x)^3 - 3(1+2x)^2(3x^2) + 3(1+2x)(3x^2)^2 - (3x^2)^3 \\
&= 1 + 3(2x) + 3(2x)^2 + (2x)^3 - 9x^2(1+4x+4x^2) \\
&\qquad + 27x^4(1+2x) - 27x^6 \\
&= 1 + 6x + 3x^2 - 28x^3 - 9x^4 + 54x^5 - 27x^6.
\end{aligned}
$$

EXERCISE 130.

Expand—

1. $(a+3)^4$.
2. $(2a-x)^5$.
3. $\left(1+\frac{x}{2}\right)^6$.
4. $(3x^{\frac{1}{2}}-2y^3)^4$.
5. $(x^{\frac{1}{3}}-2x)^5$.
6. $(x\sqrt{y}+1)^8$.
7. $(x^{-\frac{2}{3}}+\sqrt{x})^7$.
8. $\left(\frac{x}{2y}-\sqrt{xy}\right)^5$.
9. $\left(x^{-2}-\frac{y^3}{3}\right)^5$.
10. $\left(3\sqrt{\frac{a}{x^3}}-1\right)^5$.
11. $(2\sqrt{x}+\sqrt[3]{xy^2})^4$.
12. $(\sqrt[3]{a^2}-\sqrt{a^{-3}})^6$.
13. $\left(\sqrt[3]{ax^2}-3\sqrt{\frac{x}{a}}\right)^4$.
14. $\left(2\sqrt[3]{\frac{a^2}{x}}+3\sqrt{\frac{x}{a^3}}\right)^6$.
15. $(3a^{-\frac{3}{4}}\sqrt{b}-b^{-\frac{7}{2}}\sqrt[4]{a})^4$.
16. $(x^2-x+2)^3$.
17. $(2-3x+x^2)^3$.
18. $(2x^2+x-3)^4$.
19. $(a^2+2ax-x^2)^4$.
20. $(3x^2-2x-1)^4$.

Find the—

21. Sixth term of $(a-2x^3)^{11}$.
22. Eighth term of $(1+x\sqrt{y})^{13}$.
23. Seventh and eleventh terms of $(x^2-y\sqrt[3]{x})^{14}$.
24. Sixth and ninth terms of $(\frac{1}{2}a^2b-2\sqrt[3]{a})^{12}$.
25. Tenth and twelfth terms of $\left(x^{\frac{2}{3}}+\frac{y}{2\sqrt{x}}\right)^{20}$.
26. Middle term of $(3a^{\frac{3}{4}}-x\sqrt[3]{a})^{10}$.

27. Term containing x^5 in $\left(x - \frac{2}{x}\right)^{11}$.

28. Term containing x^{18} in $\left(x^2 - \frac{a}{x}\right)^{15}$.

29. Term containing x^{24} in $\left(\frac{x}{2} + \sqrt{x^3}\right)^{18}$.

30. Term not containing x in $\left(x^3 - \frac{2}{x}\right)^{12}$.

31. Term containing x in $\left(y\sqrt{x} + \sqrt[3]{\frac{y}{x}}\right)^{17}$.

FOR FRACTIONAL OR NEGATIVE EXPONENTS.

349. Examples. The binomial formula is true also when the exponent of the binomial, n, is a fraction or a negative number, provided the resulting series is convergent, though no simple elementary proof in this case can be given.

Ex. 1. Expand $(1 + 3x)^{\frac{1}{2}}$ to 4 terms.

Using formula (3), Art. 346,

$$(1 + 3x)^{\frac{1}{2}} = 1 + \tfrac{1}{2}(3x) + \frac{\frac{1}{2}(\frac{1}{2} - 1)}{1 \times 2}(3x)^2 + \frac{\frac{1}{2}(\frac{1}{2} - 1)(\frac{1}{2} - 2)}{1 \times 2 \times 3}(3x)^3 + \ldots\ldots$$

$$= 1 + \tfrac{3}{2}x - \tfrac{9}{8}x^2 + \tfrac{27}{16}x^3 + \ldots\ldots$$

Ex. 2. Expand $\dfrac{1}{\sqrt[3]{a - x}}$ to 4 terms.

$$\frac{1}{\sqrt[3]{a - x}} = (a - x)^{-\frac{1}{3}} = a^{-\frac{1}{3}}\left(1 - \frac{x}{a}\right)^{-\frac{1}{3}}$$

$$= a^{-\frac{1}{3}}\left[1 - \left(-\tfrac{1}{3}\right)\frac{x}{a} + \frac{-\frac{1}{3}(-\frac{1}{3} - 1)}{1 \times 2}\left(\frac{x}{a}\right)^2 - \frac{-\frac{1}{3}(-\frac{1}{3} - 1)(-\frac{1}{3} - 2)}{1 \times 2 \times 3}\left(\frac{x}{a}\right)^3 + \ldots\ldots\right]$$

$$= a^{-\frac{1}{3}}\left[1 + \frac{x}{3a} + \frac{2x^2}{9a^2} + \frac{14x^3}{81a^3} + \ldots\ldots\right]$$

Ex. 3. Find 6th term of $\left(x^2 - \dfrac{2}{3\sqrt[3]{x}}\right)^{-\frac{1}{2}}$.

We have [Art. 347], $x = x^2$, $a = -\dfrac{2x^{-\frac{1}{3}}}{3}$, $n = -\frac{1}{2}$, $r = 6$.

$$\therefore \text{ 6th term} = \frac{-\frac{1}{2}\cdot -\frac{3}{2}\cdot -\frac{5}{2}\cdot -\frac{7}{2}\cdot -\frac{9}{2}}{1\cdot 2\cdot 3\cdot 4\cdot 5}(x^2)^{-\frac{1}{2}-5}\left(-\frac{2x^{-\frac{1}{3}}}{3}\right)^5$$

$$= \tfrac{7}{216}\cdot x^{-11}\cdot x^{-\frac{5}{3}} = \tfrac{7}{216}x^{-\frac{38}{3}}.$$

350. Extraction of Roots of Numbers by Use of the Binomial Formula. When a number whose root is to be extracted approximates an exact power corresponding to the index of the root, the required root may often be obtained readily by use of the binomial formula.

Ex. Extract the cube root of 215 to five decimal places.

$$215 = 216 - 1 = 216(1 - \tfrac{1}{216}) = 6^3(1 - \tfrac{1}{216}).$$

$$\sqrt[3]{215} = 6(1 - \tfrac{1}{216})^{\frac{1}{3}}$$

$$= 6[1 - \tfrac{1}{3}\cdot\tfrac{1}{216} - \tfrac{1}{9}\cdot(\tfrac{1}{216})^2 - \tfrac{5}{81}(\tfrac{1}{216})^3 - \ldots\ldots]$$

$$= 6(1 - .001543 - .000002 -)$$

$$= 5.99073 +, \textit{ Root.}$$

EXERCISE 131.

Expand to five terms—

1. $(a+x)^{-2}$.
2. $(x^2-1)^{\frac{3}{2}}$.
3. $(x+3)^{\frac{1}{2}}$.
4. $(1+2x^3)^{-1}$.
5. $(x^2 - x^{\frac{1}{2}})^{-3}$.
6. $(\sqrt{x}+2)^{\frac{2}{3}}$.
7. $(x^{-1}+x\sqrt{y})^{-\frac{1}{2}}$.
8. $\dfrac{1}{(1+x)^7}$.
9. $\sqrt[4]{a-x}$.
10. $\sqrt[3]{a^3-3x}$.
11. $(\sqrt[3]{x^2}-4y^3)^{\frac{7}{2}}$.
12. $(a^{-\frac{1}{6}} + \sqrt{ax^{-3}})^{-4}$.

13. $\dfrac{1}{\sqrt[4]{x^3 - 2a}}$.

14. $\dfrac{1}{\sqrt{x^2 - 3\sqrt{x}}}$.

15. $\dfrac{1}{\sqrt[3]{a^{-1} + 3x^{\frac{1}{3}}}}$.

16. $\sqrt{x^{-\frac{2}{3}} - \dfrac{2x}{y^{-\frac{1}{2}}}}$.

17. $\left\{\dfrac{1}{\sqrt[3]{1 - \sqrt[6]{x^2y^{-3}}}}\right\}^4$.

18. $\left\{\dfrac{x}{\sqrt{y^3}} - \dfrac{2y}{\sqrt[3]{x^2}}\right\}^{\frac{5}{2}}$.

Find in the simplest form—

19. Fifth term of $(1 + x)^{\frac{7}{2}}$.

20. Eighth term of $(1 + 2x)^{-\frac{1}{2}}$.

21. Tenth term of $(a^2 - 3\sqrt{ax})^{\frac{16}{3}}$.

22. Fifth term of $\dfrac{1}{2a - 3\sqrt[3]{x}}$.

23. Fifth and tenth terms of $(x^{-1} - 2\sqrt{x})^{\frac{3}{2}}$.

24. Sixth and eleventh terms of $(x^{-3} + 3x^{-1})^{\frac{1}{3}}$.

25. Seventh and thirteenth terms of $\left(\sqrt[3]{x} - \dfrac{y}{2x}\right)^{-3}$.

26. The term containing x^{-20} in $\left(x^3 - \dfrac{1}{x}\right)^{\frac{4}{3}}$.

27. The term containing x^{-9} in $\left(x^{\frac{2}{3}} - \dfrac{2}{\sqrt{x}}\right)^{\frac{1}{2}}$.

28. The term involving x^{17} in $\dfrac{1}{\sqrt[3]{x^{-1}} - a\sqrt{x}}$.

Find the approximate value of the following to five decimal places:

29. $\sqrt{10}$. 31. $\sqrt[3]{30}$. 33. $\sqrt{94}$. 35. $\sqrt[4]{260}$.

30. $\sqrt{65}$. 32. $\sqrt[3]{128}$. 34. $\sqrt[4]{15}$. 36. $\sqrt[5]{35}$.

CHAPTER XXIX.

CONTINUED FRACTIONS.

351. A Continued Fraction is a fraction the denominator of which contains a fraction, the denominator of that fraction also containing a fraction, and so on, either for a finite or an infinite number of minor fractions.

Ex.
$$\cfrac{3}{4+\cfrac{5}{6+\cfrac{7}{8+\cdots\cdots}}}$$

Continued fractions are usually limited to those in which each numerator is unity; as,

$$\cfrac{1}{a+\cfrac{1}{b+\cfrac{1}{c+\cdots\cdots}}}$$

Continued fractions are more conveniently written and printed in the form,

$$\frac{1}{a+}\;\frac{1}{b+}\;\frac{1}{c+}\cdots\cdots$$

352. Integral and Converging Fractions. The simple fractions which compose a continued fraction are called **Integral Fractions.** Thus, in the above example, $\frac{1}{a}$, $\frac{1}{b}$, $\frac{1}{c}$, etc. are the integral fractions.

The **Converging Fractions** are the continued fractions

formed by taking one, two, three, etc. integral fractions at a time. Thus, in the above evample,

$\frac{1}{a}$ is the first convergent,

$\dfrac{1}{a+\frac{1}{b}}$ is the second convergent,

$\dfrac{1}{a+\dfrac{1}{b+\frac{1}{c}}}$ is the third convergent,

etc., etc.

358. A Common Fraction made into a Continued Fraction. By a method essentially the same as that used in finding the greatest common divisor of two numbers, a common fraction may be made into a continued fraction. For instance,

$$\frac{19}{43}=\frac{1}{\frac{43}{19}}=\frac{1}{2+\frac{5}{19}}=\frac{1}{2+\dfrac{1}{\frac{19}{5}}}$$

$$=\frac{1}{2+\dfrac{1}{3+\frac{4}{5}}}=\frac{1}{2+\dfrac{1}{3+\dfrac{1}{\frac{5}{4}}}}=\frac{1}{2+\dfrac{1}{3+\dfrac{1}{1+\frac{1}{4}}}}.$$

The process is more conveniently presented thus:

```
19)43(2
   38
    5)19(3
      15
       4)5(1
         4
         1)4(4
```

The quotients 2, 3, 1, 4, are the denominators of the integral fractions composing the continued fraction, the numerators in each case being 1.

354. Computation of Converging Fractions. We shall now obtain a method of computing the values of the successive convergents of a continued fraction.

Consider the continued fraction,

$$\frac{1}{a+}\ \frac{1}{b+}\ \frac{1}{c+}\ \ldots\ldots +\frac{1}{p+}\ \frac{1}{q+}\ \frac{1}{r+}\ \frac{1}{s+}\ \ldots\ldots$$

$$\text{1st convergent} = \frac{1}{a},$$

$$\text{2d convergent} = \frac{1}{a+\frac{1}{b}} = \frac{b}{ab+1},$$

$$\text{3d convergent} = \frac{1}{a+\frac{1}{b+\frac{1}{c}}} = \frac{bc+1}{c(ab+1)+a}.$$

An examination of the third convergent shows that it may be formed from the two preceding convergents. Thus,

Num. of 3d conv. = (num. of 2d conv.) × (3d quot.) + (num. 1st conv.),
Denom. of 3d conv. = (denom. of 2d conv.) × (3d quot.) + (denom. 1st conv.)

In general,

N. of rth con. = [*N. of* $(r-1)^{th}$ *con.*] [*rth quo.*] + [*N. of* $(r-2)^{th}$ *con.*]
D. of rth con. = [*D. of* $(r-1)^{th}$ *con.*] [*rth quo.*] + [*D. of* $(r-2)^{th}$ *con.*]

We shall now prove that if these laws hold for the formation of any convergent from the preceding convergents, they will hold for the formation of the next succeeding convergent.

Denote the convergents corresponding to the quotients p, q, r, s by $\frac{P}{P'}$, $\frac{Q}{Q'}$, $\frac{R}{R'}$, $\frac{S}{S'}$, and suppose the rth convergent to be formed according to the law.

Hence,
$$\frac{R}{R'} = \frac{Qr+P}{Q'r+P'} \quad \ldots\ldots\ldots\ldots (1)$$

An examination of the general continued fraction given above shows that the sth convergent is formed from the rth by changing r into $r+\frac{1}{s}$. Making this substitution in (1),

24

$$\frac{S}{S'}=\frac{Q\left(r+\frac{1}{s}\right)+P}{Q'\left(r+\frac{1}{s}\right)+P'}=\frac{Qrs+Q+Ps}{Q'rs+Q'+P's}$$

$$=\frac{(Qr+P)s+Q}{(Q'r+P')s+Q'}=\frac{Rs+Q}{R's+Q'}.$$

Hence, if the law is true for the rth convergent, it is true for the next. But by actual reduction the law holds for the formation of the 3d convergent; hence, by the general principle just proved it must hold for the 4th convergent; hence, for the 5th, and so on.

Ex. Find a series of converging fractions for $\frac{157}{972}$.

Forming the given fraction into a continued fraction, the quotients are,

6, 5, 4, 3, 2.

Hence, convergents are, $\frac{1}{6}$, $\frac{5}{31}$, $\frac{21}{130}$, $\frac{68}{421}$, $\frac{157}{972}$.

The first and second convergents are readily determined from the continued fraction. For the others, the following scheme may be found helpful:

$$3\text{d}=\begin{cases}4\times \underline{5}+\underline{1}=\underline{21}\\ 4\times 31+6=130\end{cases};\qquad 4\text{th}=\begin{cases}3\times \underline{21}+\underline{5}=\underline{68}\\ 3\times 130+31=421\end{cases};\ \text{etc.}$$

355. Convergents as Successive Approximations. The *first*, *third*, ... and all *odd* convergents are *larger* than the value of the entire continued fraction; the *second*, *fourth*, ... and all *even* convergents are *smaller*.

For consider the continued fraction,

$$\frac{1}{a+}\,\frac{1}{b+}\,\frac{1}{c+}\,\frac{1}{d+}\,\cdots\cdots$$

The first convergent, $\frac{1}{a}$, is larger than the entire continued fraction, since the denominator, a, is smaller than the denominator of the entire continued fraction.

The second convergent, $\cfrac{1}{a+\cfrac{1}{b}}$, is smaller than the continued fraction, since the denominator, b, is too small; hence, $\frac{1}{b}$ is too large; hence, $a+\frac{1}{b}$ is too large a denominator; hence, $\frac{1}{a+\frac{1}{b}}$ is too small. And so on alternately.

The convergents, however, approach nearer and nearer the value of the original fraction.

N. B. Should the original fraction be improper, the first convergent is an integer, and of course less than the value of the fraction. In the case of mixed numbers and improper fractions, therefore, the *odd* convergents are *less* and the *even* convergents *greater*, than the value of the entire continued fraction.

356. Degree of Approximation in Convergents. The difference between two successive convergents, $\frac{P}{P'}$ and $\frac{Q}{Q'}$, can be shown to be $\frac{1}{P'Q'}$, and therefore the difference between either of these convergents and the value of the entire continued fraction is less than $\frac{1}{P'Q'}$.

The difference between the first two convergents is

$$\frac{1}{a}-\frac{b}{ab+1}=\frac{1}{a(ab+1)}.$$

We shall now prove that if this law holds for the difference between any pair of convergents, $\frac{P}{P'}$ and $\frac{Q}{Q'}$, it will hold for the difference between the next pair, $\frac{Q}{Q'}$, $\frac{R}{R'}$.

Let the symbol $\sim$ be used to denote the *difference between.*

Let $$\frac{P}{P'}\sim\frac{Q}{Q'}=\frac{PQ'\sim P'Q}{PQ'\sim P'Q}=\frac{1}{P'Q'}\quad\ldots\ldots\ldots(1)$$

But
$$\frac{R}{R'} \sim \frac{Q}{Q'} = \frac{RQ' \sim R'Q}{R'Q'}$$
$$= \frac{(rQ + P)Q' \sim (rQ' + P')Q}{R'Q'}$$
$$= \frac{PQ' \sim P'Q}{R'Q'} = \frac{1}{R'Q'}, \text{ from (1).}$$

Hence, if the law is true for the difference between any pair of convergents, $\frac{P}{P'}$, $\frac{Q}{Q'}$, it is true for the difference between the next pair, $\frac{Q}{Q'}$, $\frac{R}{R'}$. But by actual reduction it is true for the difference between the first pair of convergents; hence, by the general principle just proved it is true for the difference between the second pair; and so on indefinitely.

Ex. In the example of Art. 354, what is the error in using the third convergent, $\frac{21}{130}$, instead of the value of the entire continued fraction, $\frac{157}{972}$?

The next convergent after $\frac{21}{130}$ is $\frac{68}{421}$; hence, the error is less than $\frac{1}{130 \times 421}$. This is called the superior limit of the error.

EXERCISE 132.

Express as continued fractions—

1. $\frac{31}{13}$. 2. $\frac{37}{11}$. 3. $5\frac{17}{24}$. 4. $\frac{45}{78}$.

Find the fourth convergent in—

5. $3 + \cfrac{1}{2 + \cfrac{1}{4 + \cfrac{1}{1 + \frac{1}{2}}}}$. 6. $1 + \cfrac{1}{1 + \cfrac{1}{3 + \cfrac{1}{2 + \frac{1}{3}}}}$.

7. $1 + \frac{1}{2+}\ \frac{1}{1+}\ \frac{1}{2+}\ \frac{1}{1+\ldots\ldots}$.

8. $\frac{1}{3+}\ \frac{1}{2+}\ \frac{1}{1+}\ \frac{1}{3+\ldots\ldots}$.

Express the following as continued fractions, and find the *fifth* convergent in each:

9. $\frac{61}{25}$. 11. $\frac{285}{168}$. 13. $\frac{251}{802}$. 15. $\frac{421}{1800}$. 17. 0.3029.

10. $\frac{108}{85}$. 12. $\frac{236}{568}$. 14. $4\frac{113}{261}$. 16. 1.59. 18. 0.5678.

Determine the superior limit of the error in taking the *fourth* convergent for the continued fraction itself in each of the examples, 9–18 inclusive.

357. Surds Expressed as Continued Fractions. A quadratic surd may be converted into an infinite continued fraction.

Ex. Convert $\sqrt{6}$ into a continued fraction.

2 is the greatest integer in $\sqrt{6}$.

Hence, $\sqrt{6} = 2 + (\sqrt{6} - 2)$

$$= 2 + \frac{\sqrt{6}-2}{1} \times \frac{\sqrt{6}+2}{\sqrt{6}+2} = 2 + \frac{2}{\sqrt{6}+2}$$

$$= 2 + \cfrac{1}{\cfrac{\sqrt{6}+2}{2}} \quad \dots\dots\dots\dots \quad (1)$$

$$= 2 + \cfrac{1}{2 + \cfrac{\sqrt{6}-2}{2}} \quad \left(\text{Since 2 is the greatest integer in } \frac{\sqrt{6}+2}{2}\right)$$

$$= 2 + \cfrac{1}{2 + \cfrac{\sqrt{6}-2}{2} \times \cfrac{\sqrt{6}+2}{\sqrt{6}+2}}$$

$$= 2 + \cfrac{1}{2 + \cfrac{1}{\sqrt{6}+2}}$$

$$= 2 + \cfrac{1}{2 + \cfrac{1}{4 + \cfrac{1}{\cfrac{\sqrt{6}+2}{2}}}} \quad \dots\dots\dots\dots \quad (2)$$

The last denominator fraction in (2), viz. $\frac{\sqrt{6}+2}{2}$, is the same as the last denominator fraction in (1); hence, continuing the process indefinitely,

$$\sqrt{6}=2+\frac{1}{2+}\;\frac{1}{4+}\;\frac{1}{2+}\;\frac{1}{4+}\cdots\cdots$$

An infinite continued fraction in which the denominators repeat themselves periodically is called a *Periodic Continued Fraction.*

358. A Periodic Continued Fraction Expressed as the Root of an Equation. A periodic continued fraction may be expressed as the root of an equation. Thus, to express

$$\frac{1}{2+}\;\frac{1}{3+}\;\frac{1}{2+}\;\frac{1}{3+}\cdots\cdots,$$

let x denote the value of the periodic continued fraction.

$$\therefore x=\cfrac{1}{2+\cfrac{1}{3+x}}.$$

Clearing of fractions,

$$2x^2+7x=3+x$$

$$x^2+3x=\tfrac{3}{2}$$

$$x=\tfrac{1}{2}(-3+\sqrt{15}).$$

The sign + is used before the radical $\sqrt{15}$, since x can have a positive value only.

EXERCISE 133.

Express each surd as a continued fraction—

1. $\sqrt{5}$. 4. $\sqrt{8}$. 7. $\sqrt{19}$. 10. $\sqrt{\frac{13}{5}}$.

2. $\sqrt{3}$. 5. $\sqrt{7}$. 8. $2\sqrt{5}$. 11. $\sqrt{\frac{11}{19}}$.

3. $\sqrt{10}$. 6. $\sqrt{14}$. 9. $3\sqrt{2}$. 12. $\frac{1}{\sqrt{33}}$.

13. $3+\sqrt{23}$. 14. $\frac{\sqrt{15}+3}{3}$. 15. $\frac{\sqrt{37}+5}{4}$.

Express each continued fraction as a surd—

16. $\frac{1}{1+}\ \frac{1}{3+}\ \frac{1}{1+}\ \frac{1}{3+}\ \ldots\ldots$

17. $\frac{1}{2+}\ \frac{1}{4+}\ \frac{1}{2+}\ \frac{1}{4+}\ \ldots\ldots$

18. $\frac{1}{1+}\ \frac{1}{1+}\ \frac{1}{1+}\ \frac{1}{1+}\ \ldots\ldots$

19. $1+\frac{1}{2+}\ \frac{1}{3+}\ \frac{1}{2+}\ \frac{1}{3+}\ \ldots\ldots$

20. $3+\frac{1}{4+}\ \frac{1}{1+}\ \frac{1}{4+}\ \frac{1}{1+}\ \ldots\ldots$

21. $4+\frac{1}{3+}\ \frac{1}{1+}\ \frac{1}{2+}\ \frac{1}{3+}\ \frac{1}{1+}\ \ldots\ldots$

Express as a continued fraction the positive root of each equation—

22. $x^2-2x=10.$ 23. $x^2--4x=8.$ 24. $5x^2-7x=2.$

25. Express each root of $3x^2-8x+1=0$ as a continued fraction.

26. The circumference of any circle is 3.1415926 times its diameter. Required the series of fractions converging to this ratio.

27. The lunar month is approximately 27.321661 days. Find a series of fractions converging to this quantity.

28. A solar year is 5 hours, 48 minutes, 49 seconds more than 365 days. Find a series of common fractions approximating nearer and nearer the ratio of this excess to a day.

CHAPTER XXX.

LOGARITHMS.

359. The **Logarithm** of a number is the exponent of that power of another number taken as the base, which equals the given number. Thus,

If 10 be the base, since $1000 = 10^3$, $\log 1000 = 3$;

if 8 be the base, since $4 = 8^{\frac{2}{3}}$, $\log 4 = \frac{2}{3}$.

if $B^l = N$, $\log_B N = l$. This is read:

log of N *to the base* $B = l$.

360. Source of Value in Logarithms. The source of new power in the use of logarithms may be illustrated by the multiplication of two numbers which are exact powers of 10, as 1000 and 100, by the use of exponents. Thus,

Since $1000 = 10^3$,

and $100 = 10^2$,

$1000 \times 100 = 10^5 = 100{,}000$, *Product.*

In like manner, if $361 = 10^{2.55751+}$,

and $29 = 10^{1.46240+}$,

we may multiply 361 by 29, by adding the exponents of the powers of 10 which equal these numbers, obtaining $10^{4.01991+}$, and then obtaining from a table of logarithms the number corresponding to this result, which will be the product. Thus, by the systematic use of exponents or logarithms, the process of multiplying one number by another is converted into the simpler process of adding two numbers (exponents).

In like manner, by the use of logarithms, the process of

dividing one number by another is converted into the simpler process of subtracting one exponent (or log) from another; the process of involution is converted into the simpler process of multiplication; and evolution, into the simpler process of division.

In the systematic use of these properties of exponents lies the source of new power in logarithms.

361. Systems of Logarithms. Any positive number except unity may be made the base of a system of logarithms.

The base used is usually denoted by placing it as a small subscript to the word log. Thus, the logarithm of n in a system whose base is a is denoted by $\log_a n$.

Two principal systems of logarithms are in use—

1. The **Common** (or **Decimal**) or **Briggian** system, in which the base is 10, used in numerical operations.

2. The **Napierian** system, in which the base is 2.7182818 +, generally used in algebraic processes, as to demonstrate properties of expressions by the use of logarithms.

COMMON SYSTEM.

362. Characteristic and Mantissa. If a given number, as 361, be not an exact power of the base, its logarithm, as 2.55751 + for 361, consists of two parts, the whole number, called the **Characteristic**, and the decimal part, called the **Mantissa.**

To obtain a rule for determining the characteristic of a given number (the base being 10), we have,

$$10000 = 10^4, \quad \text{hence,} \quad \log 10000 = 4.$$
$$1000 = 10^3, \quad \text{"} \quad \log 1000 = 3.$$

Hence, any number between 1000 and 10000 has a log between 3 and 4; that is, it consists of 3 and a fraction. Therefore, every integral number consisting of 4 figures has 3 for a characteristic.

Similarly, since $100 = 10^2$, $10 = 10^1$, $1 = 10^0$, every number between 100 and 1000, and therefore containing 3 integral figures, has 2 for a characteristic; every number between 10 and 100 (that is, every number containing 2 integral figures) has 1 for a characteristic; and every number between 1 and 10 (that is, every number containing 1 integral figure) has 0 for a characteristic.

Hence,

The characteristic of every integral or mixed number is one less than the number of figures to the left of the decimal point.

363. Characteristics of Decimal Numbers.

Since

$$1 = 10^0,$$

$$.1 = \frac{1}{10} = 10^{-1},$$

$$.01 = \frac{1}{100} = \frac{1}{10^2} = 10^{-2},$$

$$.001 = \frac{1}{1000} = \frac{1}{10^3} = 10^{-3}, \text{ etc., etc.,}$$

the logarithm of every number between .1 and 1 (as, for instance, of .3) will lie between -1 and 1; that is, will be -1, plus a positive fraction; also, the logarithm of every number between .01 and .1 (as of numbers like .0415) will lie between -2 and -1, and hence consist of -2, plus a positive fraction; and so on.

Hence,

The characteristic of a decimal number is negative, and is, numerically, one more than the number of zeros between the decimal point and the first significant figure.

The characteristic of a decimal number is written in two principal ways. Thus,

$$\log .0372 = \bar{2}.5705,$$

the minus sign being placed over the characteristic 2, to show that it alone is negative, the mantissa being positive.

We may also add and subtract 10 from the given log. Thus,

$$\log .0372 = 8.5705 - 10.$$

In practice, we use the following rule for the characteristics of decimal fractions:

Subtract the number of zeros between the decimal point and the first significant figure from 9, and annex -10 *after the mantissa.*

364. Mantissas of numbers are computed by methods which are beyond the scope of this book. After being computed they are arranged in tables, and when needed are taken from the tables.

The position of the decimal point in a number affects only the characteristic, not the mantissa.

For example, $69.72 = \frac{6972}{100} = \frac{6972}{10^2}.$

Hence, if $6972 = 10^{3.84336}$,

$$\log \frac{6972}{10^2} = \log \frac{10^{3.84336}}{10^2} = \log 10^{1.84336} = 1.84336.$$

In general,

$$\begin{aligned} \log 6972 &= 3.84336 \\ \log 697.2 &= 2.84336 \\ \log 69.72 &= 1.84336 \\ \log 6.972 &= 0.84336 \\ \log 0.6972 &= 9.84336 - 10 \\ \log .06972 &= 8.84336 - 10 \end{aligned}$$

365. Direct Use of a Table of Logarithms; that is, *given a number, to find its logarithm from the table.* We shall here insert a small table of logarithms, that the student may learn enough of their use to understand their algebraic properties. The thorough use of logarithms for purposes of computation is usually taken up in connection with the study of Trigonometry. In the given table (see pages 380, 381) the left-hand column is a column of numbers, and is headed N.

The mantissa of each of these numbers is in the next column opposite. In the top row of each page are the figures 0, 1, 2, 3, 4, 5, 6, 7, 8, 9.

N.	0	1	2	3	4	5	6	7	8	9
10	0000	0043	0086	0128	0170	0212	0253	0294	0334	0374
11	414	453	492	531	569	607	645	682	719	755
12	792	828	864	899	934	969	1004	1038	1072	1106
13	1139	1173	1206	1239	1271	1303	335	367	399	430
14	461	492	523	553	584	614	644	673	703	732
15	1761	1790	1818	1847	1875	1903	1931	1959	1987	2014
16	2041	2068	2095	2122	2148	2175	2201	2227	2253	279
17	304	330	355	380	405	430	455	480	504	529
18	553	577	601	625	648	672	695	718	742	765
19	788	810	833	856	878	900	923	945	967	989
20	3010	3032	3054	3075	3096	3118	3139	3160	3181	3201
21	222	243	263	284	304	324	345	365	385	404
22	424	444	464	483	502	522	541	560	579	598
23	617	636	655	674	692	711	729	747	766	784
24	802	820	838	856	874	892	909	927	945	962
25	3979	3997	4014	4031	4048	4065	4082	4099	4116	4133
26	4150	4166	183	200	216	232	249	265	281	298
27	314	330	346	362	378	393	409	425	440	456
28	472	487	502	518	533	548	564	579	594	609
29	624	639	654	669	683	698	713	728	742	757
30	4771	4786	4800	4814	4829	4843	4857	4871	4886	4900
31	914	928	942	955	969	983	997	5011	5024	5038
32	5051	5065	5079	5092	5105	5119	5132	145	159	172
33	185	198	211	224	237	250	263	276	289	302
34	315	328	340	353	366	378	391	403	416	428
35	5441	5453	5465	5478	5490	5502	5514	5527	5539	5551
36	563	575	587	599	611	623	635	647	658	670
37	682	694	705	717	729	740	752	763	775	786
38	798	809	821	832	843	855	866	877	888	899
39	911	922	933	944	955	966	977	988	999	6010
40	6021	6031	6042	6053	6064	6075	6085	6096	6107	6117
41	128	138	149	160	170	180	191	201	212	222
42	232	243	253	263	274	284	294	304	314	325
43	335	345	355	365	375	385	395	405	415	425
44	435	444	454	464	474	484	493	503	513	522
45	6532	6542	6551	6561	6571	6580	6590	6599	6609	6618
46	628	637	646	656	665	675	684	693	702	712
47	721	730	739	749	758	767	776	785	794	803
48	812	821	830	839	848	857	866	875	884	893
49	902	911	920	928	937	946	955	964	972	981
50	6990	6998	7007	7016	7024	7033	7042	7050	7059	7067
51	7076	7084	093	101	110	118	126	135	143	152
52	160	168	177	185	193	202	210	218	226	235
53	243	251	259	267	275	284	292	300	308	316
54	324	332	340	348	356	364	372	380	388	396
N.	0	1	2	3	4	5	6	7	8	9

N.	0	1	2	3	4	5	6	7	8	9
55	7404	7412	7419	7427	7435	7443	7451	7459	7466	7474
56	482	490	497	505	513	520	528	536	543	551
57	559	566	574	582	589	597	604	612	619	627
58	634	642	649	657	664	672	679	686	694	701
59	709	716	723	731	738	745	752	760	767	774
60	7782	7789	7796	7803	7810	7818	7825	7832	7839	7846
61	853	860	868	875	882	889	896	903	910	917
62	924	931	938	945	952	959	966	973	980	987
63	993	8000	8007	8014	8021	8028	8035	8041	8048	8055
64	8062	069	075	082	089	096	102	109	116	122
65	8129	8136	8142	8149	8156	8162	8169	8176	8182	8189
66	195	202	209	215	222	228	235	241	248	254
67	261	267	274	280	287	293	299	306	312	319
68	325	331	338	344	351	357	363	370	376	382
69	388	395	401	407	414	420	426	432	439	445
70	8451	8457	8463	8470	8476	8482	8488	8494	8500	8506
71	513	519	525	531	537	543	549	555	561	567
72	573	579	585	591	597	603	609	615	621	627
73	633	639	645	651	657	663	669	675	681	686
74	692	698	704	710	716	722	727	733	739	745
75	8751	8756	8762	8768	8774	8779	8785	8791	8797	8802
76	808	814	820	825	831	837	842	848	854	859
77	865	871	876	882	887	893	899	904	910	915
78	921	927	932	938	943	949	954	960	965	971
79	976	982	987	993	998	9004	9009	9015	9020	9025
80	9031	9036	9042	9047	9053	9058	9063	9069	9074	9079
81	085	090	096	101	106	112	117	122	128	133
82	138	143	149	154	159	165	170	175	180	186
83	191	196	201	206	212	217	222	227	232	238
84	243	248	253	258	263	269	274	279	284	289
85	9294	9299	9304	9309	9315	9320	9325	9330	9335	9340
86	345	350	355	360	365	370	375	380	385	390
87	395	400	405	410	415	420	425	430	435	440
88	445	450	455	460	465	469	474	479	484	489
89	494	499	504	509	513	518	523	528	533	538
90	9542	9547	9552	9557	9562	9566	9571	9576	9581	9586
91	590	595	600	605	609	614	619	624	628	633
92	638	643	647	652	657	661	666	671	675	680
93	685	689	694	699	703	708	713	717	722	727
94	731	736	741	745	750	754	759	763	768	773
95	9777	9782	9786	9791	9795	9800	9805	9809	9814	9818
96	823	827	832	836	841	845	850	854	859	863
97	868	872	877	881	886	890	894	899	903	908
98	912	917	921	926	930	934	939	943	948	952
99	956	961	965	969	974	978	983	987	991	996
N.	0	1	2	3	4	5	6	7	8	9

To obtain the mantissa for a number of three figures, as 364, we take 36 in the first column, and look along the row beginning with 36 till we come to the column headed 4. The mantissa thus obtained is .5611. If the number whose mantissa is sought contains four or five figures, *obtain from the table the mantissa for the first three figures, and also that for the next higher number, and subtract. Multiply the difference between the two mantissas by the fourth (or fourth and fifth) figure expressed as a decimal, and* ADD *the result to the mantissa for the first three figures.* Thus, to find the mantissa for 167.49

Mantissa for 168 = .2253
Mantissa for 167 = .2227
Difference = .0026

Since an increase of 1 in the number (from 167 to 168) makes an increase of .0026 in the mantissa, an increase of .49 of 1 in the number will make an increase of .49 of .0026 in the mantissa. But $.0026 \times .49 = .001274$ or .0013—

Hence .2227
13
Mantissa for 167.49 = .2240

Hence, to obtain the logarithm of a given number,

Determine the characteristic by Art. 362 or 363.

Neglect the decimal point, and obtain from the table (pp. 380, 381) *the mantissa for the given figures.*

Exs. Log. 52.6 = 1.7210. Log. .00094 = 6.9731 — 10.
Log. 167.49 = 2.2240. Log. .042308 = 8.6264 — 10.

EXERCISE 134.

Find the logarithms of the following numbers:

1. 37.	7. 175.	13. .0758.	19. 0.7788.	25. .08134.
2. 85.	8. 504.	14. 5780.	20. .04275.	26. .00732.
3. 15.	9. 32.9.	15. .00217.	21. 234.76.	27. 95032.
4. 6.	10. 4.75.	16. 1275.	22. 5.6107.	28. 91706.
5. 90.	11. .08.	17. 63.21.	23. 900.78.	29. 32.171.
6. 300.	12. 1.02.	18. 3.002.	24. 7781.4.	30. 328.07.

366. Inverse Use of a Table of Logarithms; that is, *given a logarithm to find the number corresponding to this logarithm, termed antilogarithm;*

From the table find the figures corresponding to the mantissa of the given logarithm;

Use the characteristic of the given logarithm to fix the decimal point of the figures obtained.

Ex. Find the antilogarithm of 1.5658.

The figures corresponding to the mantissa, .5658, are 368.

Since the characteristic is 1, there are 2 figures at the left of the decimal point.

Hence, the antilog. 1.5658 = 36.8.

In case the given mantissa does not occur in the table, *obtain from the table the next lower mantissa with the corresponding three figures of the antilogarithm. Subtract the tabular mantissa from the given mantissa. Divide this difference by the difference between the tabular mantissa and the next higher mantissa in the table.* ANNEX *the quotient to the three figures of the antilogarithm obtained from the table.*

Ex. Find antilog 2.4237.

4237 does not occur in the table, and the next lower mantissa is .4232. The difference between .4232 and .4249 is .0017.

Hence we have antilog 2.4237 = 265.29.

$$\begin{array}{r} 4232 \\ \hline 17 \,)\, 5.00 \,(\, .29 \end{array}$$

For if a difference of 17 in the last two figures of the mantissa makes a difference of 1 in the third figure of the antilog, a difference of 5 in the mantissa will make a difference of $\frac{5}{17}$ of 1 or .29 with respect to the third figure of the antilog.

EXERCISE 135.

Find the numbers corresponding to the following logarithms:

1. 1.6335.	7. 0.6117.	13. 0.4133.	19. 8.7727—10.
2. 2.8865.	8. 9.7973—10.	14. 1.4900.	20. 2.4780.
3. 2.3729.	9. 7.9047—10.	15. 3.8500.	21. 0.6173.
4. 0.5775.	10. 8.6314—10.	16. 1.8904.	22. 1.9030.
5. 3.9243.	11. 7.7007—10.	17. 2.4527.	23. 3.3922.
6. 1.8476.	12. 6.1004—10.	18. 9.6402—10.	24. 9.7071—10.

367. Properties of Logarithms. It has been shown (Arts. 199, 200) that—

$$a^m \times a^n = a^{m+n},$$

when m and n are commensurable. By the use of successive approximations approaching as closely as we please to limits, the same law may be shown to hold when m and n are incommensurable. It then follows that—

1. $\log\ ab = \log a + \log b.$
2. $\log\left(\frac{a}{b}\right) = \log a - \log b.$
3. $\log\ a^p = p \log a.$
4. $\log \sqrt[p]{a} = \frac{\log a}{p}.$

PROOF —

$$\text{Let } a = 10^m. \quad \therefore \log a = m.$$

$$b = 10^n. \quad \therefore \log b = n.$$

$$ab = 10^{m+n}. \quad \therefore \log ab = m + n = \log a + \log b \quad . \quad . \ (1)$$

$$\frac{a}{b} = 10^{m-n}. \quad \therefore \log\left(\frac{a}{b}\right) = m - n = \log a - \log b \quad . \quad . \ (2)$$

$$a^p = 10^{pm}. \quad \therefore \log a^p = pm = p \log a \quad . \quad . \quad . \quad . \quad . \ (3)$$

$$\sqrt[p]{a} = 10^{\frac{m}{p}}. \quad \therefore \log \sqrt[p]{a} = \frac{m}{p} = \frac{\log a}{p} \quad . \quad . \quad . \quad . \quad . \quad . \quad . \quad . \ (4)$$

The same properties may be proved in like manner for a system of logarithms with any other base than 10.

368. Properties Utilized for Purposes of Computation.

I. **To Multiply Numbers,**

Add their logarithms, and find the antilogarithm of the sum. This will be the product of the numbers.

II. **To Divide One Number by Another,**

Subtract the logarithm of the divisor from the logarithm of the dividend, and obtain the antilogarithm of the difference. This will be the quotient.

III. **To Raise a Number to a Required Power,**

Multiply the logarithm of the number by the index of the power. Find the antilogarithm of the product.

IV. **To Extract a Required Root of a Number,**

Divide the logarithm of the number by the index of the required root. Find the antilogarithm of the quotient.

Ex. 1. Multiply 527 by .083 by the use of logs.

$$\log 527 = 2.7218$$
$$\log .083 = 8.9191 - 10$$
$$\text{antilog } 1.6409 = 53.7+, \textit{ Product.}$$

Ex. 2. Compute the amount of \$1 at 6% for 20 years, at compound interest.

$$\text{Amount} = (1.06)^{20}$$
$$\log 1.06 = .0253$$
$$20$$
$$\text{antilog } 0.5060 = \$3.21+, \textit{ Amount.}$$

If the student will compute the value of $(1.06)^{20}$ by continued multiplication, and compare the labor involved with that in the above process by the use of logarithms, he will have a good illustration of the value of logarithms.

Ex. 3. Extract approximately the 7th root of 15.

$$\log 15 = 1.1761$$
$$\tfrac{1}{7}\log 15 = 0.1680+$$
$$\text{antilog } 0.1680+ = 1.47+, \textit{ Root.}$$

369. Cologarithm. In operations involving division it is usual, instead of subtracting the logarithm of the divisor, to add its cologarithm. The cologarithm of a number is obtained by subtracting the logarithm of the number from 10 − 10. Adding it gives the same result as subtracting the logarithm itself from the logarithm of the dividend. The use of the cologarithm saves figures, and gives a more compact and orderly statement of the work.

The cologarithm may be taken directly from the table by use of the following rule:

Subtract each figure of the given logarithm from 9 *except the last significant figure, which subtract from* 10.

Ex. 1. Find colog of 36.4.

$$\log 36.4 = 1.5611$$
$$\text{colog } 36.4 = 8.4389 - 10.$$

Ex. 2. Compute by use of logarithms $\dfrac{8.4 \times 32.4}{2\sqrt{576} \times 3.78}$.

$$\log 8.4 = 0.9243$$
$$\log 32.4 = 1.5105$$
$$\text{colog } 2 = 9.6990 - 10$$
$$\text{colog } \sqrt{576} = 8.6198 - 10$$
$$\text{colog } 3.78 = \underline{9.4225 - 10}$$
$$\text{antilog } 0.1761 = 1.5, \textit{ Result.}$$

EXERCISE 136.

Find by use of logarithms the approximate values of—

1. 75×1.4.
2. 9.8×3.5.
3. $15.1 \times .005$.
4. 831×0.25.
5. $\frac{54.7}{13.4}$.
6. $\frac{4.79}{9.51}$.
7. $\frac{0.317}{.0049}$.
8. $\frac{-78.9}{98.7}$.
9. $\frac{.00435}{.0911}$.
10. $\frac{5.74}{.039}$.
11. $4.7 \times (-0.59)$.
12. $0.48 \div (-1.79)$.
13. $\frac{-9.91}{-45.7}$.
14. $\frac{1.78 \times 19}{23.7}$.
15. $\frac{3.51 \times 67}{97.7}$.
16. $\frac{12.9}{4.7 \times 9.1}$.

17. $47.1 \times 3.56 \times .0079$.
18. $9.57 \times 59.7 \times .0759$.
19. $4.77 \times (-0.71) \div (0.83)$.
20. $\frac{47 \times 9.4}{52 \times 30.5}$.
21. $\frac{523 \times 249}{767 \times 396}$.

22. $(2.3)^3$.
23. $(6.15)^2$.
24. $(3.57)^4$.
25. $(0.96)^7$.
26. $(0.795)^6$.
27. $\sqrt{19}$.
28. $\sqrt[3]{3.29}$.
29. $\sqrt[5]{7.65}$.
30. $\sqrt[3]{1670}$.
31. $\sqrt{0.44}$.
32. $\sqrt[3]{17}$.
33. $\sqrt[3]{61}$.

34. $\sqrt[5]{27.9}$.
35. $\sqrt[7]{.00429}$.
36. $(76.5)^{\frac{2}{3}}$.
37. $(2.91)^{\frac{5}{2}}$.
38. $\sqrt[4]{10^3}$.
39. $\sqrt{30} \times \sqrt[3]{-5}$.
40. $\sqrt[3]{19} \div \sqrt{46}$.
41. $\sqrt[5]{1.47 \times 3.7}$.
42. $\sqrt[10]{13} \times \sqrt[5]{10.1}$.

43. $\sqrt[6]{.005} \times \sqrt[3]{.0765}$.
44. $2^{\frac{2}{7}} \times 7^{\frac{3}{2}}$.
45. $\sqrt[5]{\frac{11}{12}} \times \sqrt[7]{3\frac{5}{6}}$.
46. $\sqrt[4]{\frac{5}{6}} \times \sqrt[3]{-\frac{9}{5}}$.
47. $\sqrt[3]{2} \times \sqrt{3} \times \sqrt[7]{17}$.
48. $\sqrt{0.1} \times \sqrt[5]{40} \times \sqrt[3]{.01}$.
49. $(-0.1)^{13} \times (16.3)^{11}$.
50. $(.00812)^{\frac{5}{7}} \div (31.5)^{\frac{1}{30}}$.
51. $-(3.12)^3 \div \sqrt[3]{(-42.8)^2}$.
52. $\sqrt[7]{.000479} \div \sqrt[3]{.0568}$.

53. $\dfrac{\sqrt[8]{10} \times \sqrt[7]{100}}{\sqrt[21]{2000}}$.

54. $\dfrac{(-0.1)^{\frac{4}{5}} \times (0.01)^{\frac{3}{4}}}{\sqrt[4]{.000077}}$.

---

55. $\dfrac{-0.3384}{.08659}$.

56. $\dfrac{401.8}{52.37}$.

57. $\sqrt[6]{.4294}$.

58. $\sqrt[5]{-.02305}$.

59. $\sqrt[3]{\dfrac{-.03296}{7.962}}$.

60. $\dfrac{2563 \times .03442}{714.8 \times 0.511}$.

61. $\dfrac{121.6 \times 9.025}{48.3 \times 3662 \times .0856}$.

62. $\dfrac{\sqrt{5.955} \times \sqrt[3]{61.2}}{\sqrt[5]{298.54}}$.

63. $(2.6317)^{\frac{3}{4}} \times (0.71272)^{\frac{2}{5}}$.

64. $\sqrt[6]{1.6095} \div \sqrt[3]{2.945 \times \sqrt{0.7777}}$.

370. Algebraic Expressions Transformed. By the use of properties of logarithms 1, 2, 3, 4, stated in Art. 367, complex algebraic expressions may often be put into simpler and more convenient forms.

$$\text{Ex. Log}\ \frac{\sqrt{a^2 - x^2}}{(1+x)^2} = \log \sqrt{a^2 - x^2} - \log (1+x)^2$$
$$= \tfrac{1}{2} \log (a^2 - x^2) - 2 \log (1+x)$$
$$= \tfrac{1}{2} \log (a+x) + \tfrac{1}{2} \log (a-x) - 2 \log (1+x).$$

Various formulas may thus be put into a form adapted to numerical computation.

For instance, problems relating to compound interest, principal, amount, and time may be solved by the use of logarithms.

Since $a = p(1+r)^n$ (see Art. 368),

$$\log a = \log p + n \log (1+r).$$

Hence, $$\log p = \log a - n \log (1 + r)$$

$$n = \frac{\log a - \log p}{\log (1 + r)}$$

$$\log (1 + r) = \frac{\log a - \log p}{n}.$$

371. Exponential Equations. An exponential equation is one of the form $a^x = b$, where a and b are known quantities and x is unknown.

An equation of this kind can be solved by taking the logarithm of each member.

Ex. 1. Solve $(1.06)^x = 3$ (that is, find the number of years in which $1 will amount to $3 at 6% compound interest).

$$x \log 1.06 = \log 3$$

$$x = \frac{\log 3}{\log 1.06} = \frac{.4771}{.0253} \doteq 18.858+ \textit{ Years.}$$

Ex. 2. Given $0.3^x = 2$; find the value of x.

Taking the logarithm of each member of the equation,

$$x \log (0.3) = \log 2.$$

Hence, $$x = \frac{\log 2}{\log 0.3} = \frac{0.3010}{9.4771 - 10}$$

$$= \frac{0.3010}{-0.5229} = -0.575+.$$

Ex. 3. Given $a^x = b^{2x+1}$; find the value of x.

Taking the logarithm of each member of the equation,

$$x \log a = (2x + 1) \log b.$$

Whence, $$x \log a - 2x \log b = \log b$$

$$x = \frac{\log b}{\log a - 2 \log b}.$$

GENERAL PROPERTIES OF SYSTEMS OF LOGARITHMS.

372. Logarithm of Unity. In any system of logarithms the logarithm of 1 is zero.

For, taking a as the base,

$$a^0 = 1. \quad \therefore \log_a 1 = 0.$$

373. Logarithm of the Base. In any system the logarithm of the base itself is unity.

$$\text{For} \quad a^1 = a. \quad \therefore \log a = 1.$$

374. Logarithm of Zero. In any system whose base is greater than unity the logarithm of zero is minus infinity; that is, as the number approaches 0, the logarithm approaches minus infinity.

For, if $a > 1$, $\quad a^{-\infty} = \dfrac{1}{a^{\infty}} = \dfrac{1}{\infty} = 0.$

$$\therefore \log_a 0 = -\infty.$$

But in any system whose base is less than unity, the logarithm of zero is plus infinity.

$$\text{For} \quad a^{\infty} = 0, \quad \text{since } a < 1.$$

$$\therefore \log_a 0 = \infty.$$

375. Change of the Base of a System of Logarithms. Let a and b be the bases of two systems of logarithms, and n be the number. Then

$$\log_b n = \frac{\log_a n}{\log_a b} \quad \ldots\ldots\ldots\ldots \quad (1)$$

Let $\quad a^x = n$; $\quad$ hence, $x = \log_a n$,

and $\quad b^y = n$; $\quad$ " $\quad y = \log_b n$.

Also, $a^x = b^y$. $\quad \therefore a^{\frac{x}{y}} = b.$

$\therefore \log_a b = \dfrac{x}{y}$, $\quad$ or $y = \dfrac{x}{\log_a b}$.

$$\therefore \log_b n = \frac{\log_a n}{\log_a b}.$$

Hence the logarithm of a number, n, in a system whose base is a, being given, the logarithm of the same number in a system with another base, b, may be found by dividing the given logarithm by the logarithm of b in the system whose base is a.

From the above relation (1) we may also prove

$$\log_b a \times \log_a b = 1.$$

For putting $n = a$ in (1),

$$\log_b a = \frac{\log_a a}{\log_a b} = \frac{1}{\log_a b}.$$

$$\therefore \log_b a \times \log_a b = 1.$$

376. Examples.

Ex. 1. Find the logarithm of 0.7 to the base 5.

$$\text{By Art. 375, } \log_5 0.7 = \frac{\log_{10} 0.7}{\log_{10} 5}$$

$$= \frac{9.8451 - 10}{0.6990}$$

$$= \frac{-0.1549}{0.6990} = -0.2216+, \textit{ Logarithm.}$$

Ex. 2. Find the logarithm of 243 to the base 9 without the use of the tables.

Let $\log_9 243 = x$; then $9^x = 243$.

Hence, $(3^2)^x = 3^5$, or $3^{2x} = 3^5$.

Hence, $2x = 5$

$x = \frac{5}{2}$, *Logarithm.*

Ex. 3. Find the number of digits in 5^{16}.

$$\text{Log}\,(5^{16}) = 16 \log 5 = 16 \times 0.6990$$

$$= 11.1840.$$

Since the characteristic is 11, the number of digits in 5^{16} is 12.

EXERCISE 137.

Find the approximate value of x in each of the following exponential equations:

1. $40^x = 75.$
2. $20^x = 100.$
3. $5^x = 12.$
4. $7^x = 25.$
5. $1.3^x = 7.2.$
6. $5^{x+1} = 17.$
7. $3^{x-2} = 5.$
8. $12^{x+1} = 45^x.$
9. $5^{5-3x} = 2^{x+2}.$
10. $(0.4)^{-x} = 7.$
11. $20^x = 5.$
12. $27^{2x-1} = 8.$
13. $0.7^x = 0.3.$
14. $13.18^x = .0281.$
15. $0.703^x = 1.09.$

Express in terms of $\log a$, $\log b$, $\log c$, and $\log x$—

16. $\log a^2b^3.$
17. $\log ax^5.$
18. $\log b^3\sqrt{x}.$
19. $\log \sqrt[3]{a^2}\cdot\sqrt{x^3}.$
20. $\log \sqrt[4]{\dfrac{ax}{b^3}}.$
21. $\log \dfrac{a^2b\sqrt{x}}{c^5}.$
22. $\log \dfrac{\sqrt[3]{ax^2}}{b\sqrt{c}}.$
23. $\log a^{-3}b^2\sqrt{c^{-1}}.$
24. $\log \dfrac{a^4x\sqrt[3]{b}}{bc\sqrt{x}}.$

Express the value of x in terms of the logs of the known quantities in each of the following equations:

25. $a^x = 7b^x.$
26. $c^x = a^{x-1}.$
27. $13^{x+1} = a^xb.$
28. $7a^x = 3b^2.$
29. $5a^{x+1} = a^2b^{2x+1}.$
30. $6c^xb^{2x} = 11(a-b)^x.$
31. $a^2 - b^2 = (2a-1)^{3x}.$
32. $25a^{\sqrt{x}} = \sqrt{a^2-b^2}.$

Find the values of—

33. $\log_3 50.$
34. $\log_8 12.$
35. $\log_{20} 25.$
36. $\log_6 4.5.$
37. $\log_{12} 8.$
38. $\log_{4.7} 23.$
39. $\log_5 0.9.$
40. $\log_{0.3} 65.$
41. $\log_{\sqrt{2}} \sqrt{3}.$

Find without the use of the tables the base when—

42. $\log_x 9 = \frac{2}{3}$.
44. $\log_x 5 = 0.5$.
46. $\log_x 3\frac{3}{8} = -\frac{3}{2}$.
43. $\log_x 8 = \frac{3}{4}$.
45. $\log_x \frac{1}{16} = -0.8$.
47. $\log_x 3\sqrt{3} = 1.5$.

Find without the use of the tables—

48. $\log_2 16$.
50. $\log_9 \frac{1}{27}$.
52. $\log_2 .0625$.
49. $\log_8 4$.
51. $\log_{\frac{1}{32}} 16$.
53. $\log_{\sqrt{2}} 4\sqrt{2}$.

Find the number of digits in—

54. $(875)^{18}$.
55. 2^{25}.
56. 9^{90}.
57. 57^{57}.

Show that—

58. $(\frac{21}{20})^{100} > 100$.
59. $(\frac{81}{80})^{1000} > 100000$.

60. There are more than 300 zeros between the decimal point and the first significant figure in $(0.5)^{1000}$.

In the following geometrical progressions find n:

61. Given a, r, l.
63. Given a, l, s.
62. Given a, r, s.
64. Given r, l, s.

65. Given $a = 2$, $r = 5$, $l = 31{,}250$.
66. Given $a = \frac{1}{3}$, $r = 3$, $s = 364\frac{1}{3}$.
67. Given $a = 0.375$, $l = 96$, $s = 191\frac{5}{8}$.

68. Find the amount of \$1000 for 20 years at 5% compound interest.

69. Find the amount of \$875 for 18 years at 6% compound interest.

70. How long will it require for a sum of money to double itself at 5% compound interest? At 7 per centum?

71. If \$1280 amounts to \$37,770 in 50 years at compound interest, what is the rate per centum?

CHAPTER XXXI.

HISTORY OF ELEMENTARY ALGEBRA.

377. Epochs in the Development of Algebra. Some knowledge of the origin and development of the symbols and processes of Algebra is important to a right understanding of them.

The oldest mathematical treatise known is a papyrus roll, now in the British Museum, entitled "Directions for Attaining to the Knowledge of All Dark Things." It was written by a scribe named Ahmes at least 1700 B. C., and is a copy, the writer says, of a more ancient work, dating, say, 3000 B. C., or several centuries before the time of Moses. This papyrus roll contains, among other things, the beginnings of algebra as a science. Taking the epoch indicated by this work as the first, the principal epochs in the development of algebra are as follows:

1. **Egyptian**: 3000 B. C.–1500 B. C.

2. **Greek** (at Alexandria): 200 A. D.–400 A. D. Principal writer, **Diophantus**.

3. **Hindoo** (in India): 500 A. D.–1200 A. D.

4. **Arab**: 800 A. D.–1200 A. D.

5. **European**: 1200 A. D.–. Leonardo of Pisa, an Italian, published a work in 1202 A. D. on the Arabic arithmetic, but containing an account also of the science of algebra as it then existed among the Arabs. From Italy the knowledge of algebra spread to France, Germany, and England, where its subsequent development took place.

We will consider briefly the history of—

I. ALGEBRAIC SYMBOLS.
II. IDEAS OF ALGEBRAIC QUANTITY.
III. ALGEBRAIC PROCESSES.

I. HISTORY OF ALGEBRAIC SYMBOLS.

878. Symbol for the Unknown Quantity.

1. **Egyptians** (1700 B. C.) used the word "hau" (expressed, of course, in hieroglyphics), meaning "heap."

2. **Diophantus** (Alexandria, 350 A. D.?), ς′, or ς°′; plural, $\overline{\varsigma\varsigma}$.

3. **Hindoos** (500 A. D.–1200 A. D.), Sanscrit word for "color," or first letters of words for colors (as of "blue," "yellow," "white," etc.).

4. **Arabs** (800 A. D.–1200 A. D), Arabic word for "thing" or "root" (the term "root," as still used in algebra, originates here).

5. **Italians** (1500 A. D.), *Radix*, *R*, *Rj*.

6. **Bombelli** (Italy, 1572 A. D.), $\overset{1}{\smile}$.

7. **Stifel** (Germany, 1544), *A*, *B*, *C*, ...

8. **Stevinus** (Holland, 1586), ①.

9. **Vieta** (France, 1591), vowels *A*, *E*, *I*, *O*, *U*.

10. **Descartes** (France, 1637), x, y, z, etc.

879. Symbols for Powers (of x at first). **Exponents.**

1. Diophantus, δυναμις, or $\delta^{\bar{v}}$ (for square of the unknown quantity); κυβος, or $\kappa^{\bar{v}}$ (for its cube).

2. Hindoos, initial letters of Sanscrit words for "square" and "cube."

3. Italians (1500 A. D.), "census" or "zensus" or "z" (for x^2); "cubus" or "c" (for x^3).

4. Bombelli (1579), $\overset{1}{\smile}$, $\overset{2}{\smile}$, $\overset{3}{\smile}$ (for x, x^2, x^3).

5. Stevinus (1586), ①, ②, ③ (for x, x^2, x^3).

6. Vieta (1591), *A*, *A quadratus*, *A cubus* (for x, x^2, x^3).

7. Harriot (England, 1631), *a*, *aa*, *aaa*.

8. Herigone (France, 1634), a, $a2$, $a3$.
9. Descartes (France, 1637), x, x^2, x^3.

Wallis (England, 1659) first justified the use of fractional and negative exponents, though their use had been suggested before by Stevinus (1586).

Newton (England, 1676) first used a general exponent, as in x^n, where n denotes any exponent, integral or fractional, positive or negative.

380. Symbols for Known Quantities.

1. Diophantus, μοναδες (*i. e.* monads), or $\overset{\circ}{\mu}$.
2. Regiomontanus (Germany, 1430), letters of the alphabet.
3. Italians, d, from "dragma."
4. Bombelli, $\overset{\circ}{\smile}$.
5. Stevinus, ⊙.
6. Vieta, consonants, $B, C, D, F, \ldots$.
7. Descartes, a, b, c, d.

Descartes possibly used the last letters of the alphabet, x, y, z, to denote unknown quantities because these letters are less used and less familiar than $a, b, c, d, \ldots$, which he accordingly used to denote known numbers.

381. Addition Sign. The following symbols were used:

1. Egyptians, pair of legs walking forward (to the left), Λ.
2. Diophantus, juxtaposition (thus, ab, meant $a + b$).
3. Hindoos, juxtaposition (survives in Arabic arithmetic, as in $2\frac{3}{5}$, which means $2 + \frac{3}{5}$).
4. Italians, "plus," then "p" (or e, or ϕ).
5. Germans (1489), +, †, +.

382. Subtraction Sign.

1. Egyptians, pair of legs walking backward (to the right), thus, Λ; or by a flight of arrows.
2. Diophantus, ⩚ (Greek letter ψ inverted).
3. Hindoos, by a dot over the subtracted quantity (thus, $m\dot{n}$ meant $m - n$).

4. Italians, "minus," then *M* or *m*, or *de*.

5. Germans (1489), horizontal dash, —.

The signs + and − were first printed in Johann Widman's Mercantile Arithmetic (1489). These signs probably originated in German warehouses, where they were used to indicate excess or deficiency in the weight of bales and chests of goods. Stifel (1544) was the first to use them systematically to indicate the operations of addition and subtraction.

383. Multiplication Sign. Multiplication at first was usually expressed in general language. But—

1. Hindoos indicated multiplication by the syllable "*bha*," from "bharita," meaning "product," written after the factors.

2. Oughtred and Harriot (England, 1631) invented the present symbol, ×.

3. Descartes (1637) used a dot between the factors (thus, $a \cdot b$).

384. Division Sign.

1. Hindoos indicated division by placing the divisor under the dividend (no line between). Thus, $\begin{smallmatrix} c \\ d \end{smallmatrix}$ meant $c \div d$.

2. Arabs, by a straight line $\left(\text{thus, } a - b\text{, or } a \mid b\text{, or } \frac{a}{b}\right)$.

3. Italians expressed the operation in general language.

4. Oughtred, by a dot between the dividend and divisor.

5. Pell (England, 1630), ÷.

385. Equality Sign.

1. Egyptians, [symbol] [symbol] (also other more complicated symbols to indicate different kinds of equality).

2. Diophantus, general language or the symbol, ι.

3. Hindoos, by placing one side of an equation immediately under the other side.

4. Italians, æ or α; that is, the initial letters of "æqualis" (equal). This symbol was afterward modified into the form, ∞, and much used, even by Descartes, long after the invention of the present symbol by Recorde.

5. Recorde (England, 1540), =.

He says that he selected this symbol to denote equality because "than two equal straight lines no two things can be more equal."

386. Other Symbols used in Elementary Algebra.

Inequality Signs (> <) were invented by Harriot (1631).

Oughtred, at the same time, proposed ⊐, ⊐ as signs of inequality, but those suggested by Harriot were manifestly superior.

Parenthesis, ( ), was invented by Girard (1629).

The **Vinculum** had been previously suggested by Vieta (1591).

Radical Sign. The Hindoos used the initial syllable of the word for square root, "Ka," from "Karania," to indicate square root.

Rudolf (Germany, 1525) suggested the symbol used at present ($\surd$) (the initial letter, *r*, in the script form, of the word "radix," or root) to indicate square root, $\surd\surd$ to denote the 4th root, and $\surd\surd\surd$ to denote cube root.

Girard (1633) denoted the 2d, 3d, 4th, etc. roots, as at present by $\sqrt[2]{\ }$, $\sqrt[3]{\ }$, $\sqrt[4]{\ }$, etc.

The sign for **Infinity**, ∞, was invented by Wallis (1649).

387. Many **other Algebraic Symbols** have been invented in recent times, but these do not belong to elementary algebra.

Other kinds of algebra have also been invented employing other systems of the symbols.

388. General Illustration of the Evolution of Algebraic Symbols. The following illustration will serve to show the principal steps in the evolution of the symbols of algebra:

At the time of Diophantus the numbers 1, 2, 3, 4, ... were denoted by letters of the Greek alphabet, with a dash over the letters used; as, $\bar{\alpha}$, $\bar{\beta}$, $\bar{\gamma}$, ...

In the algebra of Diophantus the coefficient occupies the last place in a term instead of the first as at present.

Beginning with Diophantus, the algebraic expression, $x^2 + 5x - 4$, would be expressed in symbols as follows:

δ^υ ᾱ ϛό ε̄ ⋔ μ° δ̄	(Diophantus, 350 A. D.).
$1z\ p.5\ Rm.4$	(Italy, 1500 A. D.).
$1Q + 5N - 4$	(Germany, 1575).
$1\overset{2}{\smile}\ p.5\overset{1}{\smile}\ m.4\overset{0}{\smile}$	(Bombelli, 1579).
1② + 5① − 4⓪	(Stevinus, 1586).
$1Aq + 5A - 4$	(Vieta, 1591).
$1aa + 5a - 4$	(Harriot, 1631).
$1a2 + 5a1 - 4$	(Herigone, 1634).
$x^2 + 5x - 4$	(Descartes, 1637).

889. Three Stages in the Development of Algebraic Symbols.

1. **Algebra without Symbols** (called **Rhetorical Algebra**). In this primitive stage algebraic quantities and operations are expressed altogether in words, without the use of symbols. The Egyptian algebra and the earliest Hindoo, Arabian, and Italian algebras are of this sort.

2. **Algebra in which the Symbols are Abbreviated Words** (called **Syncopated Algebra**). For instance, "*p*" is used for "plus."

The algebra of Diophantus is mainly of this sort. European algebra did not get beyond this stage till about 1600 A. D.

3. **Symbolic Algebra.** In its final or completed state algebra has a system of notation or symbols of its own, independent of ordinary language. Its operations are performed according to certain laws or rules, "independent of, and distinct from, the laws of grammatical construction."

Thus, to express addition in the three stages we have "plus," *p*, +; to express subtraction, "minus," *m*, −; to express equality, "æqualis," *æ*, =.

Along with the development of algebraic symbolism there

was a corresponding development of ideas of algebraic quantity and of algebraic processes.

II. History of Algebraic Quantity.

390. The **Kinds of Quantity** considered in algebra are **positive** and **negative**; **particular** (or numerical) and **general**; **integral** and **fractional**; **rational** and **irrational**; **commensurable** and **incommensurable**; **constant** and **variable**; **real** and **imaginary.**

391. Ahmes (1700 B. C.) in his treatise uses *particular*, *positive* quantity, both *integral* and *fractional* (his fractions, however, are usually limited to those which have unity for a numerator). That is, his algebra treats of quantities like 8 and $\frac{1}{8}$, but not like -3, or $-\frac{2}{5}$, or $\sqrt{2}$, or $-a$.

392. Diophantus (350 A. D.) used *negative* quantity, but only in a limited way; that is, in connection with a larger positive quantity. Thus, he used $7-5$, but not $5-7$, or -2. He did not use, nor apparently conceive of, negative quantity having an independent existence.

393. The **Hindoos** (500 A. D.–1200 A. D.) had a distinct idea of independent or *absolute negative* quantity, and used the minus sign both as a quality sign and a sign of operation. They explained independent negative quantity much as is done to-day by the illustration of debts as compared with assets, and by the opposition in direction of two lines.

Pythagoras (Greece, 520 B. C.) discovered *irrational* quantity, but the Hindoos were the first to use this in algebra.

394. The **Arabs** avoided the use of negative quantity as far as possible. This led them to make much use of the process of transposition in order to get rid of negative terms in an equation. Their name for algebra was "al gebr we'l mukabala," which means transposition and reduction.

The Arabs used *surd* quantities freely.

395. In **Europe** the free use of absolute negative quantity was restored.

Vieta (1591) was principally instrumental in bringing into use *general algebraic quantity* (known quantities denoted by letters and not figures).

Cardan (Italy, 1545) first discussed *imaginary* quantities, which he termed "sophistic" quantities.

Euler (Germany, 1707–83) and Gauss (Germany, 1777–1855) first put the use of these quantities on a scientific basis.

Descartes (1637) introduced the systematic use of *variable* quantity as distinguished from constant quantity.

III. History of Algebraic Processes.

396. Solution of Equations. Ahmes solves many *simple equations of the first degree*, of which the following is an example:

"Heap its seventh, its whole equals nineteen. Find heap."

In modern symbols this is,

$$\text{Given } \frac{x}{7} + x = 19; \text{ find } x.$$

The correct answer, $16\frac{5}{8}$, is given by Ahmes.

Hero (Alexandria, 120 B. C.) solved what is in effect the *quadratic equation*,

$$\tfrac{11}{14}d^2 + \tfrac{29}{7}d = s,$$

where d is unknown, and s is known.

Diophantus solved simple equations of one, and *simultaneous equations* of two and three unknown quantities. He solved quadratic equations much as is done at present, completing the square by the method given in Art. 255. However, in order to avoid the use of negative quantity as far as possible, he made three classes of quadratic equations, thus,

$$\begin{cases} ax^2 + bx = c, \\ ax^2 + c \;\; = bx, \\ ax^2 = bx + c. \end{cases}$$

In solving quadratic equations he rejected negative and irrational answers.

He also solved equations of the form $ax^m = bx^n$.

He was the first to investigate *indeterminate equations*, and solved many such equations of the first degree with two or three unknown quantities, and some of the second degree.

The **Hindoos** first invented a *general method of solving a quadratic equation* (now known as the Hindoo method, see Art. 256). They also solved particular cases of higher degrees, and gave a general method of solving indeterminate equations of the first degree.

The **Arabs** took a step backward, for, in order to avoid the use of negative terms, they made six cases of quadratic equations; viz.:

$$ax^2 = bx, \qquad ax^2 + bx = c,$$
$$ax^2 = c, \qquad ax^2 + c = bx,$$
$$bx = c, \qquad ax^2 = bx + c.$$

Accordingly, they had no general method of solving a quadratic equation.

The Arabs also solved equations of the form $ax^{2p} + bx^p = c$, and obtained a geometrical solution of *cubic equations* of the form $x^3 + px + q = 0$.

In **Italy**, Tartaglia (1500–59) discovered the *general solution of the cubic equation*, now known as Cardan's solution. Ferrari, a pupil of Cardan, discovered the solution of *equations of the fourth degree.*

Vieta discovered many of the elementary *properties of an equation of any degree;* as, for instance, that the number of the roots of an equation equals the degree of the equation.

397. Other Processes. Methods for the **Addition, Subtraction,** and **Multiplication** of polynomial expressions were given by Diophantus.

Transposition was first used by Diophantus, though, as a process, it was first brought into prominence by the Arabs.

The **Square** and **Cube Root** of polynomial expressions were extracted by the Hindoos.

The methods for using **Radicals**, including the extraction of the square root of binomial surds and rationalizing the denominators of fractions, were also invented by the Hindoos.

The methods of using fractional and negative **Exponents** were determined by Wallis (1659) and Sir Isaac Newton.

The three **Progressions** were first used by Pythagoras 569 B. C.–500 B. C.).

Permutations and **Combinations** were investigated by Pascal and Fermat (France, 1654).

The use of **Undetermined Coefficients** was introduced by Descartes.

The **Binomial Theorem** was discovered by Newton (1655), and, as one of the most notable of his many discoveries, is said to have been engraved on his monument in Westminster Abbey.

Continued Fractions were first used by Cataldi (Italy, 1653), though none of their properties were demonstrated by him. Lord Brouncker (England, 1620–84) was the first to do this.

Logarithms were invented by Lord Napier (Scotland, 1614) after a laborious search for means to diminish the work involved in numerical computations, and were improved by Briggs (England, 1617).

The fundamental **Laws of Algebra** (the Associative, Commutative, and Distributive Laws; see Arts. 33–36) were first clearly formulated by Peacock and Gregory (England, 1830–45), though, of course, the existence of these laws had been implicitly assumed from the beginnings of the science.

Students who desire to investigate the history of Algebra in more detail should read the second part of *Fine's Number System of Algebra, Ball's Short History of Mathematics*, and *Cajori's History of Elementary Mathematics*.

CHAPTER XXXII

APPENDIX.

I. PROCESSES ABBREVIATED BY USE OF DETACHED COEFFICIENTS.

398. Multiplication by Detached Coefficients. The process of multiplying one polynomial by another (see pp. 42–4) can often be much abbreviated, and the likelihood of error diminished, by detaching the coefficients of the terms of the polynomial, performing the multiplication with respect to them, and then supplying the proper powers of the letters in the product obtained. Thus,

Ex. 1. Multiply $6x^3-5x^2-4x-3$ by $6x^2+5x-4$.

Detaching coefficients,

$$\begin{array}{rrrrrr}
6 & -5 & -4 & -3 & & \\
6 & +5 & -4 & & & \\
\hline
36 & -30 & -24 & -18 & & \\
 & 30 & -25 & -20 & -15 & \\
 & & -24 & +20 & +16 & +12 \\
\hline
36 & +0 & -73 & -18 & +1 & +12
\end{array}$$

Annexing the powers of x, $36x^5+0x^4-73x^3-18x^2+x+12$,

or $36x^5-73x^3-18x^2+x+12$, *Product.*

Ex. 2. Multiply $x^3+3a^2x-2a^3$ by $x^3-4ax^2+3a^3$.

$$\begin{array}{rrrrrrr}
1 & +0 & +3 & -2 & & & \\
1 & -4 & +0 & +3 & & & \\
\hline
1 & +0 & +3 & -2 & & & \\
 & -4 & -0 & -12 & +8 & & \\
 & & & +3 & +0 & +9 & -6 \\
\hline
1 & -4 & +3 & -11 & +8 & +9 & -6
\end{array}$$

Hence, $x^6-4ax^5+3a^2x^4-11a^3x^3+8a^4x^2+9a^5x-6a^6$, *Product.*

EXERCISE 138.

The pupil may work Exs. 12 to 23 of Exercise 9 by use of detached coefficients.

As the method is especially advantageous in multiplying polynomials with fractional coefficients, Exs. 1 to 6 of Exercise 13 should also be worked by this method.

399. Division by use of Detached Coefficients is performed similarly.

Ex. Divide $x^4 - 2x^2y^2 + 8xy^3 - 3y^4$ by $x^2 + 2xy - y^2$.

$$\begin{array}{l|l}
1+0-2+8-3 & 1+2-1 \\ \cline{2-2}
\underline{1+2-1} & 1-2+3 \\
-2-1+8 & \\
\underline{-2-4+2} & \\
3+6-3 & \\
\underline{3+6-3} &
\end{array}$$

Hence, $x^2 - 2xy + 3y^2$, *Quotient.*

400. Synthetic Division. The above process may be further abbreviated as follows:

$$\begin{array}{rr|l|l}
 & 1+0-2 & +8-3 & \underline{1\,|-2+1}, \textit{Divisor} \text{ (with signs} \\
 & -2+1 & & \text{of all the terms} \\
 & +4 & -2 & \text{except the first} \\
 & & \underline{-6+3} & \text{changed).} \\ \cline{2-2}
\textit{Quotient,} & 1-2+3 & +0+0, \textit{Remainder.} &
\end{array}$$

Hence, $x^2 - 2xy + 3y^2$, *Quotient.*

This abbreviation is made possible by noticing that if we change the sign of each term of divisor (in the division in Art. 399), we can change the successive subtractions employed into successive additions.

Further, as each successive term of the quotient is found by dividing only the first term of the remainder by the first term of the divisor, it is sufficient to add each column only as needed in order to determine the first term of each remainder, and hence the next term of the quotient. It is not necessary to multiply the first term of the divisor by each term of the quotient, since these products are not used in determining the remainders which give the successive terms of the quotient.

Thus, in the above process, having determined the first term of the quotient, 1, we multiply $-2+1$ by 1, and add the column $\begin{matrix}0\\-2\end{matrix}$; the sum, -2, divided by 1 gives -2, the second term of the quotient. We now multiply $-2+1$ by -2, set down the product, $+4-2$, in the proper place, add the column $\begin{matrix}-2\\+1\\+4\end{matrix}$, and divide the sum, 3, by 1, etc.

It is usually more convenient to set the divisor in a perpendicular column at the left. Thus,

Ex. Divide $4x^5-6x^4y+4x^3y^2-11x^2y^3+y^5$ by $2x^2-3xy-y^2$.

2	$4-6+4-11$	$+0+1$
$+3$	$6+2$	
$+1$	$0+0$	
	$+9$	$+3$
		$-3-1$
	$2+0+3-1$	$+0+0$

Hence, $2x^3+3xy^2-y^3$, *Quotient.*

EXERCISE 139.

Solve Examples 16 to 33 of Exercise 12 by the use of detached coefficients and synthetic division.

401. H. C. F. and Evolution by Detached Coefficients. In finding the H. C. F. of polynomials by the method of Arts. 119–124, and Exercise 38, the work may be abbreviated by the use of detached coefficients.

In extracting the square and cube root of polynomials (see Arts. 189, 194, Exercises 68 and 70), work may be saved in the same way.

It is to be carefully noted that the use of detached coefficients not only saves labor in all these cases, but has the further advantage of diminishing the probability of mistakes, since fewer symbols are operated with.

II. REMAINDER AND FACTOR THEOREMS. SYMMETRY.

402. Remainder Theorem. *If any polynomial of the form* $p_1x^n+p_2x^{n-1}+p_3x^{n-2}+\ldots\ldots+p_n$ *be divided by* $x-a$, *the remainder will be* $p_1a^n+p_2a^{n-1}+p_3a^{n-2}+\ldots\ldots+p_n$ *(obtained by substituting* a *for* x *in the original expression).*

Let the given expression be divided by $x-a$ till a remainder is obtained which does not contain x, and denote the quotient by Q and the remainder by R. Then

$$p_1x^n+p_2x^{n-1}+p_3x^{n-2}+\ldots\ldots+p_n\equiv Q(x-a)+R.$$

This is an identity and therefore true for all values of x.

Let $x=a$, then

$$p_1a^n+p_2a^{n-1}+p_3a^{n-2}+\ldots\ldots+p_n=Q(a-a)+R=R.$$

$$\therefore R=p_1a^n+p_2a^{n-1}+p_3a^{n-2}+\ldots\ldots+p_n.$$

Ex. Find the remainder when $2x^5+3x^4-5x^3+6x^2+8x-9$ is divided by $x-2$.

Substituting 2 for x in the given expression,

$$2\cdot 2^5+3\cdot 2^4-5\cdot 2^3+6\cdot 2^2+8\cdot 2-9, \quad \text{or } 103, \textit{ Remainder.}$$

403. Factor Theorem. *If any rational integral expression containing x become equal to zero, when a is substituted for x, then $x-a$ is a factor of the given expression.*

This follows directly from the remainder theorem when $R=0$, or it can be proved as follows:

Let E stand for the given expression. If E be divided by $x-a$ till a remainder is obtained in which x does not occur, denote the quotient by Q and the remainder by R. Then

$$E\equiv Q(x-a)+R.$$

Let $x=a$, then

$$0=Q(0)+R \text{ (since } E=0 \text{ when } x=a).$$

$$\therefore R=0,$$

Hence, $E=Q(x-a)$, or $x-a$ is a factor of E.

This principle is frequently of value in factoring expressions.

Ex. Factor $3x^3+7x^2-4$.

By trial we find that $3x^3+7x^2-4=0$, when $x=-1$.

$$\therefore x+1 \text{ is a factor of } 3x^3+7x^2-4.$$

By division. $3x^3+7x^2-4=(x+1)(3x^2+4x-4)$

$$=(x+1)(x+2)(3x-2), \textit{ Factors.}$$

It is to be noted that the only numbers that need be tried as values of x are the factors of the last term of the given expression. This follows from the fact that the last term of the dividend must be divisible by the last term of the divisor.

EXERCISE 140.

Factor by use of the factor theorem.

1. x^2-4.
2. $x^2-3x-28$.
3. x^2-3x+2.
4. x^2-a^2.
5. x^3-8a^3.
6. $a^3-b^3+3(a-b)$.
7. $(a-b)^2+3(a-b)$.
8. a^2b-ab^2.
9. a^3+5a-6.
10. $2x^2+7x-15$.
11. $2x^3-x^2-7x+6$.
12. $4x^3-4x^2-14x-6$.
13. $3x^3+8x^2+3x-2$.
14. $2x^4+x^3-14x^2+5x+6$.
15. $6x^4-13x^3-45x^2-2x+24$.
16. x^3-2x^2+1.
17. x^3-6x^2+25.
18. $x^4-28x^2+35x-90$.
19. Prove that x^n-y^n is always divisible by $x-y$.
20. Prove that x^n+y^n is divisible by $x+y$ when n is odd.
21. Show that $(1-x)^2$ is a factor of $1-x-x^n+x^{n+1}$.
22. Show that $(x-1)^2$ is a factor of $nx^{n+1}-(n+1)x^n+1$.

404. Symmetrical Expressions. An expression is symmetrical with respect to two letters when it is unaltered by an interchange of the letters.

Exs. $a+b$, ab, a^3+b^3, a^2+ab+b^2,

are each a symmetrical expression with reference to a and b.

Similarly an expression is symmetrical with respect to three or more letters, when it is unaltered by an interchange of any pair of them.

Ex. $a^3+b^3+c^3-3abc$,

is symmetrical with respect to a, b, c, since it is unaltered by substituting a for b and b for a; a for c and c for a; b for c and c for b.

Instead of complete symmetry, there are partial symmetries of different kinds which an algebraic expression may have. Thus, an expression has *cyclo-symmetry* (see Art. 144, Ex. 1) with reference to a, b, and c, if it remains unchanged after a is changed to b, b to c, and c to a.

Ex. $ab^2 + bc^2 + ca^2$ has cyclo-symmetry.

A symmetrical expression may often be denoted in an abbreviated way by writing only the typical terms of the expression with the Greek letter Σ before each one. Thus,

$$\text{for } a^2b + b^2a \quad \text{write } \Sigma a^2b;$$

$$\text{for } a^2 + b^2 + c^2 + ab + bc + ca \quad \text{write } \Sigma a^2 + \Sigma ab.$$

Similarly for a product, as $(a - b)(b - c)(c - a)$,

$$\text{write } \Pi(a - b).$$

405. Factoring Symmetrical Expressions. The factor theorem (Art. 403) is frequently of use in factoring symmetrical expressions.

Ex. 1. Factor $bc(b-c) + ca(c-a) + ab(a-b)$.

When $b = c$, the given expression reduces to

$$ca(c-a) + ac(a-c), \quad \text{or } 0.$$

Hence, $b - c$ is a factor of it.

Similarly $c - a$ and $a - b$ are factors.

$$\therefore bc(b-c)+ca(c-a)+ab(a-b) = L(b-c)(c-a)(a-b). \quad (1)$$

Where L (since the right-hand member is of the same degree with the given expression) is some number to be determined,

Since (1) is true for all values of a, b, c,

Let $a = 0, \quad b = 1, \quad c = 2.$

$$\therefore 2(1-2) = L(1-2)(2)(-1).$$

$$\therefore L = -1,$$

and the factors of the given expression are

$$-(b-c)(c-a)(a-b).$$

Ex. 2. Factor $a(b+c-a)^2+b(c+a-b)^2+c(a+b-c)^2$ $+(b+c-a)(c+a-b)(a+b-c)$.

If we put $a=0$, the given expression reduces to zero.

$$\therefore a, b, c, \text{ are factors of it.}$$

$$\therefore a(b+c-a)^2+b(c+a-b)^2+c(a+b-c)^2$$
$$+(b+c-a)(c+a-b)(a+b-c)=Labc.$$

Let $\quad a=1,\ b=1,\ c=1,$ then $L=4$.

$$\therefore 4abc \text{ are the factors required.}$$

Ex. 3. Factor $(b^3+c^3)(b-c)+(c^3+a^3)(c-a)+(a^3+b^3)$ $(a-b)$.

Evidently $b-c$, $c-a$, $a-b$, are factors; but their product is of the third degree only, while the given expression is of the fourth degree. Hence the given expression must contain another factor of the first degree, and since this factor is symmetrical as well of the first degree, it must be of the form

$$L(a+b+c).$$

$$\therefore (b^3+c^3)(b-c)+(c^3+a^3)(c-a)+(a^3+b^3)(a-b)=$$
$$L(b-c)(c-a)(a+b)(a+b+c).$$

Hence, $\quad L=1.$

In factoring symmetrical expressions, it is also useful to remember (see Ex. 35 of Exercise 12) that

$$a^3+b^3+c^3-3abc=(a+b+c)(a^2+b^2+c^2-bc-ca-ab).$$

Hence, for example,

$$x^3-8y^3+27+18xy=x^3+(-2y)^3+3^3-3x(-2y)(3)$$
$$=(x-2y+3)(x^2+4y^2+9+6y-3x+2xy).$$

EXERCISE 141.

Show that—

1. $(a+b+c)^3-(b+c-a)^3-(c+a-b)^3-(a+b-c)^3=$ $24abc$.

2. $a^3(b-c)+b^3(c-a)+c^3(a-b)=-(b-c)(c-a)(a-b)$ $(a+b+c)$.

3. $a^2(b-c)+b^2(c-a)+c^2(a-b)=-(b-c)(c-a)(a-b)$.

4. $(x+y+z)^4-(y+z)^4-(z+x)^4-(x+y)^4+x^4+y^4+z^4=12xyz(x+y+z)$.

5. $b^2c^2(b-c)+c^2a^2(c-a)+a^2b^2(a-b)=-(b-c)(c-a)(a-b)(bc+ca+ab)$.

6. $a^4(b^2-c^2)+b^4(c^2-a^2)+c^4(a^2-b^2)=-(b+c)(c+a)(a+b)(b-c)(c-a)(a-b)$.

Factor—

7. $x^3(y-z)+y^3(z-x)+z^3(x-y)$.

8. $a(b-c)^3+b(c-a)^3+c(a-b)^3$.

9. $a(b-c)^2+b(c-a)^2+c(a-b)^2+8abc$.

10. $a^3+b^3-c^3+3abc$.

11. $27-8x^3+y^6+18xy^2$.

III. PROOF OF THE BINOMIAL THEOREM FOR FRACTIONAL AND NEGATIVE EXPONENTS.

406. Proof. In Arts. 344 and 345 it has been shown that when n is a positive integer,

$$(1+x)^n=1+nx+\frac{n(n-1)}{1\times 2}x^2+\frac{n(n-1)(n-2)}{1\times 2\times 3}x^3+\dots.$$

It is evident that when n is a positive integer the number of terms in the right-hand member is limited, since all the terms after the $(n+1)$st contain $n-n$, or 0, as a factor. But if n is not a positive whole number, the number of terms must be unlimited, since no factor becomes zero.

Hence the binomial theorem for fractional and negative exponents gives an infinite series, and is valid only when this series is convergent. Hence, in the proof of this theorem for negative and fractional values of n, the *series used are limited to those cases where x has such a value as will make the series convergent and the proof is good only for such cases.*

Denote the series

$$1 + nx + \frac{n(n-1)}{1 \times 2}x^2 + \frac{n(n-1)(n-2)}{1 \times 2 \times 3}x^3 + \ \ldots \quad (1)$$

by the symbol $f(n)$.

Then, in like manner, the symbol $f(m)$ stands for

$$1 + mx + \frac{m(m-1)}{1 \times 2}x^2 + \frac{m(m-1)(m-2)}{1 \times 2 \times 3}x^3 + \ \ldots \quad (2)$$

If (1) be multiplied by (2), the product will be a series in ascending powers of x, whose coefficients will involve m and n in a way which is independent of any particular values which m and n may have (just as, for instance, $(1 + ax)(1 + bx) \equiv 1 + (a + b)x + abx^2$ is identically true, no matter whether a and b be whole numbers or fractions, positive or negative).

Hence, to determine the form which the product must have in all cases, whatever the values of m and n, we let m and n be positive integers.

$$\therefore f(n) = (1 + x)^n,$$

$$f(m) = (1 + x)^m.$$

Multiplying, $f(n) \times f(m) = (1 + x)^n \times (1 + x)^m = (1 + x)^{m+n}$

$$= 1 + (m+n)x + \frac{(m+n)(m+n-1)}{1 \times 2}x^2 + \ldots.$$

This is the *form* of the above product, and holds for all values of m and n.

$$\therefore f(m) \times f(n) = f(m + n) \ \ldots \ldots \quad (3)$$

Similarly,

$$f(m) \times f(n) \times f(p) = f(m + n) \times f(p)$$

$$= f(m + n + p),$$

And $f(m) \times f(n) \times f(p) \ldots$ to s factors $= f(m + n + p + \ldots$ to s terms).

Let $\quad m = n = p = \ldots = \frac{r}{s}$, where r and s are positive integers.

Then $\quad f\left(\frac{r}{s}\right) \cdot f\left(\frac{r}{s}\right) \ldots$ to s factors $= f\left(\frac{r}{s} + \frac{r}{s} + \frac{r}{s} + \ldots \text{to } s \text{ terms}\right)$.

$$\therefore \left[f\left(\frac{r}{s}\right)\right]^s = f(r) \ \ldots \ldots \quad (4)$$

But since r is a positive integer, $\quad f(r) = (1 + x)^r$.

Substitute for $f(r)$ in (4), $\quad (1 + x)^r = \left[f\left(\frac{r}{s}\right)\right]^s$.

Extract sth root, $\quad (1 + x)^{\frac{r}{s}} = f\left(\frac{r}{s}\right)$.

This proves the binomial theorem for any positive fractional exponent.

We can now prove that the theorem is also true for any negative exponent.

In (3) let $$m = -n.$$

$$\therefore f(-n) \times f(n) = f(-n+n) = f(0) = 1.$$

$$\therefore f(-n) = \frac{1}{f(n)} = \frac{1}{(1+x)^n} = (1+x)^{-n}.$$

$$\therefore (1+x)^{-n} = f(-n)$$

$$= 1 + (-n)x + \frac{(-n)(-n-1)}{1 \times 2}x^2 + \ldots.$$

IV. A MISCELLANEOUS EXERCISE.

EXERCISE 142.

Factor—

1. $1000x^9 - y^3$.
2. $36x^3 - 13x^2 - 40x$.
3. $36x^4 - 289x^2 + 400$.
4. $x^7 - 64x$.
5. $132x + xy - xy^2$.
6. $4x^4 + 4a^2 - a^4 - 4$.
7. $x^{\frac{2}{3}} - 9$.
8. $x^{\frac{3}{5}} + 27$.
9. $4x - y^2$.
10. $x^{2a} - y^{-6}$.
11. $a^{\frac{3}{2}} - 8b^{-1}$.
12. $25n^5 - y^{-2}$.
13. $15ax - 5ay + 12bx - 4by$.
14. $7(p-1)^2 - 27(p-1) + 18$.
15. $a^2 + 9 - 4x^2 - n^2 + 4nx - 6a$.
16. $2a^2x^2 + (a^2 + 3a)x - a^2 - 3a - 2$.
17. $3x - 8x^{\frac{1}{2}} - 35$.
18. $6x^3 - x^{\frac{3}{2}} - 15$.
19. $10x^{\frac{1}{2}} - 19x^{\frac{1}{4}} - 56$.
20. $12x^{\frac{4}{3}} + 5x^{\frac{2}{3}} - 72$.
21. $a^{\frac{1}{3}}x^{\frac{1}{2}} - 3a^{\frac{1}{3}} + 5x^{\frac{1}{2}} - 15$.
22. $30 + \sqrt{2x} - 2x$.
23. $60 - 7\sqrt{3a} - 6a$.
24. $15x - 2\sqrt{xy} - 24y$.

If $x = 2$, $y = -3$, $z = -\frac{1}{2}$, find the value of:

25. $3xy - y(x + 4z) - xz(4y + 6x) + 3yz(x + y)(y + 2z)$.
26. $(x - y - 10z)(2x + 3y + 6z) - xyz(x^3 - y^2 - 2z^2) + 12x^2y^2z^2$.
27. $(y^2 - z)(x - 5) + (x^2 + z)(y - 2z) + (2x + 5y - 10z)^2$.
28. $x^2y + xy^2 + xz^2 + x^2z + y^2z + yz^2 - 3yz(y^2 - x - 1)$.
29. $x^y + y^x + z^{xy} + (x + y)^{x-y} - \sqrt[y]{xz} - (xz + xy)^x - (y + z)x^{-x}$.

Find the H. C. F. and the L. C. M. of:

30. $a^2 + a^{\frac{3}{2}}b^{\frac{1}{2}} + a^{\frac{1}{2}}b^{\frac{3}{2}} - b^2$ and $a^2 - a^{\frac{3}{2}}b^{\frac{1}{2}} - a^{\frac{1}{2}}b^{\frac{3}{2}} - b^2$.

31. $a^2 + b\sqrt{a} - ab + \sqrt{b^3}$ and $a^2 + a\sqrt{ab} + ab^{\frac{1}{2}} - b\sqrt{b}$.

32. $a^{\frac{3}{2}} + b\sqrt{a} - 2b^{\frac{3}{2}}$ and $a^{\frac{3}{2}} + ab^{\frac{1}{2}} - b\sqrt{a} - b^{\frac{3}{2}}$.

33. $3a^2 + 7a^{\frac{3}{2}}b^{\frac{1}{2}} + a^{\frac{1}{2}}b^{\frac{3}{2}} - b^2$ and $2a^2 - 13ab - 2a^{\frac{1}{2}}b^{\frac{3}{2}} + 3b^2$.

Simplify:

34. $(\sqrt{x} - 1)(3\sqrt{x} + 2) + x^{\frac{1}{2}}(2\sqrt{x} - 5) - (2\sqrt{x} - 3)(3x^{\frac{1}{2}} - 4)$.

35. $$\frac{\dfrac{m+n}{m-n} + \dfrac{m-n}{m+n}}{\dfrac{m+n}{m-n} - \dfrac{m-n}{m+n}} - \frac{m}{n} - \frac{n}{m}.$$

36. $$\left\{ x^3 \cdot \frac{1 - \dfrac{1}{x-1}}{x - 1 - \dfrac{1}{x+1}} + \frac{x + \dfrac{x}{x-1}}{x - 1 - \dfrac{x^2}{x+1}} \right\} \div (x+1).$$

37. $8^{-\frac{2}{3}} + 25^{\frac{3}{2}} - (\tfrac{1}{3})^{-3} + 13^0 - (\tfrac{1}{32})^{-\frac{3}{5}}$.

38. $$\frac{a^{\frac{1}{3}} \sqrt[3]{8a^2 b^{-1}}}{b^{\frac{1}{3}} \sqrt{ab}} \cdot \sqrt[3]{-27a^{\frac{3}{2}}b^{-1}}.$$

39. $\sqrt{\tfrac{2}{3}} + \sqrt{\tfrac{3}{2}} - \sqrt{\tfrac{1}{6}} + \sqrt[4]{36}$.

40. $$\frac{\sqrt{2} - 2\sqrt{3} + 3\sqrt{6}}{\sqrt{2} - 2\sqrt{3} - 3\sqrt{6}}.$$

Solve:

41. $3x^2 - 2x(x+5) - (x-1)(2x+3) = (5x-1)(x+7) - 6x^2$.

42. $$\frac{3x-2}{2x-3} - \frac{7}{6} - \frac{x-1}{x+2} = -\frac{2}{3}.$$

43. $3x + 11y = 19$; $11x - 3y = 9$.

44. $2x-6y-z=5.$
$3x+y-\frac{1}{2}z=5.$
$4x-9y+z=23.$

45. $\frac{2}{x}-5y=13.$
$\frac{3}{x}+\frac{1}{2}y=4.$

46. $4x-\frac{5}{3y}=-3.$
$3x+\frac{1}{4y}=-6\frac{1}{4}.$

47. $ax+(b+1)y=c.$
$(b+1)x+ay=d.$

48. $\frac{3}{2x}+\frac{2}{3y}=-7.$
$\frac{4}{5x}+\frac{5}{6y}=2.$

49. $ax+by+cz=1.$
$bx+cy+az=1.$
$cx+ay+bz=1.$

Write the following expressions with positive exponents and reduce the results to single fractions.

50. $a+b^{-1}+2cb^{-2}.$

51. $(a-b)^{-1}+(a+b)^{-1}.$

52. $a(a-b)^{-1}-b(a+b)^{-1}.$

53. $\dfrac{x^{-1}}{1+x^{-1}}+\dfrac{1-x^{-1}}{x^{-1}}.$

54. $\dfrac{4(x-1)^{-1}+x^{-1}(x-1).}{(x-1)^{-1}-x^{-1}}$

55. $\dfrac{a^{-1}+\sqrt{a^{-1}b^{-1}}+b^{-1}}{a^{-2}+a^{-1}b^{-1}+b^{-2}}.$

56. $\dfrac{(a+b)(a-b)^{-1}-(a-b)(a+b)^{-1}}{1-(a^2+b^2)(a+b)^{-2}}$

57. $\dfrac{2xy^{-1}+1-yx^{-1}}{2xy^{-1}+x^{-1}y-3}.$

58. $(mn^{-1}+1)\div(m^2n^{-1}+n^2m^{-1}).$

59. $\dfrac{(a^{-1}+b^{-1}-c^{-1})^2}{a^{-2}b^{-2}(a+b)^2-c^{-2}}.$

60. $a(1-a^{-2})(a+a^{-1})(1+a^{-1})^{-1}(a^2+1)^{-1}.$

61. $(x^2+y^2)(x^2-y^2)^{-1}-x(x+y)^{-1}+y(y-x)^{-1}.$

62. $b(a+b)^{-1}-ab(a+b)^{-2}-ab^2(a+b)^{-3}.$

63. $(b+c)(a-b)^{-1}(a-c)^{-1}+(c+a)(b-c)^{-1}(b-a)^{-1}+$
$(a+b)(c-a)^{-1}(c-b)^{-1}.$

64. $\dfrac{(1+z)(1+z^2)^{-1}-(1+z^2)(1+z^3)^{-1}}{(1+z^2)(1+z^3)^{-1}-(1+z^3)(1+z^4)^{-1}}.$

65. $x^{p-q} \cdot x^{2q-1} \cdot x^{1-p}$.

66. $(x^{a-b})^{a+b} \div x^{-b^2}$.

67. $x^{a^2+ab} \div x^{2ab}$.

68. $\left(x^{m-1}\right)^{m+1} \cdot \left(x^m\right)^m$.

69. $z^{a-b} \cdot z^{b+a} \div (z^2)^a$.

70. $\left(x^m\right)^{n-1} \cdot \left(x^n\right)^{m+1} \cdot \left(x^m\right)^{1-2n}$.

71. $\left(x^a\right)^{a+b} \cdot \left(x^b\right)^{a-b} \div \left(x^{a+2b}\right)^a$.

72. $\left(y^{x-2}\right)^a \cdot \left(y^{a-1}\right)^x \div \left(y^{2a+x}\right)^{-1}$.

73. $x^{3m-n} \cdot x^{2n+1} \cdot x^{m+2} \div x^{2m+n+3}$.

74. $x^{p+3q} \cdot x^{q-3p} \cdot x^{p-2q} \div x^{p+q}$.

75. $x^{\frac{a}{a+1}} \cdot x^{\frac{1}{a+1}} \div \sqrt[a]{x^{-1}}$.

76. $\left\{\dfrac{a^{p-q}}{\sqrt[q]{a^{q^2-pq}}} \cdot \sqrt[p]{a^{2pq-p^2}}\right\}^{-n}$.

77. $\left[\left(x^{\frac{1}{a-c}}\right)^{a-\frac{c^2}{a}}\right]^{\frac{a}{a+c}}$

78. $\dfrac{1-a^{\frac{1}{2}}b^{\frac{1}{3}}}{b^{\frac{1}{3}}\sqrt{a}-ab} \div \dfrac{\left(1+a^{\frac{1}{2}}b^{\frac{1}{3}}\right)^{-1}}{a^{-\frac{1}{3}}b}$.

79. $\left(\dfrac{m^{p+q}}{m^q}\right)^p \div \left(\dfrac{m^q}{m^{q-p}}\right)^{p-q}$.

80. $\dfrac{a^{-2}b+5a^{-1}\sqrt{b}-66}{a^{-2}b+3a^{-1}\sqrt{b}-54}$.

Solve the following simultaneous quadratics:

81. $x^4+y^4=272$.
$x+y=6$.

82. $x^5-y^5=33$.
$x-y=3$.

83. $x^2-xy=35$.
$xy+y^2=18$.

84. $\dfrac{x}{y}+4\sqrt{\dfrac{x}{y}}=8\frac{1}{4}$.
$x-y=5$.

85. $3x^2-7xy+y^2-4x=-19$.
$7xy-y^2=-15$.

86. $x^2y+xy^2=6$.
$3x^2y^2+8xy=3$.

87. $x^{\frac{1}{3}}+y^{\frac{1}{3}}=5$.
$x+y=35$.

88. $x^{-1}+y^{-1}=4$.
$x^{-2}+y^{-2}=8\frac{1}{2}$.

89. $\dfrac{x}{a}+\dfrac{y}{2b}=\frac{5}{8}$.
$\dfrac{a}{3x}+\dfrac{5b}{4y}=5\frac{2}{3}$.

90. $x^2-xy+3y=11$.
$y^2-xy-3x=-1$.

91. $2x^2-7xy+5y=-4$.
$6x-4y=15$.

92. $\dfrac{3x}{y+3}+\dfrac{2y}{x+2}=\frac{5}{2}$.
$\dfrac{x}{2}+\dfrac{y}{3}=2$.

93. $\frac{1}{x}+\frac{1}{y}=2.$
$xy+\frac{1}{x}+\frac{1}{y}=3\frac{1}{2}.$

94. $x+y+\sqrt{xy}=14.$
$\sqrt{xy}\,(x+y)=40.$

95. $\sqrt{x^2+7}+y=6.$
$\sqrt{x^4+22y^2}+x^2=22.$

96. $\sqrt{x+y}+\sqrt{x-y}=4.$
$x^2-y^2=9.$

97. $x^2+xy+y^2=84.$
$x-\sqrt{xy}+y=6.$

98. $\sqrt{\frac{x}{y}}+\sqrt{\frac{y}{x}}=\frac{10}{3}.$
$x+y=10.$

99. $\sqrt{xy}-\sqrt{x-y}=11.$
$\sqrt{x-y}\,\sqrt{xy}=60.$

100. $y+\sqrt{x^2-1}=2.$
$\sqrt{x+1}-\sqrt{x-1}=\sqrt{y}.$

I. Transformations of Physical Formulas.

101. Given $v=at$, find the value of t in terms of a and v

102. Given $s=\frac{1}{2}at^2$, find the value of t in terms of a and s.

103. Given $s=\frac{v^2}{2a}$, find the value of v in terms of a and s.

104. Given $s=\frac{1}{2}a(2t-1)$, find t in terms of a and s.

105. Given $F=\frac{mv^2}{r}$, find each letter in terms of the other three.

106. Given $e=\frac{mv^2}{2}$, find each letter in terms of the others.

107. Given $e=\frac{wv^2}{2a}$, find each letter in terms of the others.

108. Given $t=\pi\sqrt{\frac{l}{g}}$, find l and g, each in terms of the other letters.

109. Given $C=\frac{E}{R}$, find each letter in terms of the others.

110. Given $R=\frac{gs}{g+s}$, find each letter in terms of the others.

111. Given $\frac{1}{f}=\frac{1}{p}+\frac{1}{p'}$, find each letter in terms of the others.

II. Transformations of Arithmetical Formulas.

112. Given $i = prt$, find each letter in terms of the other three.

113. Given $a = p + prt$, find each letter in terms of the other three.

III. Transformations of Algebraic Formulas.

114. Consult pages 318 and 326.

IV. Transformations of Geometrical Formulas.

115. Given $A = \frac{1}{2}bh$, find each letter in terms of the others.

116. Given $A = \frac{1}{2}h(b+b')$, find each letter in terms of the others.

117. Given $C = 2\pi R$, find each letter in terms of the others.

118. Given $A = \pi R^2$, find each letter in terms of the others.

119. Given $A = \pi RL$, find each letter in terms of the others.

120. Given $A = 4\pi R^2$, find each letter in terms of the others.

121. Given $T = \pi R(R+L)$, find each letter in terms of the others.

122. Given $T = 2\pi R(R+H)$, find each letter in terms of the others.

123. Given $V = \pi R^2 H$, find each letter in terms of the others.

124. Given $V = \frac{1}{3}\pi R^2 H$, find each letter in terms of the others.

125. Given $V = \frac{4}{3}\pi R^3$, find each letter in terms of the others.

126. Given $a+b+c = 2s$.
show that $a+b-c = 2(s-c)$; $a-b+c = 2(s-b)$, etc.

127. Also show that $1 - \frac{a^2+b^2-c^2}{2ab} = \frac{2(s-a)(s-b)}{ab}$; and that

$$1 - \frac{a^2+c^2-b^2}{2ac} = \frac{2(s-a)(s-c)}{ac};\ \text{etc.}$$

128. Also show that $1 + \frac{a^2+b^2-c^2}{2ab} = \frac{2s(s-c)}{ab}$; and that

$$1 + \frac{a^2+c^2-b^2}{2ac} = \frac{2s(s-b)}{ac};\ \text{etc.}$$

ANSWERS.

Exercise 1.

13. $5a + 7b$.
14. $6a^2 = 2(a - b)$.
15. $4(a^2 - 9b) < (7a + b^3)^2$.
16. $(x - 10y^2)(x^3 + yz) = 2ax^3$.
17. $\dfrac{9x + 2y}{3z} = 9x + \dfrac{2y}{3z^3}$.
18. $5a^3 + \dfrac{6b^2}{(x - 2y^3)^2} > \dfrac{5(a^3 + b^3)}{(x + 2y^4)^3}$.

Exercise 2.

1. 10.
2. 18.
3. 30.
4. 15.
5. 9.
6. 25.
7. 6.
11. 7.
12. 3.
13. 6.
16. 18.
17. 12.
19. 16.
24. 6.
25. 108.
29. 63.
30. 18.
31. 11.
32. 3.
33. $\frac{2}{5}$.
34. $\frac{3}{5}$.
35. 1.
36. $\frac{2}{5}$.
42. 15.
44. $\frac{42}{5}$.
45. $\frac{1}{6}$.
46. $\frac{134}{15}$.
47. 3.
48. 2.
49. 6.
58. 13.
59. 6.

Exercise 3.

1. -5.
2. -6.
3. $2x$.
7. $7x^2$.
11. $4ax$.
14. $2x$.
17. $a + b$.
18. $2x^2 - 11y^2$.
19. $2by^2$.
20. $7x^2 + 2y^2$.
24. $9x + 12xz$.
25. $2x^2y + 4xy^2$.
26. $-2x - 4y + 2z$.
27. $-xy + 2ax + y^2 - 3x^2$.
28. $-4x - y - 2z$.

Exercise 4.

1. $4ab$.
2. $-4x$.
3. $-x$.
4. $8x$.
7. $x^2 - 5x$.
15. $-1 - 2x + 2x^2 + x^3 + 3x^4$.
16. $12xy^2 - x^2y^2 - 9x^2y$.
22. $4x^n - 2x^m - x^3 + 2$.
23. $4x^3 - 2x - 2$.
24. $2x^3 + 6x^2 - 2x - 4$.

Exercise 5.

1. $5a - b$.
2. $x + 1$.
3. $1 - x$.
4. -1.
5. $2x + 1$.
6. $-x + 3y$.
7. $1 - 2x$.
8. $9x - 1$.
9. 4.
10. $4x - 1$.
11. 0.
12. $a - 1$.
13. 0.
14. 0.
15. $x + 1$.
16. 6.
17. $2c - b - d$.
18. $3x - 2x^3$.
19. $-7x^3 + x^2 - 2x - 1$.
20. $-x$.
21. 2.
22. $-2y$.
23. $-3x$.

Exercise 7.

1. $2x^4 - 3x^3 - 5x - 2$.
2. $4xyz^2 + 3xy^2z - x^2yz$.
3. $3\sqrt{3} - 3\sqrt{2} - 1$.
4. $3(x + y) + (y + z)$.
5. $1 - 5ab + 2bc - d + x$.
6. $d^2 - a + 2x + 10y$.
7. $-7ab - 2c + c^2 + 4x - 2\sqrt{y}$.
8. $4\sqrt{x} - 5x + x^2 + x^3 + x^4 - 4$.
9. 12.
10. 2; $\frac{2}{3}$.
11. 3; 13.
12. $12x - 3$.
13. $12x$.
14. $2x^2 - 4z^2$.
15. $(1 + a - 2c)x^5 - (3 + c)x^4 - (1 + a - 3c)x^3 - (2a + 5)x^2 + 2$.
17. $3a^3 - 10ab + 3a^2b^2$.
18. First, $x^3 - x^2 + x - 1$.
19. First, $-x^3 + 2x^2 - 3x + 1$.
20. $2x^3 - 2x^2y + 6xy^2 + 5y^3$.
21. $4x^3 - 2x^2y + 2xy^2 + 7y^3$.
22. $-2x^3 + 2xy^2 + 3y^3$.
23. $6x^3 - 2x^2y - 3y^3$.
24. $10x^3 - 4x^2y + 4xy^2 - y^3$.

Exercise 8.

5. $-8x^3$.
6. $15x^2$.
7. $-12a^2x^3$.
8. $42x^2y^4$.
9. $-21a^2xy$.
10. $20a^2bcd^2$.
14. $-12x^{3n}$.
15. $-35x^{2n}y^{2m}$.
16. $-3x^{n+2}y^{2n}$.
18. $6a^2x + 9ax^2$.
19. $-15x^2y + 10xy^2$.
25. $2x^{n+3} - 3x^{n+2}$.
27. $6x^{3n}y^m - 15x^{2n}y^{m+1}$.
29. $20x - 12x^2$.

Exercise 9.

1. $2x^2 - 7x - 4$.
2. $3x^2 - 7x - 6$.
3. $2x^2 - 9x - 35$.
4. $12x^2 - 25xy + 12y^2$.
5. $28x^4 + x^2y^2 - 15y^4$.
6. $30x^2y^2 + xy - 42$.
7. $32a^5c - 2ab^4c^3$.
8. $33x^5y + x^3y^3 - 14xy^5$.
9. $a^3 + b^3$.
10. $x^4 - y^4$.
11. $8x^4 - 2x^3 + x^2 - 1$.
12. $6x^3 - 19x^2y + 21xy^2 - 10y^3$.

13. $2x^5 - 5x^4 - 2x^3 + 9x^2 - 7x + 3$.
14. $3x^4y - 10x^3y^2 + 4x^2y^3 + 6xy^4 + y^5$.
15. $x^5 - 5x^4y + 10x^3y^2 - 10x^2y^3 + 5xy^4 - y^5$.
16. $4x^5 + 9x^4 - 16x^3 + 22x^2 - 21x + 6$.
17. $x^6 - x^5 - 7x^4 + 3x^3 + 17x^2 - 5x - 20$.

18. $x^6 - 6x^4y + 9x^2y^2 - y^6$.
19. $x^4 - 14x^3 + 49x^2 - 4$.
20. $a^4 + a^2b^2 + b^4$.
21. $16x^4 + 36x^2y^2 + 81y^4$.

22. $x^7 - 9x^5y^2 + 7x^4y^3 + 13x^3y^4 - 19x^2y^5 + 8xy^6 - y^7$.
23. $- x^5 + 2ax^4 + 8a^2x^3 - 16a^3x^2 - 16a^4x + 32a^5$.
24. $a^3 + b^3 + x^3 + 3ab^2 + 3a^2b$.
25. $a^2b^2 + c^2d^2 - a^2c^2 - b^2d^2$.
26. $x^{n+1} - x^{n-1} - 6x^{n-2} - 2x + 4$.
27. $x^{2n+1} - x^{2n} - 2x^{2n-1} + 3x^{2n-2} - 10x^{2n-3}$.
28. $x^{n-4} + x^{n-3} - x^{n-2} + x^{n-1} - x^n + 7x^{n+1} + 10x^{n+2}$.

Exercise 10.

1. $-x$.
2. $2x^2 - 8$.
3. x^2.
4. $-20x$.
5. $x - 2x^2 - 2x^3$.
6. $2 - x^2$.
7. $3x^2 - 6x$.
8. $-2ab - 8b^2$.
9. $a^2 - 4b^2 + 12bc - 9c^2$.
10. $y^2 - 2yz + z^2$.
11. $2x + 1$.
12. $2x + 2x^2$.
13. $x^2 + 10x - 16$.
14. 0.
15. $x^2 - 5x + 8$.
16. $3x^2 - 10xy + y^2$.
17. $x^2 - z^2$.
18. $4a^2 - ax + bx + my + cy$.
19. 0.
20. $4x^3$.
21. -12.
22. 40.
23. -3.
24. -2.
25. 7.
26. -2.
27. -1.
28. -11.
29. 1.
30. 5.
31. 29.

Exercise 11.

1. -3.
2. $-3x^2$.
3. $-2a$.
4. $5xy$.
5. $-x$.
6. $-7y^2$.
17. $x^2 - x + 1$.
18. $3x^2 - 7x + 1$.
23. $2x - 5x^2 + 3x^{2-n}$.
24. $x^4 - 2x^3 + 3x^2 + x$.

Exercise 12.

1. $3x + 1$.
2. $2x + 1$.
3. $4x - 5y$.
4. $3x + 4y$.
5. $3x + 7$.
6. $3x - 5y$.

7. $3a + 4c$.
8. $-5x + 8$.
9. $4x + y$.
10. $a + 2b$.
11. $x^2 + xy + y^2$.
12. $9x^2 - 6x + 4$.
13. $3x - 7$.
14. $25 + 20x + 16x^2$.
15. $4a^2x^2 - 2axy^2 + y^4$.
16. $x^2 - 3x + 1$.
17. $7x^2 + 8x + 1$.
18. $3a^2 - 4ax + x^2$.
19. $2y^3 - 4y^2 + y - 1$.
20. $c^4 + c^2x^2 + x^4$.
21. $2x^3 - 3x^2 + 4x - 5$.
22. $2x^3 - x + 1$.
23. $3x^3 + 4x^2y + 5xy^2 + 2y^3$.
24. $2x^4 - 3x^2y - 2y^2$.
25. $x^3 + 2x^2y + 4xy^2 + 8y^3$.
26. $x^4 - 2x^3y + 4x^2y^2 - 8xy^3 + 16y^4$.
27. $x^5 - x^4y + x^3y^2 - x^2y^3 + xy^4 - y^5$.
28. $64x^6 + 16x^4y^2 + 4x^2y^4 + y^6$.
29. $2x^2 - 5x - 1$.
30. $3x^2 - x - 5$.
31. $2x^3 - 4x^2 - x + 3$.
32. $2x^3 + 3x^2y - 4xy^2 + y^3$.
33. $3a^3 - 4a^2b + 3ab^2 - 2b^3$.
34. $x^2 + y^2 - z^2 - xz + xy - yz$.
35. $c^2 + d^2 + n^2 - cd - cn - dn$.
36. $y^4 + 2y^3 + 3y^2 + 2y + 1$.
37. $2x^4 - 4x^3 + 3x^2 - 2x + 1$.
38. $2xy - 2xz - 3yz$.
39. $x^2 - 3x + 1$.
40. $2x^3 - 3x^2 + x - 5$.
41. $2x^n - 3x^{n-1}$.
42. $4x^{3n} + 3x^{2n} - x^n$.
43. $4x^{n+1} - 3x^n + x^{n-1}$.
44. $3x^{n-1} + 2x^{n-2} - 3x^{n-3}$.

Exercise 13.

1. $\frac{3}{4}x^3 - \frac{11}{6}x + \frac{2}{3}$.
2. $\frac{3}{8}x^3 - \frac{9}{4}x^2 + \frac{5}{6}$.
3. $2x^4 - \frac{49}{36}x^3 - \frac{49}{72}x - \frac{1}{2}$.
4. $2.88x^3 + 10.86x - 19.2$.
5. $5.4x^4 - 3.3x^3 + 10.1x^2 + 1.32x - .08$.
6. $6.75x^4 + 1.2x^3y + 15.84x^2y^2 + 13.44y^4$.
7. $\frac{2}{5}x^2 - 4x + \frac{4}{5}$.
8. $\frac{4}{5}x^2 + \frac{4}{5}x - 1$.
9. $\frac{2}{5}x^2 + \frac{6}{5}xy + \frac{9}{4}y^2$.
10. $1.8x^2 - 3.2x + 0.48$.
11. $0.5x^2 - 1.8x + 3.5$.
12. $3x^2 + 4.8xy - 21.5y^2$.

Exercise 14.

1. 5.	8. $-\frac{1}{4}$.	15. $-\frac{4}{5}$.	22. 3.
2. -3.	9. 0.	16. $-\frac{6}{5}$.	23. 1.
3. -2.	10. 6.	17. $-\frac{5}{4}$.	24. 1.
4. 2.	11. -5.	18. $-\frac{5}{2}$.	25. 6.
5. $-\frac{4}{5}$.	12. 11.	19. $-\frac{7}{5}$.	26. 14.
6. $-\frac{11}{3}$.	13. 0.	20. $-\frac{3}{4}$.	27. $-\frac{11}{3}$.
7. $\frac{18}{5}$.	14. -2.	21. $\frac{5}{4}$.	28. -3.

Exercise 16.

1. 12 and 36 marbles.
2. A, \$61; B, \$39.
3. John, 36; William, 60 ex.
4. 1st, \$26; 2d, \$37; 3d, \$35.
5. 80 miles.
6. 27, 28, 29.
7. 19, 73.
8. 1st, 22; 2d, 11; 3d, 17.
9. Horse, \$67; cow, \$27.
10. A, \$97; B, \$194; C, \$194.
11. 14, 21.
12. 13, 14, 15, 16, 17.
13. 9, 15.
14. 1st, \$280; 2d, \$140; 3d, \$80.
15. Daughter, \$960; sons, each, \$1770.
16. Father, 48; son, 12 years.
17. Father, 52; son, 31; daughter, 26 years.
18. Father, 84; son, 42 years.
19. 11, 25.
20. 19, 21, 23.
21. 9, 14.
22. Eldest, \$21; rest, in order, \$13, \$9, \$7, and \$6.
23. 21, 54.
24. 21, 22.
25. 14, 16.
26. Father, 44; son, 20 years.
27. $4\frac{1}{2}$ hours; 36 miles.
28. 8 miles.
29. 36 miles.
30. 38 days.
31. A, 25; B, 15 years.
32. Father, 55; son, 25 years.
33. Silk, \$3.30; cloth, \$1.10.
34. 9.
35. 15 of each.
36. 17 of each.
37. 7 bills; 14 quarters.
38. 17 halves; 3 dimes.
39. Son, \$375; daughter, \$150.
40. 11 beggars.
41. A, \$42; B, \$37; C, \$33.
42. 5 @ 32 cts.; 7 @ 20 cts.
43. 48 feet.
44. A, \$21; B, \$68.
45. 72 pounds.
46. A, 24 miles.

Exercise 17.

1. $\frac{1}{3}x^2 - \frac{4}{3}xy$.
2. $\frac{1}{6}x^3 - \frac{1}{3}$.
3. $4x^2 - 3.5y^2$.
4. $2.6 + 0.9x - x^2$.
5. $\frac{1}{2}x^3 - \frac{1}{5}x^2 - \frac{1}{4}x + \frac{27}{8}$.
6. $0.3x^3 - 0.7x^2y + 3.3xy^2 - 2.5y^3$.
7. $\frac{1}{3}x^3 - \frac{1}{5}x + 8$.
8. $3.6x^2 + 1.29xy - 0.6y^2$.
9. $\frac{2}{3}x^4 + \frac{1}{2}a^2x^2 + \frac{2}{3}a^4$.
10. $\frac{5}{2}x^4 - \frac{17}{36}x^3 + \frac{12}{15}x - 1$.
11. $4.8a^3 + 4.55a^2b + 2.05ab^2 + 13b^3$.
12. $0.2x^3 - 1.5x^2y + 1.3xy^2 + 4.2y^3$.
13. $1.8x^5y^2 - 2.73x^4y^3 - 1.56x^3y^4 + 2.7x^2y^5$.
14. $4.8x^3 - 17.95x^2y + 18.45xy^2 - 6.3y^3$.
15. $2x^2 - 4xy + 2y^2$.
16. $x^5 - 2x^4y + 2x^3y^2 + 3x^2y^3 - 2xy^4 - 3y^5$.

17. $2 - 3x + 3x^2 - 3x^3 + 3x^4 \ldots\ldots$
18. $1 - 2x + 2x^2 - 2x^4 + 2x^5 \ldots\ldots$
19. $x^4 - x^3 + 2x^2 - 3x + 5 \ldots\ldots$
20. $1 + 2x + 7x^2 + 20x^3 + 61x^4 \ldots\ldots$

21. $\frac{3}{4}x^2 - 2xy + \frac{4}{3}y^2$.
22. $\frac{3}{8}x^3 - \frac{2}{3}x^2y + \frac{1}{2}xy^2$.
23. $6x - \frac{1}{3}y - \frac{1}{2}$.
24. $1.6x^2 - 2xy + 2.4y^2$.
25. $3.5x^2 - 3x + 1.5$.
26. $2x^3 - 2.5x^2y - 0.3xy^2 + 1.2y^3$.
27. $2z - x$.
28. 0.
29. $3x + 7y$.
30. $2a - 12c + 84d$.

31. $3 + 4x - 2y$.
32. $5a^2b + 4b^3$.
33. 2.
34. -3.
35. -1.
36. -8.
37. -1.
38. $x^5 + x - 1$.
39. $x^2 - 3$.
40. 38.

Exercise 18.

1. $n^2 + 2ny + y^2$.
2. $c^2 - 2cx + x^2$.
3. $4x^2 - 4xy + y^2$.
4. $9x^2 - 12xy + 4y^2$.
11. $x^2 - z^2$.
12. $y^2 - 9$.
14. $49x^2 - 16y^2$.
18. $4x^{2n} - 25y^{2m}$.

Exercise 19.

1. $a^2 + 2ab + b^2 - 9$.
3. $15 - 2x - x^2$.
5. $4x^2 - 9y^2 - 6y - 1$.
6. $x^4 + 6x^3 + 9x^2 - 4$.
9. $x^4 + x^2y^2 + y^4$.
11. $4x^4 - 29x^2 + 25$.
16. $a^2 + 2ab + b^2 - c^2 + 2c - 1$.
17. $x^4 + y^4 - x^4y^4 - 1$.

Exercise 20.

1. $4x^2 + y^2 + 1 + 4xy + 4x + 2y$.
2. $x^2 + 4y^2 + 4z^2 - 4xy + 4xz - 8yz$.
3. $9x^2 + 4y^2 + 25 - 12xy - 30x + 20y$.
4. $4a^2 + b^2 + 9c^2 - 4ab + 12ac - 6bc$.
9. $x^2 + y^2 + z^2 + 1 - 2xy + 2xz - 2x - 2yz + 2y - 2z$.
20. $a^4 + 4b^4 + 4ab - 1$.

Exercise 21.

1. $x^2 + 7x + 10$.
2. $x^2 - 8x + 15$.
3. $x^2 - 3x - 28$.
4. $x^2 + 4x - 32$.
5. $x^2 - 6x - 7$.
20. $a^2 + (b - 1)a - b$.
23. $x^2 + 4xy + 4y^2 - 5x - 10y - 14$.
27. $a^2 + 2ab + b^2 + a + b - 12$.

Exercise 22.

1. $2x^2 + 7x + 6$.
2. $2x^2 + x - 10$.
3. $3x^2 - 7x + 2$.
4. $5x^2 - 4x - 1$.

Exercise 23.

1. $a + x$.
2. $3 + 2x$.
9. $a + b - 2c$.
10. $2x^2 - y^2 - 1$.

Exercise 24.

1. $a^2 - 2a + 4$.
2. $x^2 + x + 1$.
3. $9x^2 + 12x + 16$.
4. $1 - 2x^2 + 4x^4$.
5. $25 + 5x^3 + x^6$.
6. $9a^4 - 3a^2y^4 + y^8$.
7. $x^4 - x^2y^2 + y^4$.
8. $a^2 - 2a + 1 + ax - x + x^2$.
9. $c^2 - c + cx + 1 - 2x + x^2$.
10. $4 + 2x + 2y + x^2 + 2xy + y^2$.
11. $x^4y^4 + x^3y^3 - 2xy + 1$.
12. $9x^4 - 15x^2y^3 + 25y^6$.

Exercise 25.

1. $a^2 + 3ab + 9b^2$.
2. $a^3 + 2a^2b + 4ab^2 + 8b^3$.
3. $x^4 - x^3 + x^2 - x + 1$.
4. $x^4 + 3x^3 + 9x^2 + 27x + 81$.
11. $1 + 4x$.
12. $3x + 5y^2$ and $3x - 5y^2$.
13. $x^2 + 9$, $x^2 - 9$, $x + 3$, $x - 3$.
14. $x - 2$.
15. $1 + x$.
16. $x^2 - 3y^2$.
17. $x^3 + y^3$, $x^3 - y^3$, $x + y$, $x - y$.
21. $x^6 + 8y^3$, $x^6 - 8y^3$, $x^4 - 4y^2$, $x^2 + 2y$, $x^2 - 2y$.
23. 1st, all integral values; 2d, all *even* integers.

Exercise 26.

1. $x^2(2x + 5)$.
2. $x(x^2 - 2)$.
3. $x(x + 1)$.
5. $7a(1 + 2a^2)$.
8. $x^2(1 - x - x^2)$.
15. $a^mb^3c^{2n}(1 + 11c)$.

Exercise 27.

1. $(2x + y)^2$.
2. $(4a - 3y)^2$.
3. $(5x - 1)^2$.
4. $(x - 10y)^2$.
5. $c(7 + 2bc)^2$.
17. $(a - b - c)^2$.

Exercise 28.

1. $(x + 3)(x - 3)$.
2. $(5 + 4a)(5 - 4a)$.
3. $(2a + 7b)(2a - 7b)$.
4. $3(x + 2y)(x - 2y)$.
5. $(10 + 9m)(10 - 9m)$.
9. $x(x^2 + 1)(x + 1)(x - 1)$.
21. $(15x^n + y)(15x^n - y)$.
24. $(x + y + 1)(x + y - 1)$.
25. $(x + y + 1)(x - y - 1)$.
29. $(4x + 2y + 1)(2y - 2x - 1)$.
30. $(11a - 8b)(9a - 2b)$.
31. $y(x^6y^4 + z^8)(x^3y^2 + z^4)(x^3y^2 - z^4)$.

Exercise 29.

1. $(5x-8y)(8y-3x)$.
2. $(3a-3b+5)(3a-3b-5)$.
3. $(a-b+1)(a-b-1)$.
4. $(3x+2y+z)(3x+2y-z)$.
5. $(x-a+y)(x-a-y)$.
6. $(a+y+x)(a+y-x)$.
7. $(a^2+x^2+y)(a^2-x^2-y)$.
8. $(x+y+1)(x-y-1)$.
9. $(1+x-y)(1-x+y)$.
10. $(c+a-b)(c-a+b)$.

Exercise 30.

1. $(c^2+cx+x^2)(c^2-cx+x^2)$.
2. $(x^2+x+1)(x^2-x+1)$.
3. $(2x^2+3x-1)(2x^2-3x-1)$.
4. $(2a^2-3ab-3b^2)(2a^2+3ab-3b^2)$.
5. $(3x^2+3xy+2y^2)(3x^2-3xy+2y^2)$.
13. $(a^2+2ab+2b^2)(a^2-2ab+2b^2)$.

Exercise 31.

1. $(1-2x)^2$.
2. $3x(2y+x)(2y-x)$.
3. $(3-x+y)(3-x-y)$.
4. $x(1+x)^2(1-x)^2$.
9. $2x(4x^2+3x+1)(4x^2-3x+1)$.
12. $(x^2-2x-1)(x-1)^2$.
17. $(x+y+1)(x+y-1)(1+x-y)(1-x+y)$.

Exercise 32.

1. $(x+3)(x+2)$.
2. $(x+2)(x-3)$.
3. $(x+3)(x-2)$.
4. $(x+11)(x-4)$.
5. $(x-5)(x-6)$.
12. $(x^2+4)(x+3)(x-3)$.
24. $(x^2-8)(x+1)(x-1)$.
31. $x(x+4)(x+3)(x-4)(x-3)$.
32. Seven factors.
37. $(x+a)(x+b)$.
38. $(x+2a)(x-3b)$.
39. $(x-a)(x-2b^2)$.
43. $2(x+1)^2(x+4)(x-2)$.

Exercise 33.

1. $(2x+1)(x+1)$.
2. $(3x-2)(x-4)$.
3. $(2x+1)(x+2)$.
4. $(3x+1)(x+3)$.
5. $(3x-5)(2x+1)$.
6. $(x+3)(2x-1)$.
7. $2x(x+4)(3x-2)$.
20. $(2x+3)(x+1)(2x-3)(x-1)$.
21. $(3x+2)(x+4)(3x-2)(x-4)$.
24. $(a^2+b^2)(5a+4b)(5a-4b)$.
27. $(5a+4b)(a+b)(5a-4b)(a-b)$.
31. $(a+b+8)(a+b-3)$.
32. $(3x-3y-2z)(x-y+3z)$.

33. $(3x^2 + 6x + 4)(x + 3)(x - 1)$. 35. $2(1 + 3x)(2 - x)$.
34. $4x(x + 4)(x + 2)(x + 1)(x - 1)$.

Exercise 34.

1. $(m - n)(m^2 + mn + n^2)$.
2. $(c + 2d)(c^2 - 2cd + 4d^2)$.
3. $(3 - x)(9 + 3x + x^2)$.
9. $x(3x + a)(9x^2 - 3ax + a^2)$.
10. $(8x - y^2)(64x^2 + 8xy^2 + y^4)$.
11. $a(1 + 7a)(1 - 7a + 49a^2)$.
12. $(a + x)(a - x)(a^2 - ax + x^2)(a^2 + ax + x^2)$.
13. $(x^2 + y)(x^2 - y)(x^4 - x^2y + y^2)(x^4 + x^2y + y^2)$.
17. $(a + b + 1)(a^2 + 2ab + b^2 - a - b + 1)$.
18. $(5 + 2b - a)(25 - 10b + 5a + 4b^2 - 4ab + a^2)$.
19. $(2 - c - d)(4 + 2c + 2d + c^2 + 2cd + d^2)$.
22. $(x + y)(x^4 - x^3y + x^2y^2 - xy^3 + y^4)$.

28. As the 24th. 29. As the 23d. 34. As the 12th. 35. As the 22d.

Exercise 35.

1. $(a + b)(x + y)$.
2. $(x - a)(x + c)$.
3. $(5y - 3)(x - 2)$.
4. $(m - 2y)(3a - 4n)$.
5. $x(a + 3)(a + c)$.
6. $y(3a - 5n)(a + b)$.
7. $x(x^2 + 2)(x + 1)$.
8. $2x(x + a)(x - a)(x - 1)$.
9. $(y^2 + 1)(y + 1)$.
16. $(x + 4)(x + 2)^2$.
17. $(a + 3)(a^2 - 3)$.
18. $(x - y)(2x + 2y - 1)$.
20. $(x - 1)(x^2 + 3x + 3)$.
23. $(3a - x)(3a + 2x)(a - x)$.
24. $(x - 2)(x + 3)(x - 1)$.
25. $(x + 3)(2x - 5)(2x - 1)$.
26. $(2x + 1)(4x - 3)(x + 1)$.
27. $(2x - 3)(4x - 3)(x - 2)$.
28. $(x + 2)(x + 1)(x - 3)$.
29. $(x + 3)(x - 2)(x - 4)$.
30. $(x - 2)(x - 1)(x - 5)$.
31. $(x - 3)(2x - 1)(3x - 1)$.

Exercise 36.

10. $3x(x^2 + x + 1)(x^2 - x + 1)(x + 1)(x - 1)$.
11. $(2a + 1)(a + 1)(2a - 1)(a - 1)$.
12. $2(x^2 + 2x + 2)(x^2 - 2x + 2)(x^2 + 2)(x^2 - 2)$.
15. $5a(x^6 + x^3 + 1)(x^2 + x + 1)(x - 1)$.
23. $(x + 2)(x - 1)(x^2 - x + 2)$.
34. $(a^2 + 5)(a + 2)(a - 2)$.
35. $(c + d - 1)(c^2 + 2cd + d^2 + c + d + 1)$.
36. $(x - y)(x - y + 2)$.
39. $(a + 3)(a + 2)(a - 3)(a - 2)$.
41. $(a + b + c)(a + b - c)(a - b + c)(a - b - c)$.
43. $(2 + n)(16 - 8n + 4n^2 - 2n^3 + n^4)$.
46. $\frac{1}{4}a(2x + y)(4x^2 - 2xy + y^2)$.

47. $(1 + x^2)(1 + x)^2(1 - x)$.
48. $6(x - 3)(4 - x)$.
51. $(x + 1)(x + 2)(x - 3)$.
53. $(x^4 + 6x^2y^2 + y^4)(x + y)^2(x - y)^2$.
54. $(x^2 + y^2 + z^2)(x + z)(x - z)$.
56. $(a + b)(a - 7b)$.
61. $(x - 1)^2(x + 2)^2$.
62. $(a + b^2)(a - b^2)(1 - x)(1 + x + x^2)$.
66. $(1 + 2abc - 3xyz)(1 - 2abc + 3xyz)$.

67. $(abc - mnp)(ax - my)$.
69. $(x - 2)(x + 5)(2x + 1)$.
71. $(2x + 5)(2x - 3)(x - 1)$.
72. $(x + 1)(x + 2)(3x - 2)$.

Exercise 37.

1. $2ab$.
2. $5x^2y$.
3. abc^2.
4. $8a^2x^2$.
5. $14m^2$.
6. $12x$.
7. $17ax^3$.
8. a^2x^2y.
9. $a + b$.
10. $x - y$.
11. $x - 3$.
12. $x(2x + 3)$.
13. $a - x$.
14. $x + 1$.
15. $x - 1$.
16. $a(2a + 1)$.
17. $x + 1$.
18. $x - 3$.
19. $4ax(a - x)$
20. $x(x - 1)$.
21. $2x - y$.
22. $x(x - 2)$.
23. $x^2 + 9$.
24. $b(1 - a^2)$.
25. $1 + a + a^2$.
26. $x - 1$.

Exercise 38.

1. $x + 1$.
2. $2x - 3$.
3. $x - 1$.
4. $3(x - 1)^2$.
5. $x(2x + 1)$.
6. $3x + 4$.
7. $2x - 1$.
8. $x + 3$.
9. $x^2 + x - 1$.
10. $x^2 - 2x + 3$.
11. $x^2 + 4$.
12. $x + 3$.
13. $x^2 + x + 1$.
14. $2x(2x - 3)$.
15. $3x - y$.
16. $x(3x - 4)$.
17. $x - 1$.
18. $x - 2$.
19. $x - 3$.
20. $2x(x - 1)$.

Exercise 39.

1. $6a^3b^3$.
2. $24x^3y^3z$.
3. $36a^2x^2y^2$.
4. $48x^3y^3$.
5. $12abc$.
6. $12a^2b^2c^2$.
7. $84x^3y^3z^2$.
8. $48a^2b^2$.
9. $42a^2b^3$.
10. $24x^3y^2$.
11. $2x(x^2 - 1)$.
12. $ab(a + b)$.

13. $14x^2(x - 3)$.
14. $(x^3 - 1)(x + 1)$.
15. $(x^2 - y^2)(x - 2y)$.
16. $6x(x + 1)(x - 1)^2$.
17. $abx^2y(x + y)(x - y)^2$.
18. $(x + 5)(x - 8)(x - 1)$.
19. $(x + 2)(2x + 3)(3x - 4)$.
20. $a^6 - b^6$.
21. $6x^2(x + 1)(x - 1)$.
22. $3ab(a + b)(a - b)$.
23. $(2x + 1)(x + 1)(2x - 1)$.
24. $6x(x^3 - 1)(x - 1)$.
25. $6x(3x + 10)(2x - 7)(x - 3)$.
26. $2x(1 + x^2)(1 + x)(1 - x)$.

27. $14x^5y^3(x+1)^3(x-1)^3$.
28. $6x^3(3x+1)(x-1)(3x-1)^2$.
29. $36x^3(2x+3)^2(2x-3)^2$.
30. $(x+a+1)(x-a-1)(x+a-1)$.
31. $(x-1)^2(x+1)^2(x+3)^2(x-3)$.

Exercise 40.

1. $(x+2)(x^2-2x+3)(x^2-2x-1)$.
2. $(3x+4)(x^2-x+1)(x^2+x-1)$.
3. $3x(2x+1)(x^2-x-1)(x^2+x+1)$.
4. $2(2x-1)(3x^2-5)(2x^2+5)$.
5. $(2x+1)^2(x-3)(x-1)(2x-1)$.
6. $(4x^2-9)(9x^2-4)$.
7. $(x-1)(3x^2+x+1)(4x^2+4x-1)$.
8. $(x+2)(3x^2-7x+5)(2x^2-x+1)$.
9. $(x+3)(x+6)(x-1)(x-2)(x-5)$.
10. $(x+1)(x+2)(2x-1)(3x-2)(2x-3)$.
11. H. C. F., a^2b^2; L. C. M., $140a^3b^3c^3d^3$.
12. H. C. F., $3(x+1)^2$; L. C. M., $18a^2b^2(x^2-1)^2(3x+2)^2(5x-2)^2$.
13. H. C. F., x^2+x-2; L. C. M., $(x^2+x-2)(x^2-x+2)(x^2+x+2)$.
14. H. C. F., $2x+1$; L. C. M., $(2x+1)^2(x-1)^2(3x^2-2x+1)$.
15. H. C. F., $x+2$; L. C. M., $4(x+2)(3x-1)^2(x-1)^2$.

Exercise 41.

1. $\frac{2a}{3x}$.
2. $\frac{4x}{5y}$.
3. $\frac{x}{2-3ax}$.
4. $\frac{3xz}{4y^2}$.
5. $\frac{1}{2a-1}$.
6. $\frac{1}{2a}$.
7. $\frac{1}{a}$.
8. $\frac{x-y}{x+y}$.
9. $\frac{2(x+1)}{3}$.
10. $\frac{5}{2(x-y)}$.
11. $\frac{a+b}{2(a-b)}$.
12. $\frac{2}{3x-4y}$.
13. $\frac{1}{2x+3y}$.
14. $\frac{7x+8y}{2x^2}$.
15. $\frac{1}{x-y}$.
16. $\frac{x+2y}{2x+3y}$.
17. $\frac{2x+y}{2x-y}$.
18. $\frac{a+b-c}{a-b-c}$.
19. $\frac{1+a-x}{x+a-1}$.
20. $\frac{2+a+b}{2-a+b}$.
21. $\frac{3x+4a}{x+a}$.

22. $\frac{x-2}{y^2}$.
23. $\frac{x+3}{x+2}$.
24. $x^2 - y^2$.
25. $\frac{x+1}{x-1}$.
26. $\frac{2x^2+3y^2}{3x^2+2y^2}$.
27. $\frac{x-5}{x^2+x-3}$.
28. $\frac{a(2x-3)}{3(x-2)}$.
29. $\frac{3x^2-2x+3}{2x^2+3x-2}$.
30. $\frac{2x^3-4x^2+2x-3}{4x^3+3x^2-18x+27}$.

Exercise 42.

1. $x - 2 + \frac{3}{x}$.
2. $2x^2 + 3 - \frac{5}{2x}$.
3. $2a^2x^2 + 1 - \frac{7+a}{5ax}$.
4. $x^2 - 4x + 5 - \frac{6}{x+1}$.
5. $x + 2y - \frac{4y^2+1}{x+y}$.
6. $3x^2 + 9 - \frac{18x+1}{x^2-3}$.
7. $x^2 - 1 + \frac{1-a}{x-1}$.
8. $x^2 - x + 2 - \frac{3(x-1)}{x^2+x-1}$.
16. $1 - x + x^2 - \frac{x^3}{1+x}$.
17. $1 - x + 2x^2 - \frac{3x^3-2x^4}{1+x-x^2}$.

Exercise 43.

1. $\frac{a^2-a+1}{a}$.
2. $\frac{x^2}{x-1}$.
3. $\frac{x^3-2x}{x-1}$.
4. $\frac{8x^3-y}{2x+1}$.
5. $\frac{a^2+ab}{a+2b}$.
6. $\frac{x^3-x}{x^2+x+1}$.
7. $\frac{a^2-x^2+a+1}{a+x}$.
8. $\frac{1-2a+a^2}{2a}$.
9. $\frac{a^2+a}{a-1}$.
10. $\frac{a^2+1}{a-2}$.
11. $\frac{x^2+xy}{x+a}$.
12. $\frac{x^3}{1+x}$.
13. $\frac{x^3-1}{x+1}$.
14. $\frac{x^4-x}{x^2+1}$.
15. $\frac{x^4}{x-1}$.
16. $\frac{1+2x^3-x^4}{1+x}$.

Exercise 44.

1. $\frac{4x}{18}, \frac{15x}{18}$.
2. $\frac{24a}{10b}, \frac{7b}{10b}, \frac{10a}{10b}$.

3. $\frac{a}{2a^2b^2}, \frac{4b}{2a^2b^2}, \frac{2ab}{2a^2b^2}$.

8. $\frac{1}{x^2-1}, \frac{3x-3}{x^2-1}$.

9. $\frac{1}{a^2-a}, \frac{2a^2-2a}{a^2-a}, \frac{3a}{a^2-a}$.

11. $\frac{x^3+x^2+x}{(x+1)(x^3-1)}, \frac{x+1}{(x+1)(x^3-1)}$.

12. $\frac{x}{x(4x^2-9)}, \frac{x(2x-3)}{x(4x^2-9)}, \frac{4x^2-9}{x(4x^2-9)}$.

18. $\frac{2x+4}{6(x^2-4)}, \frac{15x-30}{6(x^2-4)}, \frac{18}{6(x^2-4)}$.

21. $\frac{(x+1)^2}{(x+1)(x-2)(x+3)}, \frac{12(x+1)(x-2)(x+3)}{(x+1)(x-2)(x+3)}, \frac{(x-1)(x-2)}{(x+1)(x-2)(x+3)}$.

22. $\frac{2(a-b)^2}{(a^2-b^2)^2}, \frac{ab(a^2-b^2)^2}{(a^2-b^2)^2}, \frac{(a+b)^2}{(a^2-b^2)^2}, \frac{4(a^2-b^2)}{(a^2-b^2)^2}, -\frac{ab}{(a^2-b^2)^2}$.

Exercise 45.

1. $\frac{19}{6x}$.

2. $\frac{8x-9+12a}{12ax}$.

3. $\frac{15b-4c-6a}{6abc}$.

4. $\frac{4x-7}{4x}$.

5. $\frac{2a^2+ab-3b^2}{6ab}$.

6. $\frac{3a^2+b}{6a^2b}$.

7. $\frac{9+10ax^2}{12ax^3}$.

8. $\frac{2x+3}{30}$.

9. $\frac{3x+1}{24}$.

10. $\frac{2b}{a^2-b^2}$.

11. $\frac{4x}{1-x^2}$.

12. $\frac{7-6x^2}{3x^2}$.

13. $\frac{25a-20b}{12}$.

14. $\frac{8x+65}{21}$.

15. $\frac{y^3-3x^2z^3-6yz^3}{6x^2y^2z}$.

16. $\frac{a^2+b^2}{a^2-b^2}$.

17. $\frac{1}{7x-x^2-12}$.

18. $\frac{6x}{x^2-4}$.

19. $\frac{3m^2+1}{(m+1)(m-1)^2}$.

20. $\frac{x^2+2x-1}{x^2-1}$.

21. $\frac{x^3}{x^2-4}$.

22. $\frac{2x^2+3x-1}{x(x^2-1)}$.

23. 0.

24. $\frac{1+a}{9-a^2}$.

25. $\frac{1}{8x^2-2}$.

26. $\frac{4x-1}{x^2-1}$.

27. $\frac{x^2+5x+10}{(x+1)(x+2)(x+3)}$.

28. $\dfrac{5x(x+3)}{(2x+1)(2x-1)(x+1)}$.

29. $\dfrac{b^3}{(a+b)^3}$.

30. $\dfrac{3-x}{(2x+1)(2x+3)(x-1)}$.

31. $\dfrac{5x^2y-3y^3}{x(x^2-y^2)}$.

32. 0.

33. $\dfrac{x^2+4x-13}{2(x^2-1)}$.

34. 0.

35. $\dfrac{44-9x}{x^3+64}$.

36. $\dfrac{3x+2}{(x^2-1)(x-2)}$.

37. $\dfrac{x^2+90x-9}{6(x^2-9)(x-3)}$.

38. $\dfrac{1}{x(x^2+1)}$.

39. $\dfrac{x}{x^3-1}$.

40. $\dfrac{x+1}{x^2-x}$.

Exercise 46.

1. $\dfrac{5}{1-x^2}$.

2. $\dfrac{a-3b}{a^2-b^2}$.

3. $\dfrac{3xy}{4y^2-x^2}$.

4. $\dfrac{b}{a^2-b^2}$.

5. 0.

6. $\dfrac{1-6a^2}{1-4a^2}$.

7. 0.

8. $\dfrac{13}{8(1-a^2)}$.

9. $\dfrac{x}{1-x^2}$.

10. $\dfrac{6-x}{(x-2)(x-3)(x-5)}$.

11. $\dfrac{5-4b}{(a-3)(a-2)(b-2)}$.

12. $\dfrac{7a^2+19a}{12(a^2-9)}$.

13. 0.

14. $\dfrac{-7}{12x(x+1)}$.

15. $\dfrac{x}{3(x^2-9)}$.

16. $\dfrac{x^2-15x-18}{(x^2-9)(x-1)}$.

17. 0.

18. $\dfrac{17x^2-42x+39}{15(x^2-9)}$.

20. 0.

21. 1.

22. 0.

23. 1.

Exercise 47.

1. $\dfrac{2b^3x}{3acy}$.

2. $\dfrac{9y^2z}{4x^3}$.

3. $\frac{1}{4}$.

4. 1.

5. $-\dfrac{5y^{3n}z}{7}$.

6. $\dfrac{3(x+1)}{x(2x-1)}$.

7. $\dfrac{2y}{x}$.

8. $\dfrac{ab}{2a-1}$.

9. $\dfrac{x^2+2x-3}{x}$.

10. $\dfrac{a-1}{a(x+1)}$.

11. $\dfrac{2x+3}{3(3x-1)}$.

12. $\dfrac{(2x+1)^2}{(x+1)^2}$.

13. $\dfrac{a+x}{x^2(a-x)}$.

14. $\frac{a^2+a+1}{a}$.
15. $\frac{(3x-2)^2}{(2x-3)^2}$.
16. $\frac{x^2-4}{x^2-1}$.
17. $\frac{2}{x+1}$.
18. $\frac{x}{x^2-xy+y^2}$.
19. 1.
20. $\frac{1}{3ab}$.
21. 1.
22. $\frac{1}{x+y}$.
23. 1.
24. 1.
25. $-\frac{2b}{a}$.
26. $x+y$.
27. $\frac{a+x-1}{a-x+1}$.
28. 1.
29. x.
30. $\frac{a^2c+ab^2+bc^2}{a+b+c}$.
31. 1.
32. $\frac{1}{a}$.
33. $\frac{m^2+4n^2}{4mn}$.

Exercise 48.

1. $\frac{2(2-x)}{x}$.
2. $\frac{x}{2x+1}$.
3. $x+1$.
4. $\frac{b}{c}$.
5. $\frac{a-1}{a+1}$.
6. $\frac{2a-1}{a-1}$.
7. $\frac{4x^2+2x+1}{2x}$.
8. $\frac{1}{x^2}$.
9. $\frac{3}{x}$.
10. $\frac{a-1}{a+1}$.
11. $\frac{2a-1}{a}$.
12. $(a+1)^2$.
13. $-\frac{x+1}{x(x+3)}$.
14. $a+x$.
15. $\frac{a(a+1)}{a^2+1}$.
16. $\frac{1}{2x^2-1}$.
17. $\frac{x+y}{x-y}$.
18. $\frac{ac+bc-ab}{ac+bc+ab}$.
19. $\frac{x-a+1}{x+a-1}$.
20. $-\frac{1}{2}$.
21. 0.
22. $a-1$.
23. $\frac{ab-cd+1}{ab-cd-1}$.
24. $\frac{(a+b+c)^2}{2bc}$.
25. $\frac{1}{x}$.
26. $2x$.
27. $\frac{(x-y)^4}{x}$.
28. -1.
29. $\frac{1}{1+2x}$.

Exercise 49.

1. $\frac{(x-1)^2}{x^2-x-1}$.
2. $\frac{x^2-x+3}{x^2-4x+9}$.
3. $\frac{2(1-5x)}{(2x+3)(2x+1)}$.
4. $\frac{8nx}{4n^2-x^2}$.
5. $\frac{3(2x-3)}{2(3x-1)}$.
6. $\frac{3(4-3x)}{4(2-x)}$.
7. $\frac{14x}{9x^2-4}$.
8. $\frac{6}{2-x}$.
9. 0.

10. $\dfrac{x-y}{x+2y}$.
11. $\dfrac{9}{x(2-x)(x-3)}$.
12. $\dfrac{a}{a^2+1}$.
13. $\dfrac{b^2}{2a(b^2-a^2)}$.
14. $\dfrac{18x+3}{(9x^2-1)(4-9x^2)}$.
15. $\dfrac{4}{(1-x)(x-2)(x-3)}$.

16. $\dfrac{y^2}{x^2-y^2}$.
17. 1.
18. $\dfrac{1+x^4}{x(1+x^2)}$.
19. $\dfrac{1}{x^2}$.
20. 1.
21. $\dfrac{4(1-a^2)}{19a+4}$.
22. $-\dfrac{x^2}{y^2}$.
23. $-\dfrac{2x}{3}$.
24. $\dfrac{a+b}{a-b}$.
25. $5x$.
26. $-\dfrac{1}{x}$.
27. x^2+y^2.
28. $\dfrac{a+b}{a-b}$.
29. $\dfrac{x}{x+1}$.
30. $\dfrac{x}{2x-1}$.
31. $\dfrac{4}{x^2}$.
32. $\dfrac{a}{x}$.
33. $\frac{1}{4}a$.
34. $6-3a$.

35. x.
36. $\dfrac{5a-1}{2a-1}$.
37. $-\frac{4}{3}$.
38. $-x$.
39. $\dfrac{3a^2+b^2c^2}{a^2+3b^2c^2}$.
40. $\dfrac{1}{a}$.

Exercise 50.

1. 6.
2. 2.
3. 3.
4. -2.
5. 10.
6. 2.
7. 2.
8. -1.
9. 5.
10. $\frac{5}{2}$.
11. -9.
12. $-\frac{3}{2}$.
13. $\frac{16}{3}$.
14. $\frac{3}{2}$.
15. $-\frac{3}{2}$.
16. $\frac{1}{5}$.
17. $-\frac{1}{10}$.
18. $\frac{12}{7}$.
19. $-\frac{32}{5}$.
20. -2.
21. 5.
22. $\frac{11}{9}$.
23. $-\frac{3}{2}$.
24. -2.
25. -2.
26. $\frac{3}{2}$.
27. 0.
28. 1.
29. 2.
30. 5.
31. -5.
32. 3.
33. 10.
34. -0.8.
35. $\frac{24}{15}$.
36. -5.
37. $\frac{3}{16}$.
38. 5.
39. 2825.
40. .00025.
41. $-.04$.
42. 4.
43. 13.
44. $-\frac{7}{18}$.
45. $\frac{25}{17}$.
46. $\frac{41}{35}$.
47. -7.
48. 4.
49. 2.
50. $\frac{5}{4}$.
51. $-\frac{1}{2}$.
52. $-\frac{11}{12}$.
53. 0.
54. $\frac{1}{10}$.
55. -5.
56. -7.
57. -3.
58. $-\frac{24}{11}$.

Exercise 51.

1. $-\frac{4}{3}$.
2. -3.
3. $-\frac{15}{11}$.
4. 12.
5. -9.
6. $-\frac{1}{4}$.
7. $-\frac{1}{2}$.
8. -2.
9. $-\frac{3}{2}$.
10. $-\frac{2}{3}$.
11. -23.
12. -0.5.
13. -0.1.
14. -0.1.
15. $\frac{15}{11}$.
16. 8.
17. 0.4.
18. $\frac{7}{2}$.
19. 5.
20. 7.
21. 0.
22. -3.
23. $\frac{5}{2}$.

Exercise 52.

1. $3a$.
2. $\dfrac{b}{a}$.
3. $-\dfrac{c}{a}$.
4. $\dfrac{b}{a-b}$.
5. $\dfrac{a-b}{2c}$.
6. $\dfrac{3-b}{5-2a}$.
7. $\dfrac{3b+2d}{2a-c}$.
8. $\dfrac{ab}{a-b}$.
9. $\dfrac{a}{b}$.
10. $\dfrac{a^2-2}{2a-3}$.
11. $\dfrac{a-b}{a+b}$.
12. $\dfrac{a}{2}$.
13. $\dfrac{a^2-b^2}{a^2+b^2}$.
14. 0.
15. $17a$.
16. $19a^2$.
17. $\dfrac{a}{3}$.
18. $\dfrac{abcd}{ab+bc+ac}$.
19. 1.
20. $\dfrac{b}{2a+b}$.
21. $\dfrac{ab(a-b)}{a^3-2ab^2-b^3}$.
22. $\dfrac{ac^3}{ac-ab+bc}$.
23. 0.
24. $\dfrac{a^3+a}{3a^2-1}$.
25. c^2.
26. $\dfrac{ab}{2(b^2-a^2)}$.
27. $\frac{1}{2}(1-2a-a^2)$.
28. $-a$.
29. $\dfrac{ab}{2a-3b}$.
30. $\dfrac{36ab}{22b^2-15a^2}$.
31. $\dfrac{a^3-a^2-a+1}{a}$.

Exercise 53.

1. 24.
2. 45.
3. 60.
4. 63.
5. 27; 28.
6. 48; 49; 50.
7. 33 and 42.
8. $12,000.
9. 144 trees.
10. 26; 27; 28.
11. A, $32; B, $48; C, $50.
12. 16 and 81.
13. 21 and 79.
14. 14 and 54.
15. 13 years.
16. A, 21 years. B, 35 years.
17. Father, 48 years. Son, 20 years.
18. A, $960; B, $1200; C, $1080; D, $1760.
19. 70 acres.
20. 44 and 45.
22. $3\frac{3}{7}$ days.
23. $1\frac{1}{3}$ days.
24. 6 days.
25. 12 days.
26. $2\frac{2}{3}$ days.
27. 4 days.
28. 36 min.
29. $169\frac{7}{17}$ min.
30. 20 days.
32. $54\frac{6}{11}$ min. past 4. $38\frac{2}{11}$ min. past 1.
33. $32\frac{8}{11}$ min. past 6. $54\frac{6}{11}$ min. past 10.
34. $5\frac{5}{11}$ and $38\frac{2}{11}$ min. past 4. $21\frac{9}{11}$ and $54\frac{6}{11}$ min. past 7.
36. 24 hours.
37. 208 miles.
38. 6 hours.
39. 55 miles.
40. 12 mi. an hr.
41. $19\frac{11}{13}$ miles.

42. 1st, each 1½ hrs.
A, 42 mi.; B, 40 mi.
2d, each 492 hrs.
A, 1722 mi.; B, 1640 mi.

44. Hound, 150; hare, 250 leaps.

45. Hound, 72; hare, 108 leaps.

46. $63.

47. 1713 men.

48. 45 men.

49. 2160 men.

50. $46\frac{2}{3}$ bu. oats.
$53\frac{1}{3}$ bu. corn.

51. $3250 at 4%.
$1800 at 5%.

52. 10; 14; 6; 24.

53. First, 20 days; sec., 15 days.

54. 108 and 72.

55. 30 apples.

56. 334 pages.

57. 4 and 16.

58. 81 yards.

59. A, 4 days; B, 5 days; C, 6 days.

60. $5\frac{5}{11}$ and $38\frac{2}{11}$ min. past 10.

61. Dog, 600; fox, 900 leaps.

62. $19\frac{1}{11}$ days.

63. $\frac{abc}{b+c}$ miles.

64. $\frac{ab+c}{b+1}$; $\frac{a-c}{b+1}$.

65. $\frac{abc}{b-a}$ feet.

Exercise 54.

1. $156.
2. $5\frac{1}{3}$ yrs.
3. $4\frac{1}{2}$%.
4. $264.
5. 10 yrs.
6. $4\frac{1}{3}$%.
7. 8%.
8. $$333\frac{1}{3}$.
9. 25 yrs.; 16 yrs. 8 mos.
10. $540.
11. 8 months.

Exercise 55.

1. $x = 1$. $y = 1$.
2. $x = 1$. $y = -1$.
3. $x = \frac{1}{2}$. $y = -\frac{1}{3}$.
4. $x = 2$. $y = 3$.
5. $x = 2$. $y = -1$.
6. $x = 1$. $y = -\frac{1}{2}$.
7. $x = -3$. $y = 1\frac{1}{2}$.
8. $x = \frac{1}{3}$. $y = -\frac{1}{4}$.
9. $x = \frac{1}{2}$. $y = \frac{1}{3}$.
10. $x = -\frac{1}{2}$. $y = 2$.
11. $x = 3$. $y = -7$.
12. $x = 8$. $y = 9$.
13. $x = 15$. $y = 10$.
14. $x = 3$. $y = -4$.
15. $x = 10$. $y = -10$.
16. $x = 12$. $y = 18$.

Exercise 56.

1. $x = 1$. $y = 1$.
2. $x = -1$. $y = -1$.
3. $x = 2$. $y = -1$.
4. $x = -3$. $y = 0$.
5. $x = -2$. $y = \frac{1}{4}$.
6. $x = 3$. $y = -\frac{1}{4}$.
7. $x = 0$. $y = 2$.
8. $x = -2$. $y = -3$.
9. $x = -3$. $y = -4$.
10. $x = 12$. $y = 12$.
11. $x = 6$. $y = 20$.
12. $x = 15$. $y = 10$.
13. $x = \frac{5}{3}$. $y = -4$.
14. $x = \frac{7}{3}$. $y = -\frac{5}{4}$.
15. $x = 4$. $y = -3$.
16. $x = -21$. $y = -40$.

Exercise 57.

1. $x = 1.$
   $y = 1.$
2. $x = -1.$
   $y = 1.$
3. $x = 2.$
   $y = -2.$
4. $x = 3.$
   $y = -1.$
5. $x = \frac{2}{3}.$
   $y = \frac{3}{2}.$
6. $x = 3.$
   $y = -2.$
7. $x = 6.$
   $y = 6.$
8. $x = 12.$
   $y = 12.$
9. $x = 12.$
   $y = 35.$
10. $x = 7\frac{1}{2}.$
    $y = -2\frac{1}{2}.$
11. $x = \frac{1}{3}.$
    $y = \frac{1}{4}.$
12. $x = 4.$
    $y = 5.$
13. $x = 0.$
    $y = 3.$

Exercise 58.

1. $x = 5.$
   $y = 12.$
2. $x = 5.$
   $y = 2.$
3. $x = \frac{1}{4}.$
   $y = \frac{1}{6}.$
4. $x = 7.$
   $y = 5.$
5. $x = 3.$
   $y = 1.$
6. $x = 1.$
   $y = 1.$
7. $x = 3.$
   $y = 5.$
8. $x = 1.$
   $y = -1.$
9. $x = 2.$
   $y = 4.$
10. $x = -0.2.$
    $y = 0.6.$
11. $x = .015.$
    $y = .01.$
13. $x = 2.$
    $y = -3.$
14. $x = 18.$
    $y = 12.$
15. $x = 9.$
    $y = -1.$
16. $x = 17.$
    $y = 6.$
17. $x = 2.$
    $y = -1.$
18. $x = -2.$
    $y = -3.$

Exercise 59.

1. $x = 2a.$
   $y = -a.$
2. $x = -b.$
   $y = 2a.$
3. $x = \dfrac{b' - b}{ab' - a'b}.$
   $y = \dfrac{a - a'}{ab' - a'b}.$
4. $x = m + n.$
   $y = m - n.$
5. $x = \dfrac{2b + 1}{b}.$
   $y = \dfrac{a - 2}{a}.$
6. $x = a + 2b.$
   $y = 2a - b.$
7. $x = \dfrac{cn - bd}{an - bm}.$
   $y = \dfrac{ad - cm}{an - bm}.$
8. $x = \dfrac{a}{b}.$
   $y = \dfrac{b}{a}.$
9. $x = \dfrac{1}{c}.$
   $y = \dfrac{1}{d}.$
10. $x = n - m.$
    $y = n + m.$
11. $x = 3.$
    $y = \dfrac{2a + 1}{b}.$
12. $x = \dfrac{1}{a + b + c}.$
    $y = \dfrac{1}{a + b + c}.$
13. $x = a + b.$
    $y = a - b.$
14. $x = \dfrac{a - d}{b - d}.$
    $y = \dfrac{a - b}{b - d}.$
15. $x = a.$
    $y = b.$
16. $x = -a.$
    $y = b.$
17. $x = b.$
    $y = a.$
18. $x = a + 1.$
    $y = b - 1.$

Exercise 60.

1. $x = 1.$ $y = 2.$ $z = 3.$

2. $x = 2.$ $y = 3.$ $z = -4.$

3. $x = 3.$ $y = 4.$ $z = 7.$

4. $x = 3.$ $y = -2.$ $z = -4.$

5. $x = 1\frac{1}{4}.$ $y = 1\frac{1}{3}.$ $z = 1\frac{1}{4}.$

6. $x = 2.$ $y = 3\frac{1}{2}.$ $z = -4.$

7. $x = -3.$ $y = 3\frac{1}{3}.$ $z = -2.$

8. $u = 2.$ $v = 3.$ $w = 1.$ $x = 4.$

9. $x = 12.$ $y = 18.$ $z = -24.$

10. $x = a + b.$ $y = a - b.$ $z = 2a.$

11. $x = 6.$ $y = 40.$ $z = 20.$

12. $x = -a + b + c.$ $y = a - b + c.$ $z = a + b - c.$

13. $x = a - b + 1.$ $y = -a + b + 1.$ $z = a + b - 1.$

14. $x = \frac{3a - 2b}{6}.$ $y = \frac{2a + 3b}{6}.$ $z = \frac{a + b}{6}.$

Exercise 61.

1. $x = \frac{1}{3}.$ $y = -1.$

2. $x = -1.$ $y = 1$

3. $x = \frac{1}{3}.$ $y = -\frac{1}{2}.$

4. $x = -\frac{1}{2}.$ $y = \frac{1}{2}.$

5. $x = \frac{2}{3}.$ $y = \frac{3}{4}.$

6. $x = \frac{1}{6}.$ $y = -\frac{1}{6}.$

7. $x = \frac{1}{2}.$ $y = -\frac{1}{6}.$

8. $x = \frac{2n}{1 + n^2}.$ $y = \frac{2n}{1 - n^2}.$

9. $x = a.$ $y = -a.$

10. $x = \frac{1}{m}.$ $y = \frac{1}{n}.$

11. $x = \frac{a}{b}.$ $y = \frac{b}{a}.$

12. $x = 1.$ $y = 1.$

13. $x = \frac{5}{3}.$ $y = -\frac{3}{4}.$

14. $x = 1;$ $y = -\frac{1}{2};$ $z = \frac{2}{3}.$

15. $x = 2;$ $y = -\frac{1}{2};$ $z = 1.$

16. $x = \frac{1}{2};$ $y = \frac{1}{3};$ $z = \frac{1}{4}.$

17. $x = \frac{-2bc}{b + c}.$ $y = \frac{-2ac}{a + c}.$ $z = \frac{-2ab}{a + b}.$

18. $x = \frac{2a}{l + m}.$ $y = \frac{2b}{l + n}.$ $z = \frac{2c}{m + n}.$

19. $x = \frac{1}{3};$ $y = -\frac{3}{2};$ $z = 1.$

Exercise 62.

1. 9 and 14. 2. 9 and 12. 3. 2 and 8. 4. $\frac{4}{5}$.

5. Flour, 3 cts.; sugar, 5 cts.
6. $\frac{5}{8}$.
7. 49.
8. Man, \$3; boy, \$2.
9. $\frac{8}{15}$.
10. 57 pear trees; 43 apple trees.
11. Sheep, \$4; calf, \$7.
12. A, \$660; B, \$480.
13. A, 91 years; B, 30 years.
14. 84 and 60.
15. A, $4\frac{1}{2}$ miles; B, 4 miles.
16. 12 boys, \$60.
17. Length, 8 in.; breadth, 6 in.
18. A, in 24 days; B, in 48 days.
19. A, \$70; B, \$110.
20. 480 miles.

21. $\frac{4}{11}$.
22. 11 and 36.
23. $\frac{11}{4}$ and $\frac{7}{6}$.
24. $\frac{3}{7}$.
25. 23.
26. 56 and 65.
27. 24.
28. 64.
29. 253.

30. 151.
31. Silk, \$1.80; satin, \$1.50.
32. 16; 20; 24.
33. 15 gals. from 1st; 6 gals. from 2nd.
34. \$600; eldest, \$200.
35. A, \$40; B, \$50; C, \$80.
36. 8 dollars; 40 halves; 36 quarters.
37. $\frac{7}{8}$ and $\frac{5}{8}$.
38. A, \$26; B, \$14; C, \$8.
39. A, \$70; B, \$50; C, \$90.
40. 8 men; 6 women; 10 children.
41. 6 doz. at 30 cts.; 3 doz. at 40 cts.
42. 12 yards by 8 yards.
43. 15 ft. by 6 ft.

44. Fore wheel, 5 yards; hind wheel, 6 yards.

46. $6\frac{2}{3}$ miles an hour.
47. 32 miles; 5 mi. an hour.
48. $1\frac{1}{3}$ mi. an hour.
49. 24 bu. from 1st; 16 bu. from 2nd.

50. A, 60 yds. a min.; B, 80 yds. a min.
51. A, 27 mi.; 3 mi. an hr. B, 30 mi.; 5 mi. an hr.
52. A, 9; B, 12; C, 8 hrs.
54. A, $4\frac{4}{5}$ yds. a sec.; B, $4\frac{1}{5}$ yds. a sec.

55. A, $5\frac{1}{3}$ min.; B, $5\frac{1}{3}$ min.
56. A, $4\frac{4}{5}$ min.; B, $4\frac{16}{25}$ min.
57. C helped 6 days. A, in 45 days.
58. A, \$35; B, -\$26; C, \$20.

Exercise 63.

2. H. C. F. $= 2x - 3$.
3. $\frac{7}{10}$.
4. $\frac{1}{12}$.
5. $\dfrac{x^3 + 1}{x(x-1)^3}$.
6. $\dfrac{x^2 + x + 1}{x^2 - x - 1}$.
7. $\dfrac{3(4x - 15)}{5(2x - 3)}$.
8. $\dfrac{x^2 + 1}{x^2 - 1}$.
9. $\dfrac{a^2 + x^2}{a^2}$.
10. $x = -2$.
11. $x = -\frac{6}{7}$.
12. $x = -3$.
13. $x = 4$.

14. $x = \frac{3}{4}$. $y = \frac{5}{8}$.

15. $x = \frac{1}{5}$. $y = \frac{1}{6}$.

16. $x = 7$. $y = 10$.

17. $x = \dfrac{a}{a-b}$. $y = \dfrac{b}{a+b}$.

Exercise 64.

1. $x < 1$.
2. $x > 3\frac{1}{2}$.
3. $x > \dfrac{b}{a}$.
4. $x > 2$.
5. $x > 6$.
6. $x < \frac{1}{2}$.
7. $x < \dfrac{ab}{2a - 3b}$.
8. $x < 6$.
9. $x > 6$ and < 7.
10. $x < 2$ and $> 1\frac{1}{2}$.
11. 17, 18 or 19.
12. 13.

Exercise 65.

10. $27x^3y^3$.

11. $-8x^6$.

18. $\frac{81}{16}x^8y^4$.

25. $a^2 + \frac{3}{2}a + \frac{9}{16}$.

26. $x^4 + 3x^2y + \frac{9}{4}y^2$.

27. $\dfrac{x^2}{y^2} - 1 + \dfrac{y^2}{4x^2}$.

33. $\frac{9}{4}x^4 - 2x^3y + \frac{43}{36}x^2y^2 - \frac{1}{3}xy^3 + \frac{1}{16}y^4$.

Exercise 66.

1. $a^3 - 3a^2b + 3ab^2 - b^3$.
2. $x^3 + 3x^2 + 3x + 1$.
3. $1 - 4x + 6x^2 - 4x^3 + x^4$.
4. $a^3 - 6a^2 + 12a - 8$.
5. $16 + 32x^2 + 24x^4 + 8x^6 + x^8$.
6. $a^5 - 10a^4b + 40a^3b^2 - 80a^2b^3 + 80ab^4 - 32b^5$.
7. $x^5 + 15x^4 + 90x^3 + 270x^2 + 405x + 243$.
8. $a^6 - 6a^4b + 12a^2b^2 - 8b^3$.
9. $32c^5 - 80c^4d^2 + 80c^3d^4 - 40c^2d^6 + 10cd^8 - d^{10}$.
10. $a^4 - 12a^3b^2 + 54a^2b^4 - 108ab^6 + 81b^8$.
11. $343 - 441x^2 + 189x^4 - 27x^6$.
12. $x^4y^8 + 8x^3y^6 + 24x^2y^4 + 32xy^2 + 16$.
13. $8 + \dfrac{12x}{a} + \dfrac{6x^2}{a^2} + \dfrac{x^3}{a^3}$.
14. $243 - \dfrac{405c^2}{2} + \dfrac{135c^4}{2} - \dfrac{45c^6}{4} + \dfrac{15c^8}{16} - \dfrac{c^{10}}{32}$.
15. $1 - \dfrac{2c^2}{b} + \dfrac{3c^4}{2b^2} - \dfrac{c^6}{2b^3} + \dfrac{c^8}{16b^4}$.
16. $\dfrac{c^5}{32x^5} + \dfrac{5c^4}{16x^4} + \dfrac{5c^3}{4x^3} + \dfrac{5c^2}{2x^2} + \dfrac{5c}{2x} + 1$.

17. $x^6 + 3x^5 - 5x^3 + 3x - 1.$
18. $x^6 - 9x^5 + 24x^4 - 9x^3 - 24x^2 - 9x - 1.$
19. $a^8 + 4a^7c + 10a^6c^2 + 16a^5c^3 + 19a^4c^4 + 16a^3c^5 + 10a^2c^6 + 4ac^7 + c^8.$
20. $x^3 - 3x^2y + 3x^2z + 3xy^2 - 6xyz + 3xz^2 - y^3 + 3y^2z - 3yz^2 + z^3.$
21. $8x^6 - 12x^5 + 42x^4 - 37x^3 + 63x^2 - 27x + 27.$
22. $1 + 4x + 2x^2 - 8x^3 - 5x^4 + 8x^5 + 2x^6 - 4x^7 + x^8.$

Exercise 67.

1. $3xy^2.$
2. $5a^3.$
3. $12y^n.$
8. $3x^2.$
9. $5yz^3.$
10. $-\frac{1}{2}ab^2.$
15. $-8x^2.$
16. $2y^3.$
17. $-2x^3y.$

Exercise 68.

1. $x^2 - 2x + 1.$ 2. $1 - a - a^2.$ 3. $3x^2 - 2x + 1.$ 4. $5 + 3x + x^2.$

5. $n^3 - 2n^2 + 3.$
6. $2x^3 + 3x^2 - 2x - 3.$
13. $\frac{1}{2}x - 5.$
16. $x^2 + x - \frac{1}{2}.$
17. $\frac{1}{2}a^2 - \frac{1}{3}a + 6.$
18. $\frac{a}{x} + 3 + \frac{x}{a}.$
19. $\frac{2}{3}x^2 - \frac{3}{5}x + \frac{5}{2}.$
24. $1 + 2x - 2x^2 \ldots\ldots$
25. $1 - a - \frac{1}{2}a^2 \ldots\ldots$
26. $x - \frac{3}{x} - \frac{9}{2x^3} \ldots\ldots$
27. $a + \frac{2b}{a} - \frac{2b^2}{a^3} \ldots\ldots$

Exercise 69.

1. 85.
4. 325.
5. 427.
10. 90.08.
11. 14.114.
14. 0.17071.
16. 2.6457 .....
17. 3.3166 .....
18. 3.5355 .....
19. 1.8257 .....
20. 1.4529 .....
21. 0.9486 .....
22. 2.5819 .....
23. 1.2747 .....
24. 0.3415 .....
25. 0.2213 .....
26. 1.0031 .....
27. 6.0075 .....
28. 1.9318 .....
29. 1.1117 .....
30. 1.3687 .....

Exercise 70.

1. $a + 2x.$
2. $3 - a.$
3. $1 - 4x.$
4. $a^2 - a - 2.$
5. $x^2 - x + 1.$
6. $1 - 3x - 2x^2.$
7. $4x^2 - 3x - 2.$
8. $a^2 + 5a - 1.$
9. $2x^2 - 5x - 3.$
10. $2 - 3n + 3n^2.$
11. $\frac{x}{2} - \frac{1}{3}.$
12. $\frac{x^2}{2y} - \frac{2y}{3x}.$
13. $x - 1 + \frac{1}{x}.$
14. $1 + \frac{1}{a} - \frac{3}{a^2}.$
15. $x^2 + \frac{2x}{y} - \frac{3}{2y^2}.$

Exercise 71.

1. 15.	3. 124.	5. 3204.	7. 70.09.
2. 91.	4. 352.	6. 804.5.	8. 0.0503.

9. 0.997.	13. 1.542	17. 0.2147 .....
10. 4.217	14. 1.953	18. 1.021 .....
11. 1.817	15. 2.704	19. 2.0033 .....
12. 1.775	16. 0.3968	20. 2.901 .....

21. 1.730 .....   22. 0.0535 .....

Exercise 72.

1. 19.	7. $\frac{2x}{y} - \frac{y}{2x}$.	11. 1.5704 .....
2. 43.		12. $x - 3$.
3. 3.08006	8. $2 + 2x - x^2$.	13. $2a - \frac{1}{a}$.
4. 0.9457	9. 14.	
5. $1 - 3ab$.	10. 34.	14. $4x^2 - \frac{1}{2}$.
6. $x - \frac{1}{2}$.		

Exercise 73.

4. $2\sqrt[4]{a}$.	30. 81.	38. $a^{\frac{5}{2}}x^2$.	44. $a^{\frac{13}{6}}$.
6. $2a^2\sqrt[7]{b^2}$.	31. 64.	39. $3a^{\frac{1}{2}}x$.	45. $\sqrt[6]{2^{11}}$.
7. $\sqrt{a}\sqrt[3]{m^2}$.	32. 36.	40. $4x^2$.	46. $a^{\frac{11}{10}}x^{\frac{5}{4}}$.
24. 9.	33. -27.	41. $a^{\frac{17}{6}}$.	47. $2x^2$.
25. 125.	34. $\frac{16}{81}$.	42. $7a^{\frac{7}{6}}$.	48. $\frac{2a^5x^4}{c}$.
26. 8.	35. $\frac{243}{32}$.	43. $2x^{\frac{8}{3}}$.	49. $\frac{32}{225}$.
27. 16.	36. $a^{\frac{7}{6}}$.		
28. 128.	37. $2a^{\frac{4}{3}}$.		
29. 4.			

Exercise 74.

7. $3 \times 2^{-1}ac^{-2}$.	25. $\frac{75z^{\frac{1}{n}}}{cx^ny^5}$.	34. $\frac{2}{9}$.	43. $\frac{1}{2}$.
13. $\frac{7}{x^2}$.	27. $\frac{1}{8}$.	35. $\frac{1}{4}$.	44. $2a$.
14. $\frac{3a}{b}$.	28. $\frac{1}{9}$.	36. 216.	50. $3a^{\frac{1}{3}}$.
20. $\frac{3bd^3}{a^2c}$.	29. 32.	37. $\frac{81}{16}$.	51. $c^3d^{\frac{1}{3}}$.
24. $\frac{10c^{\frac{1}{3}}y^{\frac{2}{3}}}{a^2x}$.	30. 25.	38. $-\frac{3}{2}$.	52. $m^{\frac{1}{3}}$.
	31. $\frac{1}{4}$.	39. $\frac{1}{625}$.	53. $\frac{6a}{x^{\frac{1}{3}}}$.
	32. $\frac{1}{729}$.	40. $-\frac{9}{32}$.	
	33. 54.	41. $\frac{27}{343}$.	
		42. 1.	

62. $\frac{a^{\frac{1}{3}}b}{c^{\frac{1}{3}}}$.

63. $\frac{a^{\frac{1}{3}}}{3}$.

64. $\frac{1}{3}$.

65. $\frac{7a^{10}}{3x^{\frac{1}{3}}}$.

66. $\frac{x^{2m}}{y^{\frac{3m}{2}}}$.

67. $\frac{x^{m}}{y^{\frac{3}{n}}}$.

68. $x^{\frac{5-2m}{2m}}$.

Exercise 75.

12. $\frac{9}{a^4}$.

13. $\frac{x^{\frac{2}{3}}}{25}$.

14. $\frac{1}{4a^2}$.

15. $\frac{x^6}{8}$.

24. $\frac{1}{a^{\frac{15}{4}}}$.

25. $\frac{x}{y^{\frac{3}{2}}}$.

26. $\frac{x^{2a}}{c^{2b}}$.

27. $\frac{27}{125x^{\frac{7}{4}}}$.

28. $\frac{4y^{\frac{2}{3}}}{9x^{\frac{4}{3}}}$.

29. $\frac{x^{\frac{1}{36}}}{y^{\frac{1}{6}}}$.

30. $\frac{y}{x^{\frac{2}{3}}}$.

31. $\frac{2c}{a^{\frac{2}{3}}}$.

32. 1.

33. $\sqrt{a}$.

34. $\sqrt{a}$.

35. $\frac{b}{xy}$.

36. $\frac{b^2}{a^2x^4y^4}$.

37. $\frac{1}{16}$.

38. $(\frac{4}{5})^{\frac{3}{2}}$.

39. $-\sqrt[5]{\frac{3}{4}}$.

40. x^m.

41. 27.

42. $\frac{c^3}{a^4}$.

43. $a^{\frac{1}{2}}bc^6$.

44. $\frac{a^2}{\sqrt{b}}$.

45. $\frac{1}{a^3x}$.

46. $\frac{a^{\frac{4}{3}}c^2}{bd}$.

47. xyz.

48. $8x^3 - 36 + 54x^{-3} - 27x^{-6}$.

49. $x^2 - 8x^{\frac{11}{6}} + 24x^{\frac{5}{3}} - 32x^{\frac{3}{2}} + 16x^{\frac{4}{3}}$.

50. $243x^{\frac{10}{3}} + 810x^{\frac{25}{6}} + 1080x^5 + 720x^{\frac{35}{6}} + 240x^{\frac{20}{3}} + 32x^{\frac{15}{2}}$.

51. $\frac{1}{16}x^{-2} - x^{-\frac{5}{4}} + 6x^{-\frac{1}{2}} - 16x^{\frac{1}{4}} + 16x$.

52. $16x^{-2} + 160x^{-\frac{13}{6}} + 600x^{-\frac{7}{3}} + 1000x^{-\frac{5}{2}} + 625x^{-\frac{8}{3}}$.

53. $x^2 - 4x^{\frac{3}{2}}y^{-\frac{1}{2}} + 6xy^{-1} - 4x^{\frac{1}{2}}y^{-\frac{3}{2}} + y^{-2}$.

54. $1 + 4x^{-\frac{3}{2}}y^{-\frac{2}{3}} + \frac{20}{3}x^{-3}y^{-\frac{4}{3}} + \frac{160}{27}x^{-\frac{9}{2}}y^{-2} + \frac{80}{27}x^{-6}y^{-\frac{8}{3}} + \frac{64}{81}x^{-\frac{15}{2}}y^{-\frac{10}{3}} + \frac{64}{729}x^{-9}y^{-4}$.

55. 4.

56. 9.

57. $\frac{1}{9}$.

58. $\frac{1}{8}$.

59. -32.

60. 1.

61. $\frac{1}{\sqrt[n]{2}}$.

62. $\frac{1}{(-3)^n}$.

Exercise 76.

1. $2a^{\frac{2}{3}} - a + 9$.
2. $a + 1$.
3. $9x + 5x^{\frac{1}{3}}y^{\frac{2}{3}} + 6y$.
4. $4x^{\frac{2}{3}} - 1 + 12x^{-\frac{1}{3}}$.

8. $6x^3 - 7x^{\frac{5}{3}} - 19x^{\frac{4}{3}} + 5x + 9x^{\frac{2}{3}} - 2x^{\frac{1}{3}}$.

9. $2 - 4a^{-\frac{4}{3}}x^{\frac{3}{2}} + 2a^{-\frac{8}{3}}x^3$.

10. $5 - 3x^{-\frac{1}{3}}y^{\frac{1}{3}} + 12x^{-\frac{2}{3}}y^{\frac{2}{3}} + 4x^{\frac{2}{3}}y^{-\frac{2}{3}}$.

11. $5x^{\frac{2}{3}} - 3x^{\frac{1}{3}} + 1$.
12. $4x^{-1} - 3y^{-1} - 2xy^{-2}$.

13. $x^{-\frac{1}{2}} - 2x^{-\frac{1}{3}} + 3x^{-\frac{1}{6}} - 1$.

18. $x^{\frac{2}{3}}y^{-1} - 3x^{\frac{1}{3}}y^{-\frac{1}{2}} + 2 - 4x^{-\frac{1}{3}}y^{\frac{1}{2}}$.

19. $3a^{-\frac{3}{4}} - 2a^{-\frac{1}{2}}x^{\frac{1}{2}} + 4a^{-\frac{1}{4}}x - x^{\frac{3}{2}}$.

20. $2a^{\frac{1}{4}} - 3a^{-\frac{1}{12}} - a^{-\frac{5}{12}}$.
21. $x^{\frac{1}{3}} - 2x^{\frac{1}{3}}y^{\frac{1}{2}}$.
23. $a^{-\frac{1}{2}} - 2b^{\frac{1}{2}} + 3a^{\frac{1}{2}}b$.
28. $3x^{\frac{5}{2}} - 4xy^{-\frac{1}{3}} - 2x^{-\frac{1}{2}}y^{-\frac{2}{3}}$.

29. $\frac{1}{2}x^{-\frac{1}{2}}y - \frac{1}{3}x^{-\frac{1}{4}} + 3y^{-1}$.

Exercise 77.

1. $2\sqrt{3}$.
2. $3\sqrt{2}$.
3. $3\sqrt{3}$.
4. $-2\sqrt{5}$.
5. $4\sqrt{6}$.
6. $-6\sqrt{7}$.
29. $\frac{1}{3}\sqrt{6}$.
30. $\sqrt{10}$.
31. $\frac{1}{2}\sqrt{30}$.
32. $\frac{2}{5}\sqrt{30}$.
33. $\frac{5}{12}\sqrt{6}$.
34. $\frac{3}{4x}\sqrt{14a}$.

Exercise 79.

1. $\sqrt{12}$.
2. $\sqrt{45}$.
3. $\sqrt[3]{432}$.
4. $-\sqrt{20}$.
18. $-\sqrt{x-1}$.

Exercise 80.

1. $\sqrt{a}$.
2. $\sqrt{a}$.
5. $\sqrt{3a}$.
10. $\sqrt[n]{3xy^3z^5}$.

Exercise 81.

1. $\sqrt[6]{343}$; $\sqrt[6]{121}$.
2. $\sqrt[12]{625}$; $\sqrt[12]{27}$.
3. $\sqrt[12]{27}$; $\sqrt[12]{25}$.
4. $\sqrt[6]{\frac{1}{8}}$; $\sqrt[6]{\frac{4}{9}}$.
7. $\sqrt[18]{64}$; $\sqrt[18]{81}$; $\sqrt[18]{125}$.
13. $\sqrt{3}$.
14. $\sqrt[3]{15}$.
18. $\sqrt[3]{4\frac{1}{2}}$.
21. $\sqrt{3}$.

Exercise 82.

1. $5\sqrt{2}$.	13. $2\sqrt{ax}$.	19. $5ac\sqrt{b}$.	25. 0.
2. $\sqrt{2}$.	14. $\frac{3}{4}\sqrt[3]{20}$.	20. $-6b\sqrt[3]{2a}$.	26. $6\sqrt{3}-6\sqrt{2}$.
6. $2\sqrt[3]{2}$.	15. $\frac{1}{12}\sqrt[3]{6}$.	21. $8\sqrt{2}$.	27. $5\sqrt{7}$.
7. $3\sqrt[3]{5}$.	16. $4\sqrt{b}$.	22. $-19\sqrt{3}$.	28. $-5\sqrt{3}$.
11. $\frac{5}{6}\sqrt{6}$.	17. $2\sqrt[3]{2c}$.	23. $-2\sqrt{6}$.	29. $12\sqrt{6}+10\sqrt{5}$.
12. $5\sqrt{3}$.	18. 0.	24. $7\sqrt{15}$.	30. $ab\sqrt{3a}$.

Exercise 83.

1. 12.	3. $12\sqrt[3]{3}$.	5. $36\sqrt{6}$.	7. $7\sqrt{5}$.	9. $\frac{4}{3}\sqrt[3]{3}$.
2. $15\sqrt{3}$.	4. $2\sqrt[5]{3}$.	6. $30\sqrt{21}$.	8. $\frac{1}{12}$.	10. $\frac{1}{4}\sqrt[3]{3}$.

11. $a\sqrt[6]{ab^2}$.	15. $\sqrt[12]{3456}$.	19. $\sqrt[6]{\frac{3abx^2}{2y}}$.	22. $\sqrt[24]{\frac{5}{7}}$.
12. $\sqrt[6]{72}$.	16. $6\sqrt[6]{12}$.		23. $\sqrt[24]{432}$.
13. $x\sqrt[6]{864a^2x}$.	17. $\sqrt[12]{12}$.	20. $\frac{1}{2}\sqrt[20]{288}$.	24. $\frac{1}{3}\sqrt[3]{9}$.
14. $3\sqrt[6]{24}$.	18. $\sqrt[12]{\frac{4}{15}}$.	21. $\sqrt[30]{\frac{2}{7203}}$.	25. $\sqrt[12]{\frac{5}{4}}$.

26. $2\sqrt{6}-4\sqrt{3}+8\sqrt{5}$.	39. $30\sqrt{6}+54\sqrt{5}-34$.
29. $20\sqrt{2}+30-8\sqrt{15}$.	40. 2.
31. $2-4\sqrt{2}$.	41. $96-16\sqrt{3}$.
32. $2+7\sqrt{3}$.	42. $x\sqrt{6}+\sqrt{3x^2-3x}$.
33. $-6-2\sqrt{6}$.	43. $\sqrt{3x^2+3x}+x+1$.
34. $16\sqrt{15}-30$.	44. $2\sqrt{x^2-1}-6x-6$.
35. $7\sqrt{6}-12$.	46. $25-7x$.
36. $42-6\sqrt{10}+2\sqrt{6}$.	47. $2x$.
37. $2\sqrt{15}-6$.	48. $25\sqrt{3}$.
38. $-282-72\sqrt{10}$.	49. $36x^2-50x-100$.

Exercise 84.

1. 3.	14. $\frac{1}{3}\sqrt[6]{72}$.	21. $5\sqrt{7}-14$.
2. $2\sqrt{2}$.	15. 3.	22. $\sqrt{2}+3$.
3. $3\sqrt{3}$.	16. $\sqrt[6]{\frac{4}{3}}$.	23. $\sqrt{21}-5\sqrt{15}$.
10. $\frac{1}{2}\sqrt{6}$.	17. $\sqrt[6]{\frac{54}{5}}$.	24. $2\sqrt{7}+4\sqrt{6}-6\sqrt{5}$.

Exercise 85.

1. $\frac{1}{3}\sqrt{2}$.	5. $\frac{2\sqrt{7}+\sqrt{35}}{14}$.	8. $\frac{\sqrt[3]{12}-\sqrt[3]{18}}{5}$.
2. $\frac{1}{6}\sqrt{6}$.		
3. $\frac{2\sqrt{5}}{15}$.	6. $\frac{2\sqrt{3}-\sqrt{2}}{4}$.	9. $\frac{11-6\sqrt{2}}{7}$.
4. $\frac{\sqrt{6}-2\sqrt{3}}{6}$.	7. $\frac{\sqrt[3]{20}-\sqrt[3]{4}}{6}$.	10. $13+7\sqrt{3}$.

11. $\frac{13\sqrt{6}-30}{6}$.
12. $\frac{18+\sqrt{2}}{23}$.
13. $\frac{\sqrt{5}+\sqrt{30}}{5}$.
14. $\frac{6a+\sqrt{ab}-12b}{4a-9b}$.
15. $\frac{x+\sqrt{x+1}-5}{x-3}$.
16. $\frac{6a-6+5\sqrt{2a^2-a}}{14a-9}$.
17. $\sqrt{3}+\sqrt{2}$.
18. $\frac{5-\sqrt{30}-5\sqrt{6}+6\sqrt{5}}{5}$.
19. $\frac{2\sqrt{3}-\sqrt{21}}{3}$.
20. $\sqrt{x^4-1}-x^2$.
21. $\frac{\sqrt{ab+b^2}-\sqrt{ab}}{b}$.

22. 2.12132.
23. 1.05409.
24. 4.53556.
25. 0.057735.
26. 0.709929.
27. 9.00996.
28. 0.10104.
29. 2.63224.
30. 0.62034.

Exercise 86.

1. m^2.
4. 6.
7. $8x^{12}$.
10. $4a\sqrt{2x}$.

Exercise 87.

1. $2\sqrt{2}-3$.
2. $\sqrt{3}+2\sqrt{5}$.
3. $3\sqrt{3}-2\sqrt{2}$.
4. $\sqrt{6}-\sqrt{3}$.
5. $\sqrt{14}+2\sqrt{7}$.
6. $3\sqrt{5}-2\sqrt{7}$.
7. $2\sqrt{5}+\sqrt{6}$.
8. $3\sqrt{5}-4\sqrt{2}$.
9. $2\sqrt{15}-3\sqrt{3}$.
10. $\frac{1}{2}\sqrt{2}+\frac{1}{2}\sqrt{6}$.
11. $\frac{4}{3}\sqrt{3}-\frac{2}{3}\sqrt{6}$.
12. $\frac{3}{2}+\sqrt{3}$.
13. $2-\frac{1}{3}\sqrt{3}$.
14. $\sqrt{m+n}+\sqrt{m-n}$.
15. $a+3\sqrt{a^2+1}$.
16. $\sqrt{3}-1$.
17. $\sqrt{3}+\sqrt{2}$.
18. $2-\sqrt{3}$.
19. $3-\sqrt{2}$.
20. $2-\frac{1}{2}\sqrt{2}$.
21. $1+\frac{1}{3}\sqrt{6}$.
22. $1+\sqrt{2}$.

Exercise 88.

1. 8.
2. 4.
3. 3.
4. 2.
5. 2.
6. $-\frac{1}{2}$.
7. 8.
8. 14.
9. $-\frac{2}{3}$.
10. -1.
11. 12.
12. $\frac{32}{3}$.
13. 1.
14. 9.
15. $\frac{25}{9}$.
16. 64.
17. $\frac{1}{4}$.
18. $\frac{1}{2}$.
19. $\frac{49}{4}$.
20. $\frac{13}{4}$.
21. 18.
22. 5.
23. 1.
24. 9.
25. 9.
26. $\frac{(a-b)^2}{2a-b}$.
27. $\frac{4}{3}$.
28. 1.
29. 8.
30. 1.
31. $\frac{13}{8}$.
32. $\frac{1}{9}$.
33. 9.
34. 64.
35. $\frac{(a-1)^2}{4}$.
36. $\frac{1}{4a(1+b)}$.
37. $16a$.
38. $\frac{1}{3}$.
39. $\frac{1}{4}$.
40. $\frac{9a}{16}$.
41. a^2.
42. $\frac{4a}{5}$.
43. $\frac{20a^2}{25-4a^2}$.
44. -3.
45. $\frac{3}{16}$.
46. -5.

Exercise 89.

2. $\sqrt{3}$. 3. $\sqrt{5}$. 4. $\sqrt{3}$. 5. $\frac{1}{4a}\sqrt{5x}$.

6. $2 + 4\sqrt{3}$.
7. $-2\sqrt{ax}$.
8. $\frac{1}{2}\sqrt[6]{30}$; $\frac{1}{3}\sqrt[12]{105}$.
9. $3\sqrt{2} + \sqrt{3} - 3\sqrt{5} + \sqrt{10}$.
10. $\frac{4}{15}\sqrt{15} - \sqrt{6} + 2\sqrt{10}$.
11. $x - \sqrt{x-1}$.
12. $\frac{9 + 5\sqrt{3}}{6}$.
13. $\frac{2\sqrt{3} - \sqrt{2}}{5}$.

14. $\frac{54 - 38\sqrt{2}}{7}$.
15. $4 + \sqrt{15}$.
16. 2.88675.
17. 0.18362.
18. 0.218286.
19. 0.36452.
20. $2\sqrt{7}$.
21. $\sqrt[3]{23}$.
22. $3\sqrt[3]{5}$.
23. $\sqrt[4]{11}$.
24. $5 + 2\sqrt{2}$.

25. $3\sqrt{3} - 2\sqrt{2}$.
26. $4\sqrt{3} - 4\sqrt{2}$.
27. $3\sqrt{7} + 2\sqrt{11}$.
28. $\frac{x-1}{\sqrt{x}}$.
29. $\sqrt{1-x}$.
30. $\frac{3}{8}(x - \sqrt{x^2 - 36})$.
31. $2\sqrt{x^2 - y^2}$.
32. 6.

33. $\frac{1}{4}$. 34. $\frac{16}{9}$. 35. 25. 36. 4. 37. $\frac{4}{3}$.

Exercise 90.

9. $5\sqrt{-1}$.
10. $9\sqrt{-1}$.
11. $20\sqrt{-1}$.
12. $6\sqrt{-3}$.
13. $8\sqrt{-2}$.
14. 0.
15. $a\sqrt{-1} - b$.
16. $-(a + 3b)\sqrt{-1}$.
17. $-\sqrt{2}$.
18. -18.
19. 10.
20. -7.
21. $-6\sqrt{6}$.
22. $9\sqrt{2}$.
23. $28\sqrt{6}$.
24. $-10\sqrt{10}$.
25. $(y - x)\sqrt{-1}$.
26. $a(a-1)^2\sqrt{-1}$.
27. $3 + \sqrt{2}$.
28. $-6 - 5\sqrt{6}$.
29. 24.
30. $-19 - 2\sqrt{35}$.
31. $46 + 2\sqrt{-3}$.
32. $-2 - 2\sqrt{3}$.
33. $x^2 - 4x + 7$.
34. $b^2 - a^2$.
35. $x^2 - 2x + 2$.
36. $x^2 - x + 1$.
37. $\sqrt{-3}$.
38. $\sqrt{6}$.
39. $-3\sqrt{5}$.
40. $-4a\sqrt{-1}$.

41. $-\sqrt{6} + 2\sqrt{5} + 3\sqrt{-10}$.
42. $-1 + 2\sqrt{6} - a^2\sqrt{-3}$.

43. $\frac{3 + \sqrt{-2}}{11}$.
44. $\frac{1 - 4\sqrt{-3}}{7}$.
45. $1 - \sqrt{-1}$.

46. $\frac{a^2 - b^2 + 2ab\sqrt{-1}}{a^2 + b^2}$.
47. $\frac{24 + 7\sqrt{10}}{2}$.
48. $\frac{1 \pm 3\sqrt{-1}}{5}$.
49. $\frac{6 - 4\sqrt{-1}}{13}$.
50. $\frac{\sqrt{70} + 3\sqrt{-14}}{14}$.
51. $3 - \sqrt{-6}$.
52. $\sqrt{3} - \sqrt{-2}$.

53. $2\sqrt{3} + 3\sqrt{-2}$.
54. $2\sqrt{-5} - 3\sqrt{-2}$.
55. $4 - 3\sqrt{-5}$.
56. $5\sqrt{-1} + 4\sqrt{2}$.
58. $6\sqrt{-1}$.
59. $\sqrt{-1} - 11$.
60. $2 - 2\sqrt{-1}$.
62. $\frac{3}{2}\sqrt{-1} + 4$; $\frac{4}{3}\sqrt{-1} - \frac{1}{4}$.
63. $-\frac{11}{3}$.

Exercise 91.

1. ± 4.
2. ± 2.
3. $\pm \frac{1}{2}$.
4. ± 2.
5. $\pm \frac{2}{3}$.
6. $\pm \frac{3}{2}$.
7. ± 5.
8. ± 1.
9. $\pm \frac{2}{3}$.
10. $\pm a$.
11. $\pm \sqrt{\frac{b-c}{a}}$.
12. $\pm 3a$.
13. $\pm \frac{c}{2}$.
14. $\pm \frac{2b}{a}$.
15. $\pm (a+b)$.
16. ± 1.

Exercise 92.

1. $2, -16$.
2. $2, -12$.
3. $-2, 10$.
4. $6, -1$.
5. $-3, -8$.
6. $1, -\frac{7}{3}$.
7. $2, -\frac{4}{5}$.
8. $-1, \frac{7}{2}$.
9. $2, -\frac{13}{9}$.
10. $\frac{1}{2}, -\frac{5}{2}$.
11. $\frac{3}{2}, -\frac{2}{3}$.
12. $-\frac{4}{3}, \frac{4}{5}$.
13. $3, \frac{5}{3}$.
14. $5, -\frac{7}{2}$.
15. $7, -\frac{11}{2}$.
16. $\frac{5}{2}, -\frac{7}{3}$.
17. $\frac{2}{3}, -\frac{5}{6}$.
18. $1, -\frac{5}{6}$.
19. $\frac{1}{3}, -\frac{3}{2}$.
20. $2, 5$.
21. $3, -\frac{6}{5}$.
22. $3, -1$.
23. $3, -\frac{8}{5}$.
24. $3, -\frac{1}{3}$.
25. $4, -\frac{5}{8}$.
26. $1, -\frac{10}{7}$.
27. $-2, -\frac{1}{2}$.
28. $1, -\frac{10}{9}$.
29. ± 1.
30. $3, -\frac{2}{3}$.
31. $5, -\frac{15}{4}$.
32. $3, -\frac{5}{8}$.
33. $2, -5$.
34. $\frac{1}{3}, \frac{7}{2}$.
35. $5, -\frac{5}{4}$.
36. $-1 \pm \sqrt{2}$.
37. $\frac{5 \pm \sqrt{13}}{6}$.
38. $\frac{3 \pm \sqrt{5}}{3}$.
39. $\frac{-3 \pm \sqrt{29}}{10}$.
40. $\frac{6 \pm \sqrt{3}}{11}$.
41. $\frac{-5 \pm \sqrt{-7}}{4}$.
42. $\frac{7 \pm \sqrt{-11}}{6}$.
43. $\frac{1 \pm 2\sqrt{-1}}{3}$.
44. $-1, -3$.
45. $1 \pm \frac{2}{3}\sqrt{6}$.
46. $\pm \sqrt{-1}$.

Exercise 93.

1. $2a, -6a$.
2. $3b, -7b$.
3. $2c, -5c$.
4. $ab, -6ab$.
5. $\frac{3b}{2}, -\frac{4b}{3}$.
6. $\frac{5cd}{3}, -3cd$.
7. $\frac{1}{a}, -\frac{3}{2a}$.
8. $\frac{a}{c}, \frac{3a}{7c}$.
9. $\frac{1}{a}, -\frac{6}{a}$.
10. $\frac{5b}{2a}, -\frac{3b}{a}$.
11. $a, 1$.
12. $\frac{b}{2a}, -\frac{3b}{2a}$.
13. $3a, -2$.
14. $-\frac{b}{a}, \frac{5}{3a}$.
15. $-\frac{b}{a}, -\frac{a}{b}$.
16. $a, -\frac{1}{a}$.
17. $a, -\frac{a}{a+1}$.
18. $a+1, a-1$.
19. $\frac{b-2}{2}, \frac{b+2}{2}$.
20. $\frac{a}{a+b}, -\frac{b}{a+b}$.
21. $\frac{1}{ab^2}, \frac{1}{a^2b}$.
22. $\frac{a+b}{c}, \frac{c}{a+b}$.
23. $-b, \frac{ab+b^2-c}{a+b}$.
24. $1, \frac{a}{a+c}$.

Exercise 94.

1. $-1, -7$.
2. $12, -7$.
3. $\frac{5}{3}, -\frac{3}{2}$.
4. $\frac{10}{3}, -\frac{9}{2}$.
5. $\frac{3}{4}, -\frac{1}{3}$.
6. $3, \frac{1}{3}$.
7. $\frac{5}{6}, -\frac{3}{4}$.
8. $\frac{2}{3a}, -\frac{4}{a}$.
9. $\pm 2, \pm 2\sqrt{-1}$.
10. $2, -1 \pm \sqrt{-3}$.
11. $\pm 1, \pm 2$.
12. $1, \pm 1, \pm \sqrt{-1}$.
13. $\frac{1}{2}, -\frac{1}{2}, \frac{2}{3}$.
14. $-1, \frac{5}{3}$.
15. $2, -\frac{7}{5}$.
16. $\pm 2, 7, \frac{1}{7}$.
17. $1, \frac{-3 \pm \sqrt{-15}}{6}$.
18. $1, 3, -4$.
19. $-1, \pm \frac{1}{2}\sqrt{2}$.
20. $1, \frac{1}{2}, -3$.
21. $\pm 1, \frac{1 \pm \sqrt{-3}}{2}, \frac{-1 \pm \sqrt{-3}}{2}$.

Exercise 95.

1. $\pm 1, \pm 4$.
2. $\pm 1, \pm \frac{3}{2}$.
3. $1, \frac{2}{3}, \frac{-1 \pm \sqrt{-3}}{2}$, and $\frac{-1 \pm \sqrt{-3}}{3}$.
4. $16, \frac{1}{81}$.
5. $1, \frac{4}{9}$.
6. $8, -\frac{8}{27}$.
7. $1, \frac{1}{64}$.
8. $27, -\frac{1}{8}$.
9. $\frac{4}{9}, \frac{9}{16}$.
10. $1, \frac{243}{32}$.
11. $1, \frac{16}{625}$.
12. $8, -\frac{8}{125}$.
13. $1, (-\frac{7}{3})^{\frac{5}{3}}$.
14. $\frac{1}{2}, \frac{1}{162}$.
15. $4, -6$.
16. $\pm 3, \pm \frac{1}{2}\sqrt{2}$.
17. $\pm \frac{1}{3}\sqrt{3}, \pm \frac{3}{2}\sqrt{-2}$.
18. $2, \frac{2442}{32}$.
19. $2, 6, -2 \pm 2\sqrt{-2}$.
20. $6, -\frac{21}{100}$.
21. $2, -\frac{4}{3}, \frac{3 \pm \sqrt{57}}{9}$.
22. $1, -\frac{5}{2}, \frac{-3 \pm 3\sqrt{5}}{4}$.
23. $1, -8, \frac{-7 \pm 3\sqrt{21}}{2}$.
24. $2, -3, \frac{1}{2}, -\frac{3}{2}$.
25. $5, \frac{1}{3}, \frac{8 \pm 2\sqrt{37}}{3}$.
26. $1, \frac{9}{49}$.
27. $\pm \frac{2}{3}\sqrt{6}, \pm 8\sqrt{-1}$.
28. $-\frac{1}{32}, \frac{3}{4}\sqrt[3]{18}$.
29. $\frac{7}{2}, -\frac{5}{2}, 1 \pm \sqrt{5}$.
30. $\pm 1, \pm \frac{1}{25}\sqrt{310}$.
31. $\pm 2\sqrt{2}, \pm \frac{1}{2}\sqrt{-3}$.

Exercise 96.

1. $3, 2$.
2. $1, -\frac{1}{4}$.
3. $3,$* 12.
4. $6, \frac{2}{3}$.*
5. $8, \frac{8}{25}$.*
6. $9, -\frac{7}{9}$.*
7. $3, -\frac{10}{3}$.*
8. $4, -\frac{4}{7}$.
9. $2a^2, -\frac{22a^2}{3}$.
10. $a^2 + 1, \frac{81a^2 + 1}{9}$.
11. $2, -1, \frac{1 \pm \sqrt{-3}}{2}, -1 \pm \sqrt{-3}$.
12. $3, \frac{20}{3}$.
13. $5, -\frac{8}{13}$.
14. $0, 5$.
15. $8, \frac{9}{50}$.
16. $\frac{1}{3}, -\frac{11}{12}$.
17. $4, -\frac{1}{13}$.
18. $\pm \frac{2}{3}, \pm \frac{3}{4}$.
19. $\pm a, \pm 2a^2$.
20. $2, \frac{2}{3}$.

* Will not satisfy the equation as it stands.

Exercise 97.

1. $1, -6$.
2. $1, -\frac{2}{3}$.
3. $\frac{1}{3}, -\frac{4}{5}$.
4. $\frac{4}{7}, -2$.
5. $\frac{3}{4}, -\frac{1}{2}$.
6. $\frac{5}{3}, -\frac{7}{3}$.
7. $2, -\frac{8}{5}$.
8. $\pm\frac{1}{2}, \pm\frac{5}{3}$.
9. $\frac{2a}{3}, -\frac{a}{2}$.
10. $\frac{7}{4a}, -\frac{3}{a}$.
11. $\pm 3, \pm 2\sqrt{-2}$.
12. $\pm 2, \pm\frac{1}{3}$.
13. $\frac{41}{15}, \frac{14}{31}$.
14. $\frac{1}{3}, -\frac{27}{8}$.
15. $\frac{3}{125}, -\frac{27}{8}$.
16. $\frac{10}{3}, -1$.
17. $a, \frac{1}{2a}$.
18. $\frac{2}{1-a}, -1$.
19. $\frac{a}{b-a}, \frac{b}{a+b}$.
20. $\frac{a}{2b}, \frac{3b}{4a}$.

Exercise 98.

1. $1, -\frac{7}{2}$.
2. $-1, \frac{7}{4}$.
3. $-2, \frac{5}{6}$.
4. $3, -\frac{1}{4}$.
5. $\frac{5}{6}, -\frac{1}{3}$.
6. $-\frac{2}{3}, -\frac{1}{2}$.
7. $\frac{4}{3}, -\frac{9}{11}$.
8. $\frac{1}{4}, -\frac{2}{3}$.
9. $\frac{3a}{2}, -2a$.
10. $\frac{3b}{a}, -\frac{4b}{3a}$.
11. $a, \frac{1}{a}$.
12. $27, -\frac{27}{8}$.
13. $\frac{25}{4}, \frac{16}{9}$.
14. $1, \frac{4}{9}$.
15. $\pm\frac{2}{3}, \pm\frac{1}{4}$.
16. $\pm 2, \pm\frac{1}{3}\sqrt{3}$.
17. $6, \frac{12}{5}$.
18. $\frac{c(c+d)}{c-d}, \frac{d(c+d)}{d-c}$.
19. $3, \frac{3}{4}$.
20. $\frac{a}{a+3}, \frac{-3}{a+3}$.
21. 3.7320, 0.2680.
22. 1.41202, − 0.07868.
23. 4.67945, 0.32055.
24. 0.46332, − 0.86332.

Exercise 99.

1. $\frac{1}{3}, -\frac{1}{2}$.
2. $\frac{10}{9}, -\frac{5}{2}$.
3. $1, 16$.
4. $0, \pm 4$.
5. $9, \frac{16}{9}$.
6. $3, -2$.
7. $10, \frac{7}{11}$.
8. $a, \frac{1}{a}$.
9. $\frac{b}{a}, -\frac{a}{b}$.
10. $3, \frac{7}{3}$.
11. $\frac{1}{27}, -\frac{27}{8}$.
12. $0, 3, \frac{-3 \pm 3\sqrt{-3}}{2}$.
13. $1, \frac{25}{121}$.
14. $\frac{16}{3}, -3$.
15. $\frac{-1 \pm \sqrt{-1}}{2}$.
16. $-1, -3, -2 \pm \frac{1}{3}\sqrt{-10}$.
17. $15, -\frac{50}{31}$.
18. $4, \frac{1}{4}$.
19. $-1, -3, \frac{5 \pm \sqrt{-23}}{4}$.
20. $\frac{a+b}{a}, \frac{a-b}{b}$.
21. $16, 1$.
22. $-a, -b$.
23. $2, 648$.
24. $3, -\frac{3}{2}, \frac{-3 \pm \sqrt{43}}{4}$.
25. $-2, \frac{4}{3}$.

Exercise 100.

1. 5, 6. 2. 7, 8. 3. 2, 5. 4. 5. 5. 23 years.

6. 30 mi. an hr.
7. $2\frac{1}{2}$, $3\frac{1}{5}$.
8. 5, 13.
9. 36 boys.
10. 30 and 45 min.
11. 30 cents.
12. 6, 9.
13. 6, 15.
14. 7, 11.
15. 14, 3.
16. 10 boys.
17. 9 men.
18. $4\frac{1}{2}$, $7\frac{1}{2}$.
19. 18 feet.
20. 1 rod.
21. $2\frac{1}{2}$ rods.
22. 24.
23. 4 hrs., 10 min.
24. 32.
25. 5 mi. an hr.
26. 45 mi. an hr.
27. A, 20 days.
    B, 12 days.
28. 9 mi. an hr.

Exercise 101.

1. $x = 1, -13.$
   $y = 2, 16.$
2. $x = 4, -1.$
   $y = \frac{1}{2}, -2.$
3. $x = 1, 2.$
   $y = 0, -3.$
4. $x = 2, -26.$
   $y = 1, 15.$
5. $x = 4, -\frac{9}{4}.$
   $y = 1, -\frac{13}{12}.$
6. $x = -2, -\frac{34}{35}.$
   $y = 1, -\frac{7}{25}.$
7. $x = 3, -\frac{15}{4}.$
   $y = -1, \frac{17}{10}.$
8. $x = -3, 7.$
   $y = -\frac{4}{3}, 4.$
9. $x = \frac{7}{5}, 2.$
   $y = \frac{7}{5}, 1.$
10. $x = 7, 12.$
    $y = \frac{3}{2}, \frac{5}{2}.$
11. $x = 1, 10.$
    $y = -\frac{2}{3}, -\frac{5}{3}.$
12. $x = 1, \frac{39}{7}.$
    $y = -3, \frac{12}{7}.$
13. $x = 3, \frac{3}{4}.$
    $y = -2, 1.$
14. $x = 2, -\frac{7}{5}.$
    $y = 1, -\frac{28}{25}.$
15. $x = 3, \frac{17}{4}.$
    $y = 2, 3.$

Exercise 102.

1. $x = \pm 4, \mp 14.$
   $y = \pm 1, \pm 4.$
2. $x = \pm 3, \pm\frac{5}{3}\sqrt{3}.$
   $y = \mp 1, \mp\frac{1}{3}\sqrt{3}.$
3. $x = \pm 1, \pm\frac{7}{2}\sqrt{-2}.$
   $y = \pm 2, \mp\frac{3}{2}\sqrt{-2}.$
4. $x = \pm 3, \pm\frac{8}{5}\sqrt{10}.$
   $y = \pm 2, \pm\frac{1}{5}\sqrt{10}.$
5. $x = \pm 5, \pm 6\sqrt{2}.$
   $y = \pm 2, \mp 7\sqrt{2}.$
6. $x = \pm 1, \pm\frac{11}{\sqrt{91}}.$
   $y = \pm 3, \pm\frac{27}{\sqrt{91}}.$
7. $x = \pm 3, \pm\frac{5}{3}.$
   $y = \pm 5, \pm\frac{13}{3}.$
8. $x = \pm 8, \pm 3.$
   $y = \mp 5, \pm 5.$
9. $x = \pm 5, \pm\frac{13}{\sqrt{51}}.$
   $y = \mp 3, \pm\frac{8}{\sqrt{51}}.$
10. $x = \pm 1, \pm 2.$
    $y = \pm\frac{3}{2}, \pm\frac{9}{8}.$
11. $x = \pm 2, \pm\sqrt{2}.$
    $y = \pm 4, \pm 3\sqrt{2}.$
12. $x = \pm 1, \pm 14\frac{3}{4}.$
    $y = \mp 3, \pm 3\frac{7}{8}.$

Exercise 103.

1. $x = 9, 4.$
   $y = 4, 9.$
2. $x = 4, -3.$
   $y = -3, 4.$
3. $x = -3, -7.$
   $y = -7, -3.$
4. $x = 4, -5.$
   $y = -5, 4.$
5. $x = \pm 7, \pm 3.$
   $y = \pm 3, \pm 7.$
6. $x = \pm\frac{1}{2}, \pm\frac{5}{2}.$
   $y = \pm\frac{5}{2}, \pm\frac{1}{2}.$

7. $x = \frac{4}{3}, -\frac{7}{3}$.
$y = -\frac{7}{3}, \frac{4}{3}$.

8. $x = 2, 1$.
$y = 1, 2$.

9. $x = 4, -3$.
$y = -3, 4$.

10. $x = 7, -5$.
$y = -5, 7$.

11. $x = \pm\frac{3}{2}, \mp\frac{5}{2}$.
$y = \mp\frac{5}{2}, \pm\frac{3}{2}$.

12. $x = \pm 2a, \pm 3a$.
$y = \pm 3a, \pm 2a$.

13. $x = a + 1, a - 1$.
$y = a - 1, a + 1$.

14. $x = 2, 3$.
$y = 3, 2$.

15. $x = 6, 2$.
$y = 2, 6$.

16. $x = \pm\frac{1}{2}, \pm\frac{1}{3}$.
$y = \pm\frac{1}{3}, \pm\frac{1}{2}$.

17. $x = \frac{2}{3}, 2$.
$y = 2, \frac{2}{3}$.

18. $x = \frac{4}{3}, -\frac{2}{3}$.
$y = -\frac{2}{3}, \frac{4}{3}$.

19. $x = \pm\frac{2}{3}, \mp\frac{4}{3}$.
$y = \mp\frac{4}{3}, \pm\frac{2}{3}$.

20. $x = \frac{5}{2}, \frac{3}{2}$.
$y = -\frac{3}{2}, -\frac{5}{2}$.

21. $x = 5, -3$.
$y = 3, -5$.

22. $x = a - 2b, -2a - b$.
$y = 2a + b, 2b - a$.

23. $x = \pm 1 \pm \sqrt{2}$.
$y = \pm 1 \mp \sqrt{2}$.

Exercise 104.

1. $x = \pm 5$.
$y = \pm 2$.

2. $x = \frac{4}{3}, \frac{1}{2}$.
$y = \frac{2}{3}, \frac{3}{2}$.

3. $x = \pm 4, \pm\frac{7}{3}\sqrt{2}$.
$y = \pm 1 \mp \frac{4}{3}\sqrt{2}$.

4. $x = 2, -\frac{13}{5}$.
$y = 1, -\frac{14}{5}$.

5. $x = 6, -5$.
$y = -5, 6$.

6. $x = \pm 4, \pm 1$.
$y = \mp 2, \mp 3$.

7. $x = 1, \frac{5}{4}$.
$y = 3, 2$.

8. $x = \pm 1, \pm 2$.
$y = \pm 3, \pm 1$.

9. $x = 3, \frac{4}{11}$.
$y = 2, \frac{19}{11}$.

10. $x = 2, -\frac{9}{4}$.
$y = 3, -\frac{71}{16}$.

11. $x = 5, 13$.
$y = 3, \frac{5}{3}$.

12. $x = 1, \frac{9}{5}, \frac{7 \pm \sqrt{19}}{5}$.
$y = 3, \frac{5}{3}, \frac{7 \mp \sqrt{19}}{3}$.

13. $x = 2, \frac{3}{2}, 6, -\frac{5}{2}$.
$y = 3, \frac{7}{2}, -10, -\frac{3}{2}$.

14. $x = -5, -1, \frac{7 \pm \sqrt{-23}}{2}$.
$y = -2, -6, \frac{5 \mp \sqrt{-23}}{2}$.

15. $x = \frac{4}{3}, \frac{2}{3}$.
$y = -\frac{2}{3}, -\frac{4}{3}$.

16. $x = \frac{1}{3}, \frac{2}{3}, \frac{-3 \pm \sqrt{33}}{18}$.
$y = \frac{2}{3}, \frac{1}{3}, \frac{-3 \mp \sqrt{33}}{18}$.

17. $x = 2, 4, \frac{-3 \mp \sqrt{33}}{3}$.
$y = \frac{4}{3}, \frac{2}{3}, \frac{-3 \pm \sqrt{33}}{9}$.

18. $x = 3, -2, -2 \pm \sqrt{5}$.
$y = -2, 3, -2 \mp \sqrt{5}$.

19. $x = \pm 4, \pm\frac{2}{3}\sqrt{35}$.
$y = \pm 1, \pm\frac{1}{3}\sqrt{35}$.

20. $x = 2, 1, \frac{3 \pm \sqrt{-55}}{2}$.
$y = 1, 2, \frac{3 \mp \sqrt{-55}}{2}$.

21. $x = 3, -2, \frac{1 \pm 3\sqrt{-3}}{2}$.
$y = -2, 3, \frac{1 \mp 3\sqrt{-3}}{2}$.

22. $x = 3, -1, 1 \pm \sqrt{-10}$.
$y = -1, 3, 1 \mp \sqrt{-10}$.

23. $x = 1,\ 4,\ \frac{5 \pm \sqrt{-159}}{2}$; $\quad y = -4,\ -1,\ \frac{-5 \pm \sqrt{-159}}{2}$.

24. $x = \pm 2,\ \pm 1,\ \pm 2\sqrt{-1},\ \pm\sqrt{-1}$.
$y = \pm 1,\ \pm 2,\ \mp\sqrt{-1},\ \mp 2\sqrt{-1}$.

25. $x = 3,\ -2,\ 1,\ -3$.
$y = 2,\ -3,\ 3,\ -1$.

26. $x = 3,\ 4,\ -2 \pm \sqrt{3}$.
$y = 4,\ 3,\ -2 \mp \sqrt{3}$.

27. $x = \pm 2a,\ \pm 2b$.
$y = \pm b,\ \pm a$.

28. $x = \frac{1}{2},\ \frac{1}{3}$.
$y = \frac{1}{3},\ \frac{1}{2}$.

29. $x = 5,\ \frac{5}{27}$.
$y = -1,\ \frac{25}{27}$.

30. $x = 3a,\ -9a$.
$y = 2a,\ -7a$.

31. $x = 3,\ 4$.
$y = -1,\ 0$.

32. $x = -\frac{1}{3},\ \frac{41}{147}$.
$y = -\frac{1}{2},\ \frac{41}{28}$.

33. $x = -3,\ -4,\ 6 \pm \sqrt{43}$.
$y = -4,\ -3,\ 6 \mp \sqrt{43}$.

34. $x = \frac{1}{a},\ \frac{4}{a}$.
$y = -\frac{1}{b},\ -\frac{4}{b}$.

35. $x = 6a,\ \frac{7a}{6}$.
$y = \frac{3b}{2},\ \frac{35b}{4}$.

36. $x = a + b,\ \frac{b(a-b)}{a}$.
$y = a + b,\ \frac{a(b-a)}{b}$.

37. $x = 1,\ \frac{10}{7},\ 2,\ 5$.
$y = \frac{2}{a},\ \frac{5}{7a},\ -\frac{1}{a},\ -\frac{10}{a}$.

38. $x = 64,\ 1$.
$y = 1,\ 64$.

39. $x = 16,\ 9,\ (\frac{27}{2})^{\frac{2}{3}}$.
$y = 9,\ 16,\ (-\frac{27}{2})^{\frac{2}{3}}$.

40. $x = \frac{3}{2},\ \pm\sqrt{-1}$.
$y = \frac{13}{24},\ 0$.

41. $x = \pm 4,\ \pm 2,\ \pm\sqrt{-\frac{47}{7}} \pm \sqrt{-\frac{15}{7}}$.
$y = \pm 2,\ \pm 4,\ \pm\sqrt{-\frac{47}{7}} \mp \sqrt{-\frac{15}{7}}$.

42. $x = 8,\ 2$.
$y = 2,\ 8$.

43. $x = 4,\ 3,\ 6,\ 2$.
$y = \frac{1}{2},\ 2,\ 1,\ 3$.

Exercise 105.

1. 3, 7.
2. 3, 8.
3. 7 by 12 ft.
4. 7 by 11 rods.
5. 15 by 18 rods.
6. 28.
7. 9 by 18 in.
8. 40 yds.
9. 18c. @ \$12. or 16c. @ \$14.
10. $\frac{3}{5}$.
11. 40 and 60 mi.
12. Row 6 mi.; stream, 2 mi. an hour.
13. 21 and 13 ft.
14. 10 and 12 ft.
15. \$5.
16. $\frac{5}{2}$ and $\frac{4}{5}$.

Exercise 106.

11. $x^2 - 5x + 6 = 0$.
23. $x^2 - 2x - 1 = 0$.
24. $x^2 + 6x + 6 = 0$.
25. $2x^2 - 4x + 1 = 0$.
26. $2x^2 - 2x + 1 = 0$.
28. $4x^2 + a^2 + 4c^2b = 4ax$.

Exercise 107.

1. Real; uneq.
2. Real; uneq.
3. Real; eq.
4. Imag.
15. $\frac{1}{3}$.
16. $\frac{1}{5}$.
17. $\frac{3}{4}$.
18. $\frac{25}{8}$.
19. $-\frac{16}{5}$.
20. ± 10.
21. $-\frac{1}{2}$.
22. $\pm\frac{4}{5}$.
23. -2.
24. $3, -4$.
25. $-3, -\frac{1}{3}$.
26. $-1, \frac{17}{7}$.

Exercise 108.

1. $\frac{4}{3}$.
2. $\frac{7}{3}$.
3. $\dfrac{3a - 2b}{4a - 3b}$.
4. 3; 2.
5. $6a^2b$.
6. $2\frac{3}{5}$.
7. $a^2 - x^2$.
8. $\dfrac{3x + 4}{3x + 5}$.
9. $\frac{1}{2}\sqrt{3}$.
10. $6bc$.
11. $3y$.
12. $\frac{1}{7}$.
13. $\dfrac{a}{a - 1}$.
14. $\dfrac{25}{x}$.
15. 36.
16. $(a - 1)^2$.
17. $\dfrac{a - 1}{a(a + 1)}$.
18. $2, -\frac{2}{3}$.
19. $5, -4$.
20. $-7, -\frac{11}{4}$.
21. $\pm 3, -2$.
22. $0, -4$.
23. $0, 1, -\frac{5}{3}$.
24. 0, 5, 20.
25. 0, 5.
26. $3, -1$.
27. $3a^2$.
28. $x = -3;\ y = -4$.
29. $\begin{cases} x = a - 1, & \dfrac{a - 1}{a + 1} \\ y = a + 1, & 1. \end{cases}$
42. 4 and 10.
43. 5 and 11.
44. 13 and 19.
45. 4 and 8.

Exercise 109.

1. $x = 1, 5$.
   $y = 14, 7$.
2. $x = 3$.
   $y = 2$.
3. $x = 15, 8, 1$.
   $y = 1, 6, 11$.
4. $x = 9$.
   $y = 4$.
5. $x = 3$.
   $y = 7$.
6. $x = 8$.
   $y = 7$.
7. $x = 5, 12, 19$, etc.
   $y = 2, 7, 12$, etc.
8. $x = 7, 37, 67$, etc.
   $y = 1, 14, 27$, etc.
9. $x = 3, 14, 25$, etc.
   $y = 2, 18, 34$, etc.
10. $x = 22, 57$, etc.
    $y = 10, 23$, etc.
11. 99, 75, 51, 27, 3.
    8, 32, 56, 80, 104.
12. 243, 126, 9.
    78, 195, 312.
13. $\frac{17}{5}, \frac{12}{5}, \frac{7}{5}, \frac{2}{5}$.
    $\frac{9}{12}, \frac{21}{12}, \frac{33}{12}, \frac{45}{12}$.
14. Sheep, 3, 14, 25.
    Calves, 16, 10, 4.
15. 12 ways.*
16. 11 ways.*
17. 159; an indefinite No.

* Including zero values.

Exercise 110.

1. 15.
2. $2\frac{1}{4}$.
3. 5.
4. 8.
5. ± 2.
6. $\frac{3}{5}$.
7. $\frac{3}{4}\sqrt{2}$.
8. 35.
9. 4.
10. 8.
11. $x = -2y$.
12. $x^2 - 2x + 2 = y^2 - 2y$.
13. $9, -\frac{1}{3}$.
14. $\frac{3}{2}x^2, -\frac{5}{x}$.
15. $\frac{13}{2}$.
16. $\frac{1}{10}$.
17. $w = \frac{3}{2} - 4x^3 + \frac{3}{2x^3}$.
18. $1029\frac{1}{3}$ ft.
19. 13 in.
20. 12 in.
21. $157\frac{1}{7}$.
22. 4.
23. $9(\pm\sqrt{6} - 2)$.

Exercise 111.

1. 31.
2. -25.
   $s = -81$.
3. -46.
   $s = -364$.
4. $-\frac{23}{3}, -13$.
5. $54, 94\frac{1}{2}$.
6. $\frac{1}{4}, 0$.
   $s = 3\frac{3}{4}$.
7. -7.2.
   $s = -37.8$.
8. 164.
9. -69.
10. -189.
11. $148\frac{3}{4}$.
12. -300.
13. $573\frac{1}{3}$.
14. -165.
15. $\frac{(17 - 5b)bc}{4}$.
17. $-77\sqrt{3}$.

Exercise 112.

1. $a = 4$; $s = 286$.
2. $a = -5\frac{2}{3}$; $s = 209$.
3. $a = 5$; $d = 4$.
4. $a = 11$; $d = -3$.
5. $a = 5\frac{1}{2}$; $d = -2\frac{1}{2}$.
6. $a = \frac{3}{2}$; $d = -\frac{1}{3}$.
7. $a = -\frac{2}{3}$; $d = \frac{5}{12}$.
8. $a = 3\frac{2}{3}$; $d = -\frac{1}{6}$.
9. $n = 21$; $d = 1$.
10. $n = 21$; $d = -2$.
11. $n = 18$; $d = -\frac{1}{12}$.
12. $n = 16$; $d = \frac{5}{6}$.
13. $a = 7$; $n = 6$.
14. $a = -5$; $n = 10$.
15. $a = \frac{7}{2}$; $n = 7$.
16. $a = -\frac{1}{3}$; $n = 16$.
17. 12.
18. 8.
19. 10.
20. 4 or 9.

Exercise 113.

1. $d = -2$.
2. $d = \frac{1}{4}$.
3. $d = \frac{1}{6}$.
4. $d = -\frac{1}{12}$.
5. $d = \frac{5}{6}$.
6. $-1\frac{11}{15}$.
7. x.
8. $\frac{x}{x^2 - y^2}$.

Exercise 114.

1. 5, 7.
2. $4\frac{1}{2}$, 3.
3. $-\frac{11}{6}, -\frac{14}{9}$.
4. $5c - 7b,\ 4c - 6b$.
5. n^2.
6. $\frac{7n(n+1)}{2}$.
7. 102d term.
8. $8, 7\frac{1}{3}$ . . . . .
9. $3, -2, -7, -12$.
10. 1, 3, 5, 7.
11. $24d$.

12. 1, 4, 7 . . . . . 13. 12. 14. -5, $-4\frac{1}{3}$ . . . . .

15. $\begin{cases} 11,\ 6,\ 1,\ -4,\ -9. \\ -\frac{73}{7},\ -\frac{18}{7},\ \frac{37}{7},\ \frac{92}{7},\ 21. \end{cases}$

Exercise 115.

1. 486.
2. 192.
3. $-\frac{8}{27}$. $s = 33\frac{19}{27}$.
4. 16. $s = 55$.
5. $\frac{8}{81}$. $s = \frac{422}{243}$.
6. 32.
7. -63.
8. $\frac{463}{27}$.
9. $-\frac{4095}{256}$.
10. $\frac{4095}{8192}$.
11. $\frac{40}{3}(3 + \sqrt{3})$.
12. $31(\sqrt{6} + \sqrt{3})$.
13. $21\sqrt{2} + 28$.

Exercise 116.

1. 2. $s = 728$.
2. 5. $s = -425$.
3. 16. $s = \frac{2059}{4}$.
4. 45. $s = \frac{6305}{243}$.
5. $-\frac{5}{27}$. $s = -\frac{24605}{27}$.
6. $\frac{8}{9}$. $s = \dfrac{65 + 19\sqrt{6}}{9}$.
7. -4.
8. $\frac{3}{8}$.
9. $-\frac{3}{4}$.
10. $-\frac{1}{3}$.
11. 5.
12. 6.
13. 5.
14. 6.
15. 5.
16. 4.
17. 6.

Exercise 117.

1. $r = \frac{1}{2}$.
2. $r = \frac{2}{3}$.
3. $r = -2$.
4. $r = -4$.
5. $r = -\frac{1}{3}$.
6. $r = \sqrt{2}$.
7. $\pm\frac{3}{4}$.
8. ± 5.

Exercise 118.

1. 3.
2. $\frac{4}{5}$.
3. $-\frac{27}{5}$.
4. $\frac{45}{8}$.
5. $\frac{125}{48}$.
6. $\frac{1}{4}$.
7. $\frac{27}{16}$.
8. $6(2 + \sqrt{2})$.
9. $\frac{1}{2}(3\sqrt{2} + 4)$.
11. a.
12. $\frac{7}{11}$.
13. $\frac{169}{333}$.
14. $5\frac{24}{111}$.
15. $3\frac{173}{330}$.
16. $1\frac{193}{370}$.
17. $3\frac{237}{1100}$.
18. $1\frac{3}{110}$.
19. $1\frac{1}{37}$.
20. $\frac{2506}{3333}$.

Exercise 119.

1. $\frac{2}{3}$, $\frac{2}{5}$ . . . . .
2. $\frac{3}{16}$, $\frac{1}{5}$ . . . . .
3. $\frac{20}{9}$, $\frac{10}{3}$ . . . . .
4. 96, 48 . . . . .
11. 5, 15, 45.
12. $\begin{cases} 7,\ 14,\ 28. \\ 63,\ -21,\ 7. \end{cases}$
13. 3, 6, 12.
14. 1, 3, 9, 27.
15. $\begin{cases} 5,\ 8,\ 11. \\ 15,\ 8,\ 1. \end{cases}$
16. \$15, 30, 60, 120.
17. $2\sqrt{2} + 3$.
18. $\begin{cases} 2,\ 4,\ 8,\ 12. \\ \frac{25}{2},\ \frac{15}{2},\ \frac{9}{2},\ \frac{3}{2}. \end{cases}$
19. $-24\frac{1}{2}$, $\frac{9}{4}$.
20. \$64, 80, 100, 125.
21. 27, 36, 48 yrs.
22. $\begin{cases} 2,\ 4,\ 6,\ 9. \\ 2,\ \frac{1}{4},\ -\frac{1}{2},\ 9. \end{cases}$

Exercise 120.

1. $\frac{4}{45}$.
2. $-\frac{6}{25}$.
3. $\frac{1}{39}$.
4. 1, $\frac{4}{5}$, $\frac{2}{3}$, $\frac{4}{7}$.
5. $\frac{6}{7}$, $\frac{2}{3}$, $\frac{6}{11}$, etc.
6. $-\frac{10}{7}$, $-\frac{10}{11}$, etc.
7. $\frac{48}{17}$.
8. -36.
9. $\dfrac{x^2 - 1}{x^2 + 1}$.
10. $\dfrac{xy}{2x - y}$, $\dfrac{xy}{3x - 2y}$.
11. 12 and 18.
12. $-\frac{1}{3}$, $-\frac{2}{5}$, $-\frac{1}{2}$ . . . . .
13. 8 and 32.

Exercise 121.

1. 360.
2. 167,960.
3. 2520.
4. 462.
5. 969.
6. 24,024.
7. 2,441,880.
8. 43,758.
9. 15,120.
10. 1,961,256.
11. 7,893,600.
12. 286.
13. 40,320.
14. 4536.
15. 1st, 60,480; 2d, 181,440.
16. 2520.
17. 21.
18. 56.
19. 8,877,690.

Exercise 122.

1. 60.
2. 3780.
3. 453,600.
4. 34,650.
5. 1,663,200.
6. 1,135,134,000.
7. 168,168.
8. 23,000.
9. 7560.
10. 3,531,528.
11. 15,840.
12. 302,400.
13. 1,596,000.
14. 420.
15. 36.
16. 2880.
17. 8640.
18. 1200.
19. 4845.
20. 210.
21. 5040.
22. 31.
23. 72,000.
24. 27,720.

Exercise 123.

1. 2.
2. -15.
3. -4.
4. -8.
5. $b = c(c + a)$.
6. $c = d(b - ad)$.
7. $\frac{b}{d} - d - a = 1$.
8. $m - a = \frac{d - bn + n^2}{mn} = \frac{dm - cn}{n^2}$.

Exercise 124.

1. $1 + x - x^2 + x^3 - x^4 \ldots\ldots$
2. $2 + x + 2x^2 + 4x^3 + 8x^4 \ldots\ldots$
3. $1 - 6x + 7x^2 - 13x^3 + 20x^4 \ldots\ldots$
4. $3 - 7x^2 + 14x^4 - 28x^6 + 56x^8 \ldots\ldots$
5. $1 - 2x + 2x^3 - 2x^4 + 2x^6 \ldots\ldots$
6. $x + 5x^2 + 7x^3 - x^4 - 23x^5 \ldots\ldots$
7. $1 - x - x^2 - \frac{1}{2}x^4 + \frac{1}{4}x^5 \ldots\ldots$
8. $1 + \frac{1}{3}x + \frac{1}{3}x^2 - \frac{5}{9}x^3 - \frac{11}{27}x^5 \ldots\ldots$
9. $\frac{1}{2} - 2x + \frac{25}{4}x^2 - \frac{79}{4}x^3 + \frac{499}{8}x^4 \ldots\ldots$
10. $\frac{2}{3} - \frac{11}{9}x + \frac{41}{27}x^2 - \frac{107}{81}x^3 + \frac{353}{243}x^4 \ldots\ldots$
11. $\frac{1}{2} - \frac{3}{4}x^2 + \frac{11}{4}x^3 + \frac{5}{8}x^4 - 2x^5 \ldots\ldots$
12. $x^{-2} - 3x^{-1} + 3 - 3x + 3x^2 \ldots\ldots$
13. $2x^{-1} + 2 - x - 3x^2 - 2x^3 \ldots\ldots$
14. $2x^{-1} + \frac{1}{2} - \frac{3}{4}x + \frac{9}{8}x^2 - \frac{27}{16}x^3 \ldots\ldots$
15. $x^{-2} - 2x^{-1} + 1 - \frac{1}{2}x + \frac{1}{4}x^2 \ldots\ldots$
16. $x^{-2} - x^{-1} - 2x + 2x^2 - 4x^3 \ldots\ldots$
17. $x^{-2} - x^{-1} + x - x^3 + x^5 \ldots\ldots$

Exercise 125.

1. $1 - x - \frac{1}{2}x^2 - \frac{1}{2}x^3 - \frac{5}{8}x^4 \ldots\ldots$
2. $1 + 2x - x^2 + 2x^3 - \frac{7}{8}x^4 \ldots\ldots$
3. $2 - \frac{1}{4}x + \frac{7}{64}x^2 + \frac{211}{512}x^3 \ldots\ldots$
4. $a - \frac{x^2}{2a} - \frac{x^4}{8a^3} - \frac{x^6}{16a^5} \ldots\ldots$
5. $1 + \frac{1}{3}x - \frac{1}{9}x^2 + \frac{5}{81}x^3 \ldots\ldots$
6. $1 - x - 3x^2 - \frac{17}{3}x^3 \ldots\ldots$

Exercise 126.

1. $\frac{4}{x}-\frac{3}{x-3}$.
2. $\frac{2}{x-2}+\frac{1}{x+1}$.
3. $\frac{5}{x-3}-\frac{3}{x+3}$.
4. $\frac{1}{x}-\frac{4}{2x+1}+\frac{3}{2x-1}$.
5. $x+3-\frac{3}{2x}+\frac{4}{x-3}$.
6. $\frac{1}{x}+\frac{3}{2x+1}-\frac{2}{x-1}$.
7. $x-6+\frac{4}{4x-3}-\frac{3}{3x+2}$.
8. $x^2-x+1+\frac{8}{3(x-1)}-\frac{5}{3(x+2)}$.
9. $\frac{1}{2x}-\frac{5}{3x-2}+\frac{3}{3x-1}$.
10. $\frac{3}{x-1}-\frac{1}{x+1}+\frac{2}{2x-3}-\frac{6}{2x+3}$.
11. $\frac{3a}{x+a}-\frac{2a}{x-4a}$.
12. $x+2a+\frac{a^2}{2x-a}-\frac{a^2}{x+a}$.
13. $x-2+\frac{4}{x-2\sqrt{2}}+\frac{4}{x+2\sqrt{2}}$.
14. $\frac{3}{x-2+\sqrt{2}}+\frac{3}{x-2-\sqrt{2}}-\frac{6}{x}$.
15. $\frac{1+\sqrt{2}}{2x-5+\sqrt{2}}+\frac{1-\sqrt{2}}{2x-5-\sqrt{2}}$.

Exercise 127.

1. $\frac{1}{x}-\frac{1}{x^2}+\frac{3}{x+1}$.
2. $\frac{1}{x+1}-\frac{2}{(x+1)^2}+\frac{1}{(x+1)^3}$.
3. $-\frac{3}{x+2}+\frac{12}{(x+2)^2}-\frac{10}{(x+2)^3}$.
4. $\frac{3}{x}+\frac{5}{x^2}-\frac{9}{3x-4}$.
5. $-\frac{1}{3(1-2x)}-\frac{5}{3(x-2)}-\frac{4}{(x-2)^2}$.
6. $\frac{7}{2x-1}+\frac{12}{(2x-1)^2}+\frac{5}{(2x-1)^3}$.
7. $\frac{2}{x}-\frac{3}{x^2}-\frac{1}{x^3}-\frac{10}{5x-1}$.
8. $\frac{5}{x-1}-\frac{2}{(x-1)^2}+\frac{3}{x+2}-\frac{13}{(x+2)^2}$.
9. $\frac{1}{x-2}+\frac{6}{(x-2)^2}+\frac{12}{(x-2)^3}+\frac{3}{(x-2)^4}$.
10. $\frac{1}{x}-\frac{2}{x^2}-\frac{10}{(2x+3)^2}+\frac{48}{(2x+3)^3}$.
11. $\frac{2}{2x+3}-\frac{6}{(2x+3)^2}-\frac{1}{x-1}+\frac{4}{(x-1)^2}$.
12. $\frac{1}{x}-\frac{2}{x^2}-\frac{1}{x+1}+\frac{4}{(2x-5)^2}$.

Exercise 128.

1. $\frac{2}{x} - \frac{2x-5}{x^2+2}$.

2. $\frac{3}{x-2} - \frac{3x-4}{x^2+2x+4}$.

3. $\frac{10x-7}{3(x^2+x+1)} - \frac{4}{3(x-1)}$.

4. $\frac{3x+5}{x^2+1} - \frac{3}{x} - \frac{5}{x^2}$.

5. $\frac{x-4}{x^2+x+1} - \frac{x-6}{x^2-x+1}$.

6. $\frac{x+5}{x^2-x+3} + \frac{2x-15}{(x^2-x+3)^2}$.

7. $\frac{2x+3}{3(2x^2-3)} - \frac{2x-2}{(2x^2-3)^2} - \frac{1}{3x}$.

8. $\frac{2x-1}{x^2+1} - \frac{x+2}{(x^2+1)^2} - \frac{3}{2x} - \frac{1}{2(x-1)}$.

Exercise 129.

1. $x = y - y^2 + y^3 - y^4 \dots\dots$
2. $x = y + 3y^2 + 13y^3 + 67y^4 \dots\dots$
3. $x = y - 2y^2 + 5y^3 - 14y^4 \dots\dots$
4. $x = y + \frac{1}{2}y^2 + \frac{1}{6}y^3 + \frac{1}{24}y^4 \dots\dots$
5. $x = y - \frac{1}{2}y^2 + \frac{1}{3}y^3 - \frac{1}{4}y^4 \dots\dots$
6. $x = 2y + \frac{8}{3}y^2 + \frac{28}{9}y^3 + \frac{444}{135}y^4 \dots\dots$
7. $x = \frac{1}{2}y - \frac{3}{8}y^2 + \frac{5}{16}y^3 - \frac{35}{128}y^4 \dots\dots$
8. $x = y - \frac{1}{2}y^3 + \frac{1}{12}y^5 + \frac{5}{18}y^7 \dots\dots$
9. $x = y + \frac{y^3}{3} + \frac{2y^5}{15} + \frac{17y^7}{315} \dots\dots$
10. $x = -\frac{1}{2}(y-1) + \frac{3}{8}(y-1)^2 - \frac{5}{16}(y-1)^3 \dots\dots$
11. $x = (y-1) - \frac{1}{2}(y-1)^2 + \frac{1}{3}(y-1)^3 - \frac{1}{4}(y-1)^4 \dots\dots$

Exercise 130.

1. $a^4 + 12a^3 + 54a^2 + 108a + 81$.
2. $32a^5 - 80a^4x + 80a^3x^2 - 40a^2x^3 + 10ax^4 - x^5$.
3. $1 + 3x + \frac{15x^2}{4} + \frac{5x^3}{2} + \frac{15x^4}{16} + \frac{3x^5}{16} + \frac{x^6}{64}$.
4. $81x^2 - 216x^{\frac{3}{2}}y^2 + 216xy^4 - 96x^{\frac{1}{2}}y^6 + 16y^8$.
5. $x^{\frac{5}{3}} - 10x^{\frac{7}{3}} + 40x^3 - 80x^{\frac{11}{3}} + 80x^{\frac{13}{3}} - 32x^5$.
6. $x^8y^4 + 8x^7y^{\frac{7}{2}} + 28x^6y^3 + 56x^5y^{\frac{5}{2}} + 70x^4y^2 + 56x^3y^{\frac{3}{2}} + 28x^2y + 8xy^{\frac{1}{2}} + 1$.
7. $x^{-\frac{14}{3}} + 7x^{-\frac{7}{2}} + 21x^{-\frac{7}{3}} + 35x^{-\frac{7}{6}} + 35 + 21x^{\frac{7}{6}} + 7x^{\frac{7}{3}} + x^{\frac{7}{2}}$
8. $\frac{x^5y^{-5}}{32} - \frac{5}{16}x^{\frac{9}{2}}y^{-\frac{7}{2}} + \frac{5}{4}x^4y^{-2} - \frac{5}{2}x^{\frac{7}{2}}y^{-\frac{1}{2}} + \frac{5}{2}x^3y - x^{\frac{5}{2}}y^{\frac{5}{2}}$.
9. $x^{-10} - \frac{5}{3}x^{-8}y^3 + \frac{10}{9}x^{-6}y^6 - \frac{10}{27}x^{-4}y^9 + \frac{5}{81}x^{-2}y^{12} - \frac{1}{243}y^{15}$.

10. $243a^{\frac{5}{2}}x^{-\frac{15}{2}} - 405a^2x^{-6} + 270a^{\frac{3}{2}}x^{-\frac{9}{2}} - 90ax^{-3} + 15a^{\frac{1}{2}}x^{-\frac{3}{2}} - 1.$

11. $16x^2 + 32x^{\frac{11}{6}}y^{\frac{2}{3}} + 24x^{\frac{5}{3}}y^{\frac{4}{3}} + 8x^{\frac{3}{2}}y^2 + x^{\frac{4}{3}}y^{\frac{8}{3}}.$

12. $a^4 - 6a^{\frac{11}{6}} + 15a^{-\frac{1}{3}} - 20a^{-\frac{5}{2}} + 15a^{-\frac{14}{3}} - 6a^{-\frac{41}{6}} + a^{-9}.$

13. $a^{\frac{4}{3}}x^{\frac{8}{3}} - 12a^{\frac{1}{2}}x^{\frac{5}{2}} + 54a^{-\frac{1}{3}}x^{\frac{7}{3}} - 108a^{-\frac{7}{6}}x^{\frac{13}{6}} + 81a^{-2}x^2.$

14. $64a^4x^{-2} + 576a^{\frac{11}{6}}x^{-\frac{7}{6}} + 2160a^{-\frac{1}{3}}x^{-\frac{1}{3}} + 4320a^{-\frac{5}{2}}x^{\frac{1}{2}} + 4860a^{-\frac{14}{3}}x^{\frac{4}{3}}$
$+ 2916a^{-\frac{41}{6}}x^{\frac{13}{6}} + 729a^{-9}x^3.$

15. $81a^{-3}b^2 - 108a^{-2}b^{-2} + 54a^{-1}b^{-6} - 12b^{-10} + ab^{-14}.$

16. $x^6 - 3x^5 + 9x^4 - 13x^3 + 18x^2 - 12x + 8.$

18. $16x^8 + 32x^7 - 72x^6 - 136x^5 + 145x^4 + 204x^3 - 162x^2 - 108x + 81.$

20. $81x^8 - 216x^7 + 108x^6 + 120x^5 - 74x^4 - 40x^3 + 12x^2 + 8x + 1.$

21. $-14,784a^6x^{10}.$

22. $1716x^7y^{\frac{7}{2}}.$

23. $3003x^{16}y^6.$

$1001x^{\frac{34}{3}}y^{10}.$

24. $-198a^{\frac{47}{3}}b^7.$

$7920a^{\frac{32}{3}}b^4.$

25. $\frac{20995}{64}x^{\frac{17}{6}}y^9.$

$\frac{20995}{256}x^{\frac{1}{2}}y^{11}.$

26. $-61,236a^{\frac{65}{12}}x^5.$

27. $-1320x^6.$

28. $1365a^4x^{18}.$

29. $\frac{4341}{16}x^{24}.$

30. $-112,640.$

31. $24,310xy^{11}.$

Exercise 131.

1. $a^{-2} - 2a^{-3}x + 3a^{-4}x^2 - 4a^{-5}x^3 + 5a^{-6}x^4 \ldots\ldots$

2. $x^3 - \frac{3}{2}x + \frac{3}{8}x^{-1} + \frac{1}{16}x^{-3} + \frac{3}{128}x^{-5} \ldots\ldots$

3. $x^{\frac{1}{2}} + \frac{3}{2}x^{-\frac{1}{2}} - \frac{9}{8}x^{-\frac{3}{2}} + \frac{27}{16}x^{-\frac{5}{2}} - \frac{405}{128}x^{-\frac{7}{2}} \ldots\ldots$

4. $1 - 2x^2 + 4x^4 - 8x^6 + 16x^8 \ldots\ldots$

5. $x^{-6} + 3x^{-\frac{15}{2}} + 6x^{-9} + 10x^{-\frac{21}{2}} + 15x^{-12} \ldots\ldots$

6. $x^{\frac{1}{3}} + \frac{2}{3}x^{-\frac{1}{6}} - \frac{4}{9}x^{-\frac{2}{3}} + \frac{40}{81}x^{-\frac{7}{6}} - \frac{160}{243}x^{-\frac{5}{3}} \ldots\ldots$

7. $x^{\frac{1}{2}} - \frac{1}{2}x^{\frac{5}{2}}y^{\frac{1}{2}} + \frac{3}{8}x^{\frac{9}{2}}y - \frac{5}{16}x^{\frac{13}{2}}y^{\frac{3}{2}} + \frac{35}{128}x^{\frac{17}{2}}y^2 \ldots\ldots$

8. $1 - 7x + 28x^2 - 84x^3 + 210x^4 \ldots\ldots$

9. $a^{\frac{1}{4}} - \frac{1}{4}a^{-\frac{3}{4}}x - \frac{3}{32}a^{-\frac{7}{4}}x^2 - \frac{7}{128}a^{-\frac{11}{4}}x^3 - \frac{77}{2048}a^{-\frac{15}{4}}x^4 \ldots\ldots$

10. $a - a^{-2}x - a^{-5}x^2 - \frac{5}{3}a^{-8}x^3 - \frac{10}{3}a^{-11}x^4 \ldots\ldots$

11. $x^{\frac{7}{3}} - 14x^{\frac{5}{3}}y^3 + 70xy^6 - 140x^{\frac{1}{3}}y^9 + 70x^{-\frac{1}{3}}y^{12} \ldots\ldots$

12. $a^{\frac{2}{3}} - 4a^{\frac{4}{3}}x^{-\frac{3}{2}} + 10a^2x^{-3} - 20a^{\frac{8}{3}}x^{-\frac{9}{2}} + 35a^{\frac{10}{3}}x^{-6} \ldots\ldots$

13. $x^{-\frac{1}{2}} + \frac{1}{2}ax^{-\frac{5}{2}} + \frac{5}{8}a^2x^{-\frac{9}{2}} + \frac{15}{16}a^3x^{-\frac{13}{2}} + \frac{195}{128}a^4x^{-\frac{17}{2}} \ldots\ldots$

14. $x^{-1} + \frac{3}{2}x^{-\frac{5}{2}} + \frac{27}{8}x^{-4} + \frac{135}{16}x^{-\frac{11}{2}} + \frac{2835}{128}x^{-7} \ldots\ldots$

15. $a^{\frac{1}{3}} - a^{\frac{4}{3}}x^{\frac{1}{3}} + 2a^{\frac{7}{3}}x^{\frac{2}{3}} - \frac{14}{3}a^{\frac{10}{3}}x + \frac{35}{3}a^{\frac{13}{3}}x^{\frac{4}{3}} \ldots\ldots$

16. $x^{-\frac{1}{3}} - x^{\frac{4}{3}}y^{\frac{1}{2}} - \frac{1}{2}x^3y - \frac{1}{2}x^{\frac{14}{3}}y^{\frac{3}{2}} - \frac{5}{8}x^{\frac{19}{3}}y^2 \ldots\ldots$

17. $1 + \frac{4}{3}x^{\frac{1}{3}}y^{-\frac{1}{2}} + \frac{14}{9}x^{\frac{2}{3}}y^{-1} + \frac{140}{81}xy^{-\frac{3}{2}} + \frac{455}{243}x^{\frac{4}{3}}y^{-2} \ldots\ldots$

18. $x^{\frac{5}{2}}y^{-\frac{15}{4}} - 5x^{\frac{5}{6}}y^{-\frac{5}{4}} + \frac{15}{2}x^{-\frac{5}{6}}y^{\frac{5}{4}} - \frac{5}{2}x^{-\frac{5}{2}}y^{\frac{15}{4}} - \frac{5}{8}x^{-\frac{25}{6}}y^{\frac{25}{4}} \ldots\ldots$

19. $\frac{35}{128}x^4$.

20. $-\frac{429}{16}x^7$.

21. $\frac{1040}{81}a^{-\frac{17}{6}}x^{\frac{8}{3}}$.

22. $\frac{81}{32}a^{-5}x^{\frac{4}{3}}$.

23. $\frac{3}{8}x^{\frac{9}{2}}$; $\frac{143}{128}x^{12}$.

24. $\frac{22}{3}x^9$; $-\frac{55913}{81}x^{19}$.

25. $\frac{7}{16}x^{-9}y^6$; $\frac{91}{4096}x^{-17}y^{13}$.

26. $\frac{44}{6561}x^{-20}$.

27. $-\frac{429}{128}x^{-9}$.

28. $a^{30}x^{17}$.

29. 3.16228.

30. 8.062258.

31. 3.10723.

32. 5.03968.

33. 9.695359.

34. 1.96799.

35. 4.01553.

36. 2.036168.

Exercise 132.

1. $2 + \frac{1}{2+}\ \frac{1}{1+}\ \frac{1}{1+}\ \frac{1}{2}$.

2. $3 + \frac{1}{2+}\ \frac{1}{1+}\ \frac{1}{3}$.

3. $5 + \frac{1}{1+}\ \frac{1}{2+}\ \frac{1}{2+}\ \frac{1}{3}$.

4. $\frac{1}{1+}\ \frac{1}{1+}\ \frac{1}{1+}\ \frac{1}{1+}\ \frac{1}{1+}\ \frac{1}{1+}\ \frac{1}{5}$.

5. $\frac{36}{11}$.

6. $\frac{16}{9}$.

7. $\frac{11}{8}$.

8. $\frac{11}{37}$.

9. $2 + \frac{1}{2+}\ \frac{1}{3+}\ \frac{1}{1+}\ \frac{1}{2}$. 5th conv. $= \frac{61}{25}$.

10. $1 + \frac{1}{3+}\ \frac{1}{1+}\ \frac{1}{2+}\ \frac{1}{3+}\ \frac{1}{2}$. 5th conv. $= \frac{47}{37}$.

11. $1 + \frac{1}{2+}\ \frac{1}{3+}\ \frac{1}{1+}\ \frac{1}{3+}\ \frac{1}{1+}\ \frac{1}{3}$. 5th conv. $= \frac{49}{34}$.

12. $\frac{1}{2+}\ \frac{1}{2+}\ \frac{1}{1+}\ \frac{1}{1+}\ \frac{1}{2+}\ \frac{1}{5+}\ \frac{1}{1+}\ \frac{1}{2}$. 5th C. $= \frac{13}{31}$.

13. $\frac{1}{3+}\ \frac{1}{5+}\ \frac{1}{8+}\ \frac{1}{6}$. Only 4 convergents.

14. $4 + \frac{1}{2+}\ \frac{1}{3+}\ \frac{1}{4+}\ \frac{1}{2+}\ \frac{1}{1+}\ \frac{1}{2}$. 5th conv. $= \frac{297}{67}$.

15. $\frac{1}{4+}\ \frac{1}{3+}\ \frac{1}{1+}\ \frac{1}{1+}\ \frac{1}{1+}\ \frac{1}{2+}\ \frac{1}{3+}\ \frac{1}{4}$. 5th conv. $= \frac{11}{47}$.

16. $1 + \frac{1}{1+}\ \frac{1}{1+}\ \frac{1}{2+}\ \frac{1}{3+}\ \frac{1}{1+}\ \frac{1}{1+}\ \frac{1}{2}$. 5th conv. $= \frac{27}{17}$.

17. $\frac{1}{3+}\ \frac{1}{3+}\ \frac{1}{3+}\ \frac{1}{6+}\ \frac{1}{1+}\ \frac{1}{2+}\ \frac{1}{1+}\ \frac{1}{10}$. 5th conv. $= \frac{73}{241}$.

Exercise 133.

1. $2 + \frac{1}{4+}\ \frac{1}{4+} \cdots\cdots$

2. $1 + \frac{1}{1+}\ \frac{1}{2+} \cdots\cdots$

3. $3 + \frac{1}{6+}\ \frac{1}{6+} \cdots\cdots$

4. $2 + \frac{1}{1+}\ \frac{1}{4+} \cdots\cdots$

5. $2 + \frac{1}{1+}\ \frac{1}{1+}\ \frac{1}{1+}\ \frac{1}{4+} \cdots\cdots$

6. $3 + \frac{1}{1+}\ \frac{1}{2+}\ \frac{1}{1+}\ \frac{1}{6+} \cdots\cdots$

7. $4 + \frac{1}{2+}\ \frac{1}{1+}\ \frac{1}{3+}\ \frac{1}{1+}\ \frac{1}{2+} \cdots\cdots$

8. $4 + \frac{1}{2+}\ \frac{1}{8+}\ \frac{1}{2+} \cdots\cdots$

9. $4 + \frac{1}{4+}\ \frac{1}{8+}\ \frac{1}{4+} \cdots\cdots$

10. $1 + \frac{1}{1+}\ \frac{1}{1+}\ \frac{1}{1+}\ \frac{1}{1+}\ \frac{1}{2+} \cdots\cdots$

11. $\frac{1}{1+}\ \frac{1}{3+}\ \frac{1}{5+}\ \frac{1}{2+}\ \frac{1}{28+}\ \frac{1}{2+} \cdots\cdots$

12. $\frac{1}{5+}\ \frac{1}{1+}\ \frac{1}{2+}\ \frac{1}{1+}\ \frac{1}{10+} \cdots\cdots$

13. $7 + \frac{1}{1+}\ \frac{1}{3+}\ \frac{1}{1+}\ \frac{1}{8+} \cdots\cdots$

14. $2 + \frac{1}{3+}\ \frac{1}{2+}\ \frac{1}{3+} \cdots\cdots$

15. $2 + \frac{1}{1+}\ \frac{1}{3+}\ \frac{1}{2+}\ \frac{1}{1+} \cdots\cdots$

16. $\frac{\sqrt{21} - 3}{2}$.

17. $\sqrt{6} - 2$.

18. $\frac{\sqrt{5} - 1}{2}$.

19. $\frac{\sqrt{15} - 1}{2}$.

20. $\frac{\sqrt{2} + 5}{2}$.

21. $\frac{\sqrt{37} + 11}{4}$.

22. $4 + \frac{1}{3+}\;\frac{1}{6+}\cdots\cdots$ 23. $5 + \frac{1}{2+}\;\frac{1}{6+}\cdots\cdots$

24. $1 + \frac{1}{1+}\;\frac{1}{1+}\;\frac{1}{1+}\;\frac{1}{4+}\;\frac{1}{9+}\;\frac{1}{4+}\cdots\cdots$

25. $\begin{cases} 2 + \frac{1}{1+}\;\frac{1}{1+}\;\frac{1}{6+}\;\frac{1}{1+}\;\frac{1}{1+}\;\frac{1}{1+}\;\frac{1}{1+}\;\frac{1}{6+}\cdots\cdots \\ \frac{1}{7} + \frac{1}{1+}\;\frac{1}{1+}\;\frac{1}{1+}\;\frac{1}{1+}\;\frac{1}{6+}\cdots\cdots \end{cases}$

26. $3, \frac{22}{7}, \frac{333}{106}, \frac{355}{113}\cdots\cdots$

27. $27, \frac{82}{3}, \frac{765}{28}, \frac{3907}{143}\cdots\cdots$

28. $\frac{1}{4}, \frac{7}{29}, \frac{8}{33}, \frac{31}{128}, \frac{39}{161}, \frac{655}{2704}, \frac{694}{2865}\cdots\cdots$

Exercise 134.

4. 0.7782.
5. 1.9542.
11. 8.9031 − 10.
15. 7.3365 − 10.
16. 3.1055.
17. 1.8008.
18. 0.4774.
19. 9.8914 − 10.
20. 8.6309 − 10.

Exercise 135.

1. 43.
2. 770.
4. 3.78.
8. 0.627.
9. .00803.
17. 283.6,
18. 0.4367.
19. .05925.

Exercise 136.

1. 105.
2. 34.3.
3. .0755.
4. 207.71.
5. 4.082.
6. 0.5036.
7. 64.7.
8. − 0.7995.
9. 0.04775.
10. 147.1.
11. − 2.773.
12. − 0.2681.
13. 0.2168.
14. 1.427.
15. 2.407.
16. 0.3016.
17. 1.324.
18. 43.36.
19. − 4.08.
20. 0.2785.
21. 0.4287.
22. 12.16.
23. 37.82.
24. 162.5.
25. 0.7518.
26. 0.2526.
27. 4.359.
28. 1.487.
29. 1.502.
30. 11.86.
31. 0.6633.
32. 2.571.
33. 3.936.
34. 1.946.
35. 0.459.
36. 18.02.
37. 14.44.
38. 5.624.
39. − 9.365.
40. 0.3933.
41. 1.403.
42. 2.052.
43. 0.1755.
44. 22.58.
45. 1.19.
46. − 1.162.
47. 3.271.
48. 0.1424.
49. − 2.158.
50. .02864.
51. − 2.483.
52. 0.873.
53. 1.792.
54. .01567.
55. − 3.908.
56. 7.672.
57. 0.8686.
58. − 0.4704.
59. − 0.1606.
60. 0.2415.
61. .0725.
62. 3.076.
63. 1.805.
64. 0.7876.

Exercise 137.

1. 1.17. 2. 1.54. 3. 1.54. 4. 1.65.

5. 7.52. 6. 0.76. 7. 3.47. 8. 1.88.

9. 1.206. 10. 2.12. 11. 0.537. 12. 0.81.

13. 3.37. 14. − 1.38. 15. − 0.24. 16. $2\log a + 3\log b$.

17. $\log a + 5\log x$. 18. $3\log b + \frac{1}{2}\log x$.

20. $\frac{1}{4}\{\log a + \log x - 3\log b\}$.

22. $\frac{1}{3}\log a + \frac{2}{3}\log x - \log b - \frac{1}{2}\log c$.

25. $\dfrac{\log 7}{\log a - \log b}$.

26. $\dfrac{\log a}{\log a - \log c}$.

27. $\dfrac{\log b - \log 13}{\log 13 - \log a}$.

28. $\dfrac{\log 3 - \log 7 + 2\log b}{\log a}$.

29. $\dfrac{\log a + \log b - \log 5}{\log a - 2\log b}$.

30. $\dfrac{\log 11 - \log 6}{2\log b + \log c - \log(a-b)}$.

31. $\dfrac{\log(a+b) + \log(a-b)}{3\log(2a-1)}$.

32. $\left\{\dfrac{\frac{1}{2}\log(a+b) + \frac{1}{2}\log(a-b) - \log 25}{\log a}\right\}^2$.

33. 3.56. 34. 1.19. 35. 0.71. 36. 0.84. 37. 0.83.

38. 2.03. 39. −.065. 40. − 3.46. 41. 1.58. 42. 27.

43. 16. 44. 25. 45. 32. 46. $\frac{4}{3}$. 47. 3.

48. 4. 49. $\frac{4}{3}$. 50. − $\frac{1}{2}$. 51. − $\frac{4}{3}$. 52. − 4.

61–64. See page 326, formulas 17–20.

65. 7. 66. 8.

67. 9. 68. \$2654.

69. \$2497. 70. 14.2 and 10.24 yrs.

Exercise 142.

25. 0. 26. $44\frac{1}{4}$. 27. $\frac{1}{3}$.

28. $-27\frac{3}{4}$. 29. 25. 34. $-x + 11\sqrt{x} - 14$.

35. $-\dfrac{m^2+n^2}{2mn}$. 36. $\dfrac{2x^2}{2-x^2}$.

37. $-6\frac{3}{4}$. 38. $-\dfrac{6a}{b^{\frac{3}{2}}}$. 39. $\frac{5}{3}\sqrt{6}$.

40. $\dfrac{27\sqrt{6} + 99\sqrt{2} - 48\sqrt{3} - 176}{94}$.

$\frac{2}{9}$.

$\frac{5}{17}$.

$\frac{6}{5}, \frac{7}{8}$.

$3, -\frac{2}{3}, 5$.

$1, -2$.

$-\frac{67}{36}, -\frac{3}{8}$.

$-\frac{1}{10}, \frac{1}{12}$.

$\frac{ab^2+b+2c}{b^2}$.

$\frac{2a}{a^2-b^2}$.

$\frac{a^2+b^2}{a^2-b^2}$.

$\frac{x^2}{x+1}$.

$(x+1)^2$.

55. $\frac{ab}{a-\sqrt{ab}+b}$.

56. $\frac{2(a+b)}{a-b}$.

57. $\frac{x+y}{x-y}$.

58. $\frac{m}{m^2-mn+n^2}$.

59. $\frac{ac+bc-ab}{ac+bc+ab}$.

60. $\frac{a-1}{a}$.

61. 0.

62. $\frac{b^3}{(a+b)^3}$.

63. 0.

64. $\frac{1+z^4}{z(1+z^2)}$.

65. x^q.

66. x^{a^2}

67. x^{a^2-ab}.

75. $x+\frac{1}{\sqrt[a]{x}}$.

76. $\frac{1}{a^{np}}$.

77. x.

78. $\frac{b^{\frac{3}{2}}}{a^{\frac{5}{3}}}$.

79. m^{pq}

80. $\frac{\sqrt{b}+11a}{\sqrt{b}+9a}$.

$x=4, 2, 3\pm\sqrt{-55}$.
$y=2, 4, 3\mp\sqrt{-55}$.

84. $x=9, 5\frac{20}{117}$.
$y=4, \frac{20}{117}$.

$x=1, 2, \frac{3\pm\sqrt{19}}{2}$.
$y=-2, -1, \frac{-3\pm\sqrt{19}}{2}$.

85. $x=2, -\frac{2}{3}$.
$y=-1, 15; \frac{-7\pm 2\sqrt{46}}{3}$.

$x=\pm 7, \pm\frac{5}{3}\sqrt{2}$.
$y=\pm 2, \mp\frac{2}{3}\sqrt{2}$.

86. $x=3, -1, 9\pm\frac{11}{3}\sqrt{6}$.
$y=-1, 3, 9\mp\frac{11}{3}\sqrt{6}$.

$x=8, 27$.
$y=27, 8$.

$x=\frac{3}{5}, \frac{2}{5}$.
$y=\frac{2}{3}, \frac{3}{5}$.

$x=\frac{a}{2}, \frac{5a}{68}$.
$y=\frac{b}{4}, \frac{75b}{68}$.

$x=5, \frac{13}{4}$.
$y=7, -\frac{7}{4}$.

91. $x=\frac{1}{2}, \frac{59}{17}$.
$y=-3, \frac{22}{55}$.

92. $x=2, 2\frac{4}{5}$.
$y=3, 1\frac{4}{5}$.

93. $x=2, \frac{2}{3}$.
$y=\frac{2}{3}, 2$.

94. $x=2, 8, 2\pm 4\sqrt{-6}$.
$y=8, 2, 2\mp 4\sqrt{-6}$.

95. $x=\pm 3, \pm\sqrt{-7}$.
$y=2, 6$.

96. $x=5$. $y=\pm 4$.

97. $x=2, 8$. $y=8, 2$.

98. $x=1, 9$.
$y=9, 1$.

99. $x=9, 25$.
$y=25, 9$.

100. $x=1, \frac{5}{3}$.
$y=2, \frac{3}{5}$.

Lightning Source UK Ltd.
Milton Keynes UK
UKHW010955030323
417983UK00006B/391